S0-BZS-775

Origin and Development of Living Systems

Origin and Development of Living Systems

J. BROOKS

Exploration and Production Research Division,
BP Research Centre, Sunbury-on-Thames,
Middlesex, England

G. SHAW

School of Chemistry, University of Bradford,
Bradford, Yorkshire, England

1973

ACADEMIC PRESS – London and New York

ACADEMIC PRESS INC. (LONDON) LTD.
24/28 Oval Road,
London NW1

United States Edition published by
ACADEMIC PRESS INC.
111 Fifth Avenue
New York, New York 10003

Library of Congress Catalog Card Number: 72–7712
ISBN: 0–12–135740–6

PRINTED IN GREAT BRITAIN BY
BUTLER AND TANNER LTD.
FROME AND LONDON

Preface

During the latter half of the 1960s we embarked on a research project which was concerned initially with the chemical nature of the outer wall material of pollen grains and related microspores. The wall material contains the substance sporopollenin which is extremely resistant to chemical and especially biological decay to such an extent that it survives in ancient sedimentary deposits for thousands of millions of years. What began essentially as an exercise in organic chemistry soon developed and expanded in a manner which we could not originally have foreseen. We found that our studies were leading us into almost every major discipline of science. We had to become biologists (growing lilies and other plants and micro-organisms and injecting them with chemicals and radioactive chemicals), biochemists (investigating the products formed in our biological materials and their associated enzyme systems), physicists and radiochemists (investigating the labelled materials and using techniques such as neutron activation analysis for trace metal determinations), geologists, palynologists and geochemists (examining numerous rock samples, including some of the oldest of Precambrian sediments, for their entrapped spore coats or related organic residues).

Outside our own specialist field of organic chemistry we had to read widely and enlist the advice, always freely and patiently given, of many colleagues in other disciplines. We finally emerged from this bath of knowledge with a feeling that we could usefully present some of the various pieces of information we had gleaned about the living system in particular but also about the environment in which it had originated and developed. In so doing we hope that we have in this book presented a faithful unbiased account of modern thinking on these subjects but make no apologies for introduction of our own ideas and interpretation of results where we think appropriate.

We were particularly anxious to point out the vast mine of information which unquestionably exists in the ancient Precambrian sediments and which desperately cries out for many more workers. We found an enormous lack of knowledge of such matters especially about the organic and inorganic geochemistry which has been and requires to be carried out, not only among chemists but among most scientists. If this book in any way stimulates workers from almost any scientific discipline to think about possible ways in which their own particular subject might be used to wrest some of the secrets from these ancient rocks we believe that this exercise will not have been in vain.

The book is suitable for those who have a preliminary general science education. Although the contents include some degree of specialized work, especially, perhaps in the geochemical field, the main aim had been to provide a general framework of relevant basic science against which the living system and its environment can be examined.

We consider that the broad span of subjects covered should be included in University or College Foundation or General Courses in science and we believe that the book will be of special value to students taking such courses.

October 1972

J. B.*
G. S.

* Present address: H. Foster and Co. (Stearines) Ltd., Aire Place Mills, Kirkstall Road, Leeds LS3 1JL.

Acknowledgements

We would like to express our debt and gratitude to numerous colleagues and friends who have helped in many ways by discussion and correspondence during the preparation of this book. In particular we are especially grateful to Dr. Marjorie Muir (Royal School of Mines, Imperial College, London) for stimulating discussion and for various photographs used in the text. We also wish to thank Dr. Morris Viljoen (Johannesburg Consolidated Investment Company Limited, Randfontein), Professor John G. Ramsey (Royal School of Mines, Imperial College, London), Professor D. J. L. Visser (University of Pretoria), Dr. Tom Cross (University of Bradford), Dr. Robert Hutchison (British Museum, Natural History), Mike Collett (Exploration and Production Research Division, BP Research Centre, Sunbury-on-Thames, Middlesex) and many others for helpful comments and permission to use photographs and other data.

We also thank members of the Drawing Office of the Exploration and Production Research Division, BP Research Centre, Sunbury-on-Thames, for preparation of many of the figures used in the book.

One of us (J. B.) wishes to comment that the views expressed in this book are those of the two authors and should not necessarily be taken as representing in any way those of the British Petroleum Company Limited, nor for that matter of any of the persons named above.

Contents

1

The Universe and the Living System. Suitable Subjects for Investigation

With reference to somewhat exotic topics, specifically extra-sensory perception, Koestler has commented:

> "... academic science reacts to the phenomenon of extra-sensory perception much as the Pigeon League reacted to the Medicean Stars; and it seems to me for no better reason. If we have to accept that an electron can jump from one orbit into the other without traversing the space between, then why are we bound to reject out of hand the possibility that a signal of nature no more puzzling than Schroedinger's electron-waves should be emitted and received without sensory intervention?"

He concludes:

> ". . . there seems to be no justification in refusing to investigate 'empirical phenomena because they do not fit into that already abandoned philosophy."*

This is a not uncommon attitude held by many towards science and scientific research workers. The truth is different. Scientists, and of course there are always exceptions, do not take readily to subjects such as extra-sensory perception, not for lack of interest or imagination or because they are necessarily bound to any specific philosophy but for much more mundane reasons, namely the inability to devise experiments which might lead to useful objective results within a reasonable space of time. Contrary to some public opinion scientists are human with reputations to make, careers to think about. They are consequently

* Reprinted by permission of A. D. Peters and Company.

under pressure to produce results and must in the main, therefore, follow routes which have a reasonable hope of success within a reasonable time sequence. This means that it is most unlikely that there will ever be any serious substantial study by any large body of research workers of subjects such as extra-sensory perception until someone comes along with a precise objective and demonstrably successful method for looking at such subjects. He will of course have had most of the fun. This does of course lead to criticism of much of science and of many scientists as mere technicians only interested in expanding the obvious or already proven; in producing better fibres or cars or bombs. This is in fact true. Much of the heralded rapid advance of science over the past two or three decades is at least in part a myth. Thus the oft heard statement that science has produced more knowledge in the past twenty years than in the previous two hundred needs qualification. It may be true that there have been more publications in the past twenty years than in the past two hundred and equally true that the rate of enhancement of original discovery has increased enormously, but it is doubtful whether the *rate* of discovery of *new* fundamental laws of nature is greater now than it was in the nineteenth century and it may even be less.

What we are saying in part here is that there comes a time when a subject becomes ripe for investigation. Until that particular time comes no matter how bright or clever one might be it may be to no avail. Answers to specific problems frequently require methodology and knowledge without which the problems must remain unsolved.

There are perhaps two fundamental problems which when solved would allow us to have complete and absolute knowledge, namely the origin and nature of that which we call life and the nature and origin of the universe, and indeed these may well turn out to be the same problem since the subjects are so mutually interdependent. These are pretty exotic problems by any standards, but even though the subjects may even prove difficult to define, let alone the solutions found, they do nevertheless, unlike subjects such as extra-sensory perception, represent real problems for which there must be answers. In recent years there has been a marked quickening of interest in both these topics. This is not so much due to any previous lack of interest in the subjects since throughout the history of man there have been many who would speculate on these matters, but rather to an awareness that the time may now be ripe for a detailed investigation into some of the problems. In particular, during the past two decades the basic discoveries of chemistry, physics and biology have come together in a quite remarkable way to give an insight into the nature of the living system which allows, at least by previous standards, a sort of precise definition of that system. During the same period of time there have been equally rapid advances in techniques for the investigation and structure determination of micro-amounts of organic and bio-chemicals in many sources. These techniques allow an examination to be made of the organic and microfossil residues residing in the oldest of Precambrian sedimentary rocks which were laid down at the very beginning of the earth's biological history. There has been a slow but nevertheless inexorable increase in the study of the geology of these ancient Precambrian sediments. The critic might think that the practising geologist spends too much time with recent deposits, especially those with a commercial mineral content, and indeed many of the more academic studies have relied heavily for funds on their potential commercial applications. However it was at Barberton in the East Rand that the first organic residues (Fig. 1.01) in the most ancient Fig-Tree series Precambrian sediments were discovered in 1961 by J. G. Ramsey (now Professor of Geology at the Imperial College, London). He writes:

"My first visit to Barberton arose out of an invitation I received from Professor Gevers of the Geology Department of Witwatersrand, Johannesburg, 1961. I gave a course in structural geology and during my stay I was very attracted by all I heard about the geological problems of Barberton. When I went there I was not disappointed for the exposures were excellent and the structural pattern very interesting. During the course of my mapping programme I had a good look at the old sediments. Even though the rocks are so old and have suffered a complex deformation history they often show beautifully preserved sedimentation structures. Fine details of cross-bedding, ripple drift bedding, bottom structures, grooves and flutes in the greywacke, imbricate conglomerates and desiccation cracks can be seen at many localities. I was interested to see that the morphology of these structures was very like that in much more recently deposited sediments. I am sure that you could take the outcrops of Fig Tree greywackes, place them amongst the Alpine flysch, and no one would notice anything queer. Likewise, the cross bedded and ripple marked shales of the Moodies System are identical to many shales found in the Alpine Upper Tertiary Molasse. I feel that these observations point to the fact that physical conditions at the surface during the deposition of the Barberton sediments were not unlike those existing today. Probably the gravitational constant also had about the same value as it does today. The other surprise I turned up was the peculiar organic-like forms in the Fig Tree greywackes of Sheba mine. I remember the discovery very clearly, looking into a rather gloomy cross-cut deep in the mine to find an 'oolite' structure in the wall rocks. Although I am not a sedimentologist I realised that oolitic greywackes are not common (do they exist at all?). However the objects, whatever they were, looked 'nicely' (I am a structural geologist!) deformed so I collected several specimens of ellipsoidal objects which clearly indicated that the slaty cleavage at this locality coincided with the plane normal to the greatest shortening in the rock (i.e. perpendicular to the short axes of the ellipsoids). On my return to Johannesburg the specimens I had collected were sliced—I looked at them with interest for they clearly were not oolites, in fact no calcareous material was to be seen in the rock. Also the internal

Fig. 1.01. The first organic residues found in the Fig-Tree sediments. (a) and (b) Photomicrographs of the organic objects identified by Ramsey in 1961. (c) Photomicrograph of thin section of Fig-Tree sediment, from the same locality as Ramsey's specimens, showing organic remains, later identified as algal structures. (Photograph taken by Dr. Marjorie Muir. × 100.)

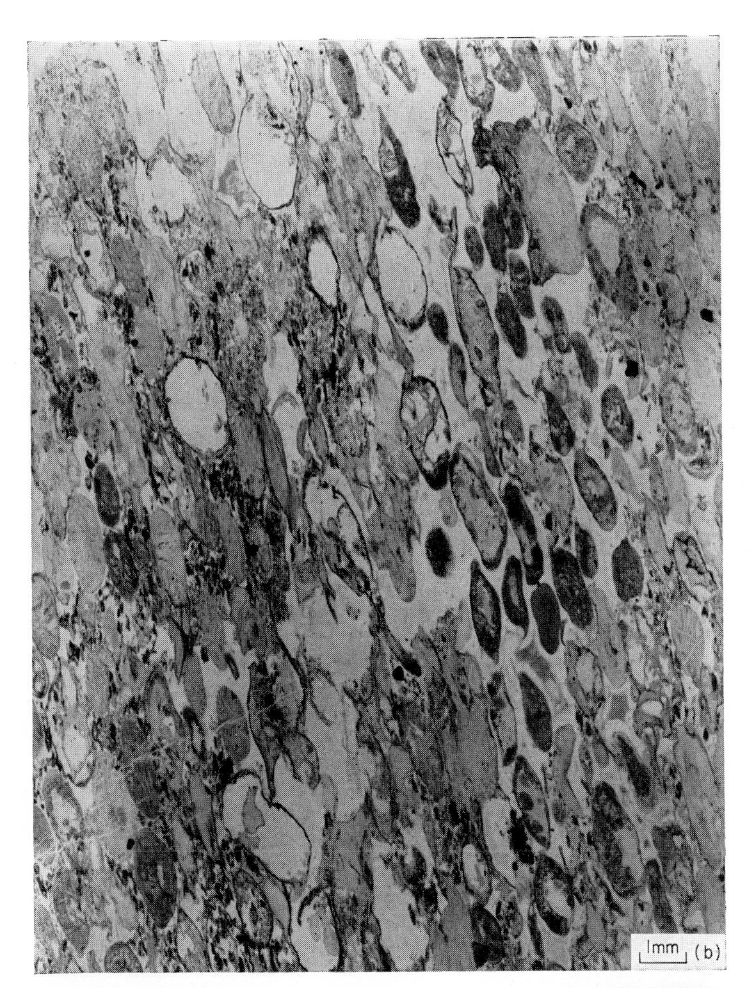

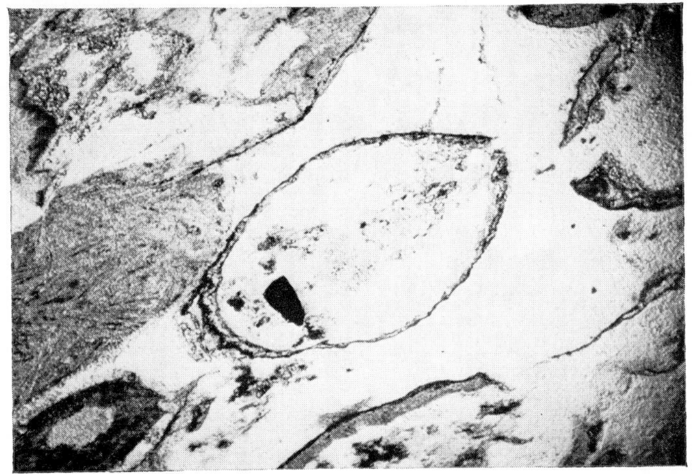

(c)

(a)

Fig. 1.02. Apollo 14 lunar photography of the moon surface. The Apollo 14 Craft was launched from Kennedy Space Centre, U.S.A. at 4.03 pm on January 31 and was recovered from the Pacific Ocean at 4.05 pm on February 9, 1971. (a) One of the Apollo 14 Astronauts photographing a field of boulders located just north-west of the Lunar Module during the Apollo 14 extravehicular activity. (b) The following is a quote by *Astronaut Mitchell* made during the postflight press conference describing the area shown in the photograph: "Here we have progressed up to the rim of Cone Crater. This is not the most numerous of the field nor the largest boulders in the field. However, this boulder, to give you a field for size, is

structure was clearly not as one usually sees in oolites, and many showed a chambered appearance looking distinctly organic. When I returned to London I showed them to several palaeontologists (keeping the age of the rocks dark) and everyone said—yes, organic, but not easily identified specifically, probably algal. Then I revealed the age (at that time believed to exceed 2900 million years now known to be greater than 3100 million years)—they were shocked, to say the least, but none retracted their original opinion. I wrote to Johannesburg to tell them about it, but they did not seem interested (they made no reference in their reply). So I went ahead and published an account (mostly details of the tectonic structure) but included photographs of the supposed organic objects and a short

(b)

about two and a half to three feet high, almost, in diameter. And a thing to point out to you: This photograph is looking almost due north from the most easterly part of our traverse, and the rim of Cone Crater is right there beyond that rock. It is less that 100 feet away. And you cannot see a thousand-foot crater right there. The primary geological objective of the mission was to sample these rocks that are on the rim of Cone Crater, because presumably, as the theory tells us, the largest rocks and the most deep rocks will be found right on the rim of the crater from which they were excavated. And since, to sample the Fra Mauro Formation, we would like to obtain samples from the very bottom of Cone Crater, if possible, those rocks, therefore, would be the oldest and quite representative of the early crust of the moon." Photograph No. 71–H–506–351 issued by National Aeronautics and Space Administration, Washington, C.D. 20546, U.S.A.

discussion of them (Univ. of Witwatersrand, Econ. Geol. Research Unit, Information Circular 14, Structural Investigations in the Barberton Mountain Land, E. Transvaal, June 1963). This was subsequently published in full in the Transactions of the Geological Society of S. Africa (vol. 66, 1963, pp. 353–398, plates I–IV). You can find the references to the objects on p. 361 and Plate I). The research unit must have eventually come around to the ideas of organisms in these rocks for they invited Professor Pflug (Giessen, W. Germany) to look at them, and Barghoorn also visited the area in 1964." (J. G. Ramsey, April, 1972)

(a)

Fig. 1.03. Moon rocks and boulders. (a) Close-up view of a large, interesting boulder in the boulder field located just north-west of the Lunar Module, photographed by one of the Apollo 14 astronauts. There is an apparent contrast in structure between the top of the rock

In a world which professes a considerable interest in its origins, which is forever seeking the meaning for the existence of living things, which spends enormous amounts of money investigating the outer planets with the avowed object of seeking to understand the life process, it is quite incredible how little money has been spent and how little interest there has been in these most remarkable ancient sediments which abound in quantity in Southern Africa. The examination of sediments such as the Fig-Tree and Onverwacht has barely started. They clearly contain information about the very beginnings of life on this planet and that information, although residing perhaps most obviously in the organic remains, must also be entrapped equally in the many inorganic constituents of the rocks. Only the ingenuity of man is required to unlock the secrets. These rocks must contain most of the evidence necessary to enable research workers to draw up ultimately a detailed

(b)

and the lower part. (b) A lunar boulder, with a hammer and small collection bag on the top, to give some indication of size. This view of several boulders clustered together was taken by astronauts of the Apollo 14 mission. Photographs No. 71–H–359 and 71–H–361 issued by N.A.S.A. Washington D.C. 20546, U.S.A.

and precise history of living matter on this planet. At the same time studies of this type should with any luck give firm indications about the true nature and origin of life on this planet at least, and this information in turn may lead to a greater understanding of the role, function and origin of living systems in the universe at large. They may moreover give an insight into extant living systems which can perhaps only be gleaned from looking backwards into the dim past. Certainly there can be no doubt that the book of this particular piece of knowledge has been well written and now only needs to be read and decoded; the tools are available and we believe that the time is now ripe.

There have also been startling advances in our knowledge of the constituents of the universe (Fig. 1.02), of the fundamental matter of which it is composed and the ways in which this matter may be converted into atoms and molecules. Much of the knowledge has come from both astronomical and sub-atomic studies and rational theories are now available which explain both the mechanism of formation of the galaxies and their contained stars, of the elements which compose these bodies and of the energy interconversions which operate at all levels of matter. However it would be over-optimistic if not foolish to suggest that these studies were so far advanced that the road to a rapid increase in knowledge in this field lay clearly ahead. The search for the fundamental particle of all matter has produced some two hundred particles and the number grows daily and somehow one sees the particles stretching infinitely into the distance like droplets shaken from some celestial globule. There does however seem to be order in the particles' nature; they may be grouped into distinct systems and they may be classified and catalogued in a quantitative way, but it is a characteristic weakness of most mathematical concepts that they rarely tell us anything about the reality of materials, only about their behaviour.

We have endeavoured in the chapters which follow to outline the various states of matter from the simplest of particles to the complexities of the living system, and wherever possible to point out the equilibria which appear to exist between all these states and to suggest practical methods for further experimental studies in some of these fields.

SUGGESTED FURTHER READING Chapter 1

Books

Brooks, J., Grant, P., Muir, M. D., Shaw, G. and van Gizjel, P. (Eds) (1971). "Sporopollenin". Academic Press, London.

Cloud, P. E. (Ed.) (1970). "Adventures in Earth History". Freeman, San Francisco.

Faegri, K. and Iversen, J. (1964). "Textbook of Pollen Analysis". Blackwell, Oxford.

Gilluly, J. and Waters, A. C. (1968). "Principles of Geology". Freeman, San Francisco.

Koestler, A. (1959). "The Sleepwalkers". Hutchinson, London.

Raup, D. M. and Stanley, S. M. (1971). "Principles of Paleontology". Freeman, San Francisco.

Rickman, H. P. (1967). "Living with Technology". Zenith Books, New York.

Singer, S. F. (Ed.) (1971). "Is There an Optimum Level of Population?". McGraw-Hill, New York.

Tschudy, R. H. and Scott, R. A. (Ed.) (1969). "Aspects of Palynology". Wiley, New York.

Watson, J. D. (1968). "The Double Helix". Weidenfeld and Nicolson, London.

Articles

Cloud, P. E. (1971). This Finite Earth. *In* "Focus: The Environment". A and S The Review. Indiana University Press.

Polanyi, M. (1967). Life Transcending Physics and Chemistry. *Chem. Engng News*. August 21, 55–66.

Ramsey, J. R. (1963). "Structural Investigations in the Barberton Mountain Land, E. Transvaal". *University of Witwatersrand, Econ. Geol. Research Unit, Information Circular No. 14.*

Shaw, G. (1970). Sporopollenin. *In* "Phytochemical Phylogeny" (Ed. J. Harbone). Academic Press, London.

2

Dating methods

GEOLOGICAL AGE DETERMINATION

The first records of attempts to determine geological ages seem to be those of Xenophanes of Colophon (570–480 B.C.) and the Greek historian Herodotos (484–424 B.C.). Xenophanes noticed that changes in fossil fauna reflected evidence of geological changes. Herodotos (*ca* 450 B.C.), observed that the annual overflow of the River Nile spread a thin layer of sediment over its valley and inferred that the Nile Delta had grown by such annual increments of river-mud deposits during many thousands of years. This hypothesis was confirmed in 1854, when the foundation of the collosal statue of Rameses II at Memphis was discovered under 9 ft of river-laid deposit. Since the statue is known historically to be about 3200 years old, the rate of sediment deposition on top of the statue must therefore have averaged about 3·5 in. each century. Application of the deposition rate estimates to a Nile delta strata 40 ft below the surface at Memphis, where burnt bricks were found, suggested that humans had inhabited the region about 13,500 years ago.

Attempts to define and classify geological time into finite periods were made by Lehmann (1767), who assumed that parts of the original crust of the earth still existed and did not contain evidence for organic life; he designated such rocks at the Primitive Class. Fossil-containing rocks were assumed to have formed by secondary processes from the Primitive Class and were called the Secondary Class, and more recent rocks which contained evidence of abundant organic life were called the Third Class. In 1841 Lyell classified European fossiliferous rocks into Periods and Groups in a form similar to that used at the present time, and these classifications were elaborated by Dana (1880), who divided the major sequences of geological time into Periods and Epoch sub-divisions.

From measurements of the rate of denudation and accumulation of sediments, Williams (1893) introduced the term Geochronology. In this system the standard period of time is represented by the Eocene and defined as a geochrone. Geological time is sub-divided according to the thickness of sediments.

The time scale used by Williams was:

Cambrian	→ Carboniferous	45 geochrones
Triassic	→ Cretaceous	9 geochrones
Eocene	→ Recent	3 geochrones

From these estimates of geological time and rates of deposition of Post-Cambrian sediments, Goodchild (1897) combined values for the rate of sedimentation of the most abundant rocks and obtained a figure of 704 million years for the age of the Post-Cambrian. The rates of sedimentation used were:

1 foot of limestone deposited in 25,000 years
1 foot of shale deposited in 3000 years
1 foot of sandstone deposited in 1500 years

There have been studies on many other deposits for which the rate of decomposition can be determined. The postglacial varved clays show both summer and winter layers, which like the growth rings in trees are seasonal additions. Some of the older sedimentary formations also appear to be seasonally layered as is well shown in the banded anhydrites of the Permian in Western Texas and the Eocene Green River Formation formed by deposition at Lake Uintah in Colorado and Utah and Lake Gosiute in Wyoming. Studies by Bradley on deposition processes of the Green River Formation suggest that the 2600 feet of shale required 6·5 million years to accumulate.

It was hoped that quantitative measurements of the amounts of all the sedimentary rocks of the earth's crust and their rates of deposition from marine and non-marine environments would indicate the degree of erosion and ultimately the age of the geological deposits. Unfortunately, there is no indication that measured modern rates of sedimentation can be assumed to be an average for all geological time. The amounts of sedimentary rocks formed since the beginning cannot therefore be accurately determined since large amounts of these deposits formed during one geological formation will have been destroyed by erosion and by re-working into another formation. The rocks may also have been substantially modified by heating and in the case of most Precambrian formations this may have altered them beyond recognition. The rate of sedimentation therefore is an unreliable measure of geological time, but it does at least indicate that the earth's crust is very ancient.

A totally different attempt to date the earth from the sodium content of the oceans was made by Joly (1899) and Lane (1929). These studies assumed:

(1) that originally the oceans were fresh water and contained no salt (sodium chloride);
(2) that all the salt produced from the earth's crust and carried into the sea remains in solution;
(3) that the rate of addition of salt to the sea remained constant throughout geological time.

The salinity of all the oceans is nearly uniform, hence it was reasoned that given accurate values for the total salt content of the oceans and for the current annual increment of salt, it would be possible to determine the age of the earth's crust.

The results gave a value for the age of the earth of approximately 100 million years, which in 1899 and for some time afterwards seemed reasonable. Further considerations, however, showed that this minimum age was subject to and based on many factors and assumptions such as:

(1) salt entrapped in sedimentary rocks;
(2) contribution from volcanic sources of oceanic salt;
(3) that the primeval oceans were fresh water;
(4) salt is continually mined (14 million tons annually) and artificially returned to rivers and thus to the oceans.
(5) the rate of erosion at the present time may not be representative of geological time.

There are so many uncertainties involved in assessing the salinity history of the oceans that this method of age determination must be regarded as unreliable.

Lord Kelvin attempted to estimate the age of the earth using physical principles and observations. In a series of reports between 1862 and 1897, he estimated that the time required for the earth to have cooled sufficiently to support organic life could not be greater than 100 million years and he finally concluded that the earth was between 20 and 40 million years old.

Radioactivity

A major step forward in geochronology was the discovery of radioactivity at the turn of the present century and this marked the beginning of a new era in geological studies.

Many atoms that occur in nature have unstable nuclei which decay spontaneously to a lower energy state. These atoms are present in radioactive elements and the process of decay is called radioactivity. When a radioactive atom (called the parent atom, P) decays, it changes to another kind of atom (called the daughter atom, D). In alpha decay, the nucleus of the parent atom loses 2 protons and 2 neutrons and the mass number of the element is reduced by 4 and the atomic number decreases by two. In beta decay, the nucleus emits a high-speed electron, one of the neutrons changes into a proton and the atomic number increases by one. In electron capture the nucleus captures an electron which combines with a proton to form a neutron and the atomic number decreases by one. Beta decay and electron capture leave the mass number unchanged.

Since radioactive decay involves only the nucleus of the atom, the rate of disintegration is independent of all chemical and physical conditions, such as temperature, pressure and chemical reactions. This means that each of the radioactive particles of a nuclide has the same chance of decaying or surviving whatever their age may be and whatever their chemical histories. The process of radioactive decay is statistically random.

When a radioactive nuclide undergoes a process of nuclear disintegration, the rate of radioactive decay is proportional to the number (N) of reactant nuclei present:

$$-\frac{dN}{dt} = \lambda N \qquad (1)$$

where λ is a constant representing the probability that an atom will disintegrate in unit time.

The equation relating explicitly the number of radioactive parent atoms remaining after time t is given by integration of equation (1).

$$N = N_0 e^{-\lambda t} \qquad (2)$$

or

$$N_0 = N e^{\lambda t} \qquad (3)$$

where N_0 is the original number of atoms when $t = 0$.

Changes due to rates of decay are usually quoted in terms of half-life, that is the time required for the radioactivity of a substance to be reduced to half its initial value (i.e. $N/N_0 = \frac{1}{2}$).

Thus, from equation (2):

$$N/N_0 = e^{-\lambda t}$$

and given

$$N/N_0 = \tfrac{1}{2}$$

$$\tfrac{1}{2} = e^{-}$$

If τ represents the "half-life" of the radioactive compound then

$$\tfrac{1}{2} = e^{-\lambda\tau}$$
$$2 = e^{\lambda\tau}$$
$$\therefore \quad \tau = \log_e{}^{2/\lambda}$$
$$\tau = \frac{0{\cdot}6931}{\lambda} \tag{3}$$

Equation (3) shows that the half-life (τ) of a radioactive compound is inversely proportional to its decay constant (λ) and is independent of the initial amount of the compound.

Various techniques have been developed to detect and measure the amounts of radiation emitted from radioactive compounds. These methods give a value A which is a measure of the rate of loss of the radioactivity.

$$A \propto -\frac{\mathrm{d}N}{\mathrm{d}t}$$

Therefore from equation (2)

$$A = A_0 e^{-\lambda t}$$

where A is the activity of disintegrating material at time t and A_0 the initial radioactivity.

In geological dating, time must be measured backwards and this has led to the use of the following basic geochronological equation for radioactive dating:

$$\frac{D}{P} = e^{\lambda t} - 1$$

where:

D is the present number of atoms of the daughter nuclide in the sample of rock, mineral or sediment formed since time t;
P is the present number of atoms of parent nuclide in the sample of rock, mineral or sediment;
λ is the decay constant of the parent nuclide;
t is the time measured from the present.
The apparent age (t) of the sample is therefore:

$$t = \frac{1}{\lambda}\log_e\left(1 + \frac{D}{P}\right) \tag{4}$$

The geochronological limits to the use of this equation depend on the relationship between the apparent and true ages of the examined material. The apparent age and true age are generally considered to be equal provided that:

(i) the rock, mineral or sediment has been present in a closed system where there has been no net migration of parent or daughter nuclides into or out of the system during its lifetime;
(ii) the sample is representative of the system being studied;
(iii) the constant λ is accurately known;
(iv) there is no contamination with parent and daughter nuclides during the chemical and physical processes and analysis of the sample;

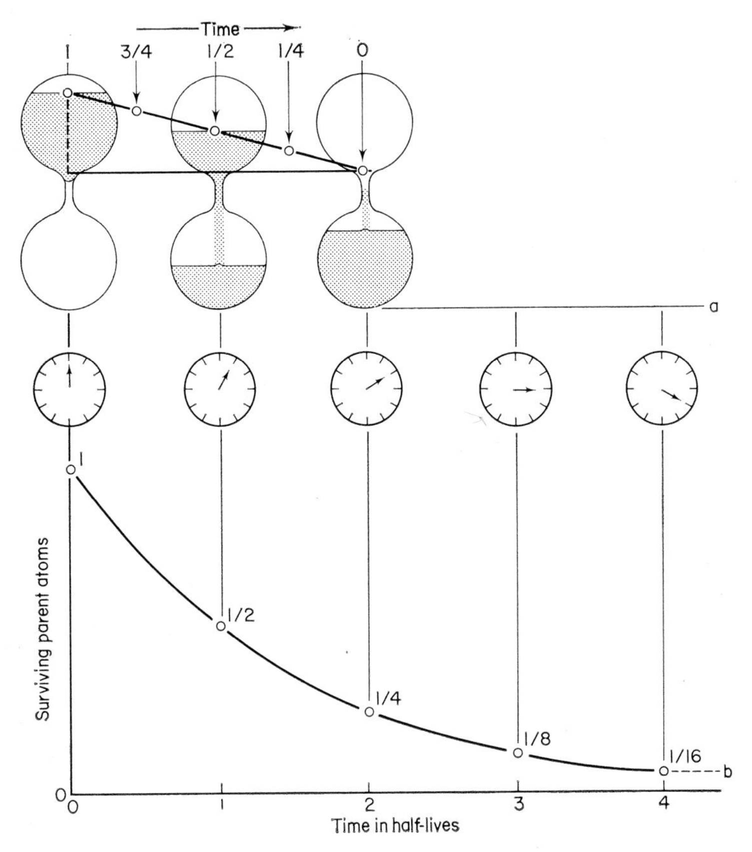

Fig. 2.01. Half-life decay of radioactive material. (a) Normal straight-line relationship for the depletion of everyday processes—where depletion is directly proportional to time. (b) Exponential decay of radioactive material which approaches the zero line (i.e. number of surviving parent atoms = zero) asymptotically. The end of the half-life interval is the beginning of a new interval.

(v) the analysis of parent and daughter nuclide concentrations in the sample is accurate.

Radioactive decay is exponential, so if half a radioactive parent compound ($N_0/2$) remains after one day, half this amount ($N_0/4$) will remain after two days, ($N_0/8$) after three days and so on. The number of remaining parent atoms N_t after "n half-lives" will be $N_0/2^n$. Fig. 2.01 shows the number of surviving radioactive parent nuclides as a function of time and indicates that in radioactive decay the end of one half-life is the beginning of a new one.

At the time of crystallization of a rock, mineral or sediment the radioactive component (P) is concentrated in a particular mineral. With the passage of time the daughter (D) material is continually produced and accumulates. The rate of change ($P \rightarrow D$) can be determined. This is analogous to the gradual running down of a radioactive clock and the amount of D present indicates the time that has passed. It is important that the rate of change ($P \rightarrow D$) is not very great compared with the age of the sample, otherwise the amount of D produced may be too small to be measured even by the most accurate chemical and physical methods.

APPLICATION OF RADIOMETRIC STUDIES TO DATING THE PAST

If the present concentrations of the parent (P) and daughter (D) nuclide can be determined by chemical and physical methods then

Table 2.01. Naturally occurring radioactive elements

Parent P	% Abundance	Daughter D	Decay	Half-life $t_{\frac{1}{2}}$ (in years)
^{238}U	99·27	^{206}Pb	$8\alpha + 6\beta$	$4{\cdot}51 \times 10^9$
^{235}U	0·72	^{207}Pb	$7\alpha + 4\beta$	$7{\cdot}13 \times 10^8$
^{232}Th	100·00	^{208}Pb	$6\alpha + 4\beta$	$1{\cdot}39 \times 10^{10}$
^{187}Re	62·93	^{187}Os	β	$\sim 5{\cdot}0 \times 10^{10}$
^{176}Lu	2·59	^{176}Tb ^{176}Hf	electron capture β	$2{\cdot}4 \times 10^{10}$
^{147}Sm	15·09	^{143}Nd	α	$1{\cdot}25 \times 10^{16}$
^{138}La	0·089	^{138}Ba ^{135}Ce	electron capture β	$\sim 7{\cdot}0 \times 10^{10}$
^{115}In	95·77	^{115}Sn	β	$6{\cdot}0 \times 10^{14}$
^{87}Rb	27·85	^{87}Sr	β	$4{\cdot}70 \times 10^{10}$
^{40}K	0·0119	^{40}Ar ^{40}Ca	electron capture β	$1{\cdot}33 \times 10^9$

knowing the rate (λ) at which D has been formed, an age can be determined.

Of the many radioactive nuclides that occur in nature (Table 2.01) only four (^{235}U, ^{328}U, ^{87}Rb and ^{40}K) (Table 2.02) have proved useful

Table 2.02 The chief methods of radiometric age determination

Parent Nuclide P	Daughter Nuclide D	Half-life $t_{\frac{1}{2}}$ (in years)	Minerals normally studied
Uranium ^{238}U	Lead ^{206}Pb	$4{\cdot}51 \times 10^9$	Zircon Uraninite Pitchblende
^{235}U	^{207}Pb	$7{\cdot}13 \times 10^8$	Zircon Uraninite Pitchblende
Rubidium ^{87}Rb	Strontium ^{87}Sr	$4{\cdot}70 \times 10^{10}$	Muscovite Biotite Lepidolite Microcline Glauconite Whole metamorphic rock
Potassium ^{40}K	Argon ^{40}Ar	$1{\cdot}33 \times 10^9$	Muscovite Biotite Hornblende Glauconite Sanidine Whole volcanic rock

in radiometric age determinations of ancient rocks. The other naturally occurring radioactive nuclides are not generally used for radiometric dating because:

(1) they have too slow a rate of decay;
(2) they have a very rapid rate of decay (these materials however have found special applications in dating very recent sediments);
(3) their occurrence is rare.

Uranium–Lead Method

The major Uranium isotopes are ^{238}U, ^{235}U and ^{234}U and all naturally occurring uranium contains both ^{238}U and ^{235}U radioactive nuclides. The natural abundances of these isotopes are:

$$^{238}U = 99{\cdot}2739 \pm 0{\cdot}0007\%$$
$$^{235}U = 0{\cdot}7204 \pm 0{\cdot}0007\%$$
$$^{234}U = 0{\cdot}0057 \pm 0{\cdot}0002\%$$

In addition most uranium minerals also contain radioactive thorium (^{232}Th; $t_{\frac{1}{2}} = 1{\cdot}39 \times 10^{10}$ years). Radioactive ^{238}U, ^{235}U and ^{232}Th decay to the stable isotopes of lead, ^{206}Pb, ^{207}Pb and ^{208}Pb respectively and are generally used in geochronological radiometric studies, as the standard method to which other methods may be compared. Thus $^{235}\text{U} \rightarrow {}^{207}\text{Pb} + 7\,{}^{4}\text{He}$ ($t_{\frac{1}{2}} = 0{\cdot}71 \times 10^9$ years) and $^{238}\text{U} \rightarrow {}^{206}\text{Pb} + 8\,{}^{4}\text{He}$ ($t_{\frac{1}{2}} = 4{\cdot}51 \times 10^9$ years).

Uranium–lead radiometric dating was originally applied to uranium minerals such as pitchblende (a complex silicate) and uraninite. However, these minerals are of rare occurrence and their study was limited. The development of sophisticated and delicate analytical techniques has enabled accurate determinations of trace amounts of uranium and lead in minerals to be made. The increased potential of uranium–lead studies resulted from the discovery of the mineral zircon ($ZrSiO_4$), which is of widespread occurrence in igneous rocks, and contains trace amounts (1000 ppm) of uranium. The advantages of using zircon are threefold:

(1) Zircons are ubiquitous in igneous rocks, whereas uraninite and pitchblende are generally restricted to pegmatite dikes;
(2) zircons are less subject than are common uranium minerals to chemical changes during geological alterations;
(3) there is little likelihood of lead incorporation into zircon mineral crystals when they were formed, or of lead being added from underground solutions, since the lead atom which is about 50% larger than the zirconium atom, could not fit readily into the tightly packed lattice, and also lead is normally bivalent whereas zirconium is tetravalent, and this would lead to difficulties in chemical bonding.

Age determinations of rocks using uranium and thorium radioactive decay into lead can be calculated directly from the general law of radioactive decay (equation (4)) in which uranium and thorium constitute the parent atoms *P* and lead the daughter atoms *D*. The increase of daughter atoms ^{206}Pb, ^{207}Pb and ^{208}Pb can be determined using equations similar to those for the increase of ^{206}Pb.

$$\underset{\text{determined}}{^{206}\text{Pb}} = \underset{\text{original}}{^{206}\text{Pb}} + {}^{238}\text{U}\,(e^{\lambda t} - 1)$$

dividing by the non-radiogenic ^{204}Pb

$$\underset{\text{determined}}{\frac{^{206}\text{Pb}}{^{204}\text{Pb}}} = \underset{\text{original}}{\frac{^{206}\text{Pb}}{^{204}\text{Pb}}} + \frac{^{238}\text{U}}{^{204}\text{Pb}}\,(e^{\lambda t} - 1) \qquad (5)$$

A plot of equation (5) with $^{206}Pb/^{204}Pb$ as ordinate values and $^{238}U/^{204}Pb$ as abscissa gives a line relationship, whose slope represents $(e^{\lambda t}-1)$. Thus, radiometric ages can be calculated from:

$$\frac{^{206}Pb}{^{238}U} = e^{\lambda 238 t}$$

$$\frac{^{207}Pb}{^{235}U} = e^{\lambda 235 t}$$

$$\frac{^{207}Pb}{^{206}Pb} = \frac{e^{\lambda 235 t}}{k e^{\lambda 238 t}} \qquad \left(k = \frac{^{238}U}{^{235}U} \text{ today}\right)$$

$$\frac{^{208}Pb}{^{232}Th} = e^{\lambda 232 t}$$

where k is the present value $\frac{^{238}U}{^{235}U}$ when the radiation is due to uranium $k = 2632$ ($k = 2013$ for thorium and $k = 2485$ when both U and Th are present in equal amounts). The age (A) of a rock, mineral or sediment is then calculated from:

$$A = \frac{Pb^{x} \times k}{\alpha} \tag{7}$$

where:

A is the age in millions of years;
Pb^{x} is the amount of lead in parts per million;
k is a constant representing the rate at which α-particles are emitted from the uranium or thorium;
α is the observed count of α-particles per mg of sample per hour.

If the sample has remained in a closed geological system throughout its history, then all the uranium and thorium radiometric ages should be identical or *concordant*. If the ages do not agree, they are *discordant*.

Discordant Ages

Daughter atoms may escape from minerals which have been heated or otherwise disturbed, if poorly included in the crystal lattice whilst the radioactive parent atoms are more easily accommodated. This means that discordant radiometric ages tend to be less than the correct values. The discordant ages also show some apparent regularities which can be related to fundamental primary processes which have operated in different degrees and at uniform rates since the mineral was first formed. Examination of very old rocks suggest that a continuous diffusion of lead from mineral systems has occurred through time. It has been shown for example that discordant uranium–lead ratios for ancient rocks from five continents are similar, which suggests a constant loss of lead. It is not yet understood how important and widespread these diffusion losses are, but it is clear that lead may be lost from mineral systems by various processes.

Radiometric age determinations are made by selecting a fresh sample of zircon-bearing igneous rock which is crushed until the mineral grains separate and these are removed and freed of matrix and fragments of other minerals. A small sample (usually about 50 mg) of pure zircon is used as a control. The amount of lead in the mineral sample is determined using a mass spectrometer and the number of α-particles emitted per hour is measured by a Geiger counter.

Common Lead Method

Studies of lead isotopes indicate that three types can be identified in geological samples:

(1) *Primeval lead* ^{204}Pb assumed to have been present during the early stages of formation of the earth and to have a homogeneous distribution so that lead from any given area had the same isotopic composition.

(2) *Radiogenic lead* (^{206}Pb, ^{207}Pb, ^{208}Pb). Formed by the radioactive decay of uranium or thorium.

(3) *Common lead* differs from primeval lead in containing additional radiogenic lead. Of the ^{204}Pb, ^{206}Pb, ^{207}Pb and ^{208}Pb isotopes only the 204 isotope has not been produced in significant amounts by radioactive processes since the solidification of the earth.

Since the formation of the earth it is calculated: (1) that about 20% of ^{232}Th has decayed to ^{208}Pb; (2) 50% of ^{238}U has decayed to ^{206}Pb and (3) 99% of ^{235}U has decayed to ^{207}Pb. If we consider a section of rock, mineral or sediment from part of the earth's crust then it should contain a proportional share of primeval lead, uranium and thorium. Since at the time of the earth's crust formation there was no radiogenic lead present, ^{206}Pb, ^{207}Pb and ^{208}Pb must have been formed by radioactive decay of uranium and thorium. When common lead is removed from a single, constantly evolving U–Th–Pb environment it is called a single-stage lead and examination of its isotopic composition provides the basis of the common lead method of dating.

Unfortunately numerous geological events can confuse these observations. If a single stage lead deposit passes through crustal rocks and becomes contaminated by the addition of lead from other environments which have a totally different radiogenic history, the resulting lead will give ages which are much younger. Other leads which may have been geologically isolated at an early time could have been reworked and redeposited much later. If a sample contains no extraneous radiogenic lead, its calculated age would be that of the initial material, although the age of the associated deposit determined independently would be much younger.

The main limitation of the common lead method for geological dating is that geological events that have disturbed and altered a mineral cannot be recognized during the isotopic studies without independent geological evidence. Thus a wrong geological history interpretation of the sediment will lead to serious misinterpretations of the common lead method.

Rubidium–Strontium Method

Radioactive rubidium, (^{87}Rb) which constitutes 28% of naturally-occurring rubidium, decays by β-decay emission to strontium (^{87}Sr). Rb–Sr ages are determined from the isotope ratios in rubidium-bearing minerals such as mica and potassium feldspars and the results can therefore be compared directly with potassium–argon (K–Ar) determination on the same samples. Rb–Sr dating methods have been extensively used, because in addition to igneous rocks, they can be applied both to metamorphic rocks and to glauconite minerals in sedimentary rocks. Rb–Sr dating is not so widely used however as the K–Ar method.

The age of minerals which are Rb-rich and contain no common strontium can be calculated directly by determination of the total strontium content (i.e. ^{87}Sr), but in the majority of minerals common strontium is present in significant amounts and special analytical methods have been developed to determine the amount of radiogenic strontium in the presence of common ^{87}Sr.

The radioactive decay ($^{87}\text{Rb} \rightarrow {}^{87}\text{Sr}$) is a very slow process ($t_{\frac{1}{2}} = 50 \times 10^9$ years) compared with the age of the earth ($4{\cdot}5 \times 10^9$ years) and the law of radioactive decay (equation (2)) can be written as:

$$N = N_0{}^{-\lambda t}$$

where:

N_0 is the number of ^{87}Rb atoms decayed
N is the number of ^{87}Sr atoms formed
t is the age in years

Since ^{87}Rb and ^{87}Sr have the same mass number then:

$$\underset{\text{(age in years)}}{t} = \frac{\%\ \text{radiogenic}\ ^{87}\text{Sr}}{\%\ ^{87}\text{Rb} \times \lambda} \tag{8}$$

Strontium is an abundant element in nature and common strontium usually occurs in rubidium-bearing minerals and if it remains undetected will distort the calculated age to a greater value. Fortunately, original strontium always contains some ^{86}Sr, which is not produced by radioactive decay. Thus the relative amount of ^{86}Sr can be used to estimate the original ^{87}Sr present in the mineral. In these analyses two types of related rock sample are required:

(1) a rock which is extremely rich in Sr and poor in Rb provides the ^{87}Sr/^{86}Sr ratio for common Sr in the system.
(2) a rock which is very rich in Rb provides the ^{87}Rb/^{87}Sr ratio.

From the amount of ^{86}Sr in the Rb-rich rock sample, the amount of ^{87}Sr can be determined in the original and in the radiogenic samples.

Rb–Sr dating has been applied with success to metamorphic rocks, and unless there is geological evidence to the contrary, it is generally assumed that during a metamorphic event there is a complete homogenization of strontium isotopes. In such a case, any ^{87}Sr that may be present is diluted by the common rock strontium and the new ratio ^{87}Sr/^{86}Sr will be slightly greater than that prior to the metamorphic event. The time at which the metamorphosis occurred is given by the amount of ^{87}Sr that has subsequently been produced by the decay of ^{87}Rb.

If sediments could be dated using the Rb–Sr method it would be possible to produce an accurate geological time scale in relation to fossil remains, but at the moment the dating of whole rock sediments and metasediments is in a preliminary stage. Radiogenic ^{87}Sr present in sediments is distributed between that contained in detrital grains and that formed in authigenic minerals developed in sediments at the time of deposition or soon afterwards. Thus, coarse-grained sediments would tend to retain their own radiogenic ^{87}Sr, while fine-grained sediments such as shales, clays and mudstones would be subject to a more homogeneous distribution of Sr isotopes. As the sediments are produced from a source rock, chemical and physical alterations may lead to the selective removal of Sr and Rb during transportation. Under these conditions the Sr-isotope abundance and concentration of Sr in the source material and the resultant sedimentary deposits, may be very different. An interesting example shows that whereas Sr may tend to co-precipitate with calcium carbonate, Rb is geochemically much more mobile and during sedimentations is more likely to be fixed by absorption on fine-grained particles or trapped in the pore spaces of sediments.

Although there are numerous problems bearing on radiogenic Rb–Sr studies on sedimentary rocks which tend to give higher ages, if the sediment is young in origin or where there has been a selective enrichment of Rb relative to Sr during weathering, transportation and sedimentation, the maximum radiogenic age may be close to the real age.

Some igneous and metamorphic rocks contain no minerals with particularly high Rb to Sr ratios and other samples are so fine grained that such minerals cannot generally be separated for study. In these cases the whole rock method has been developed in which several rock samples are analysed without separating the mineral components. Whole pieces of rock from different parts of the rock body normally differ in Rb content, and the $^{87}Sr/^{86}Sr$ ratio of each sample is plotted against its $^{87}Rb/^{86}Sr$ ratio.

Hypothetically, when the rock originally crystallized, different parts of it, regardless of Rb concentration, would have the same $^{87}Sr/^{86}Sr$ ratio and this would produce a horizontal line relationship. Gradually during the radiogenic history of the rocks, ^{87}Rb would be lost in proportion to the Rb concentration in each part, and this corresponds to the amount of radiogenic ^{87}Sr gained. As the ratio $^{87}Sr/^{86}Sr$ changed in each part of the rock, the slope of the line increases progressively indicating a measure of the age of crystallization of the rock samples. The intercept of the line at the ordinate indicates the isotopic composition of common strontium at the beginning, which can be applied in equation (8) for the age determination of the rocks.

Potassium–Argon Method

All naturally occurring potassium contains three isotopes ^{39}K, ^{40}K and ^{41}K of which ^{40}K ($t_{\frac{1}{2}} = 1{\cdot}3 \times 10^9$ years) is naturally radioactive. Because potassium is abundant and widespread in the earth's crust, ^{40}K has great age-dating potential for many different types of rocks. ^{40}K decays to two daughter products, calcium (^{40}Ca) 89% and argon (^{40}Ar) 11%. The dual decay of ^{40}K theoretically provides two separate dating methods, but in practice, however, the K–Ca method is of no use since calcium is extremely abundant and widespread and 97% of common calcium is ^{40}Ca. Argon as an inert gas is never bound chemically in the lattice of a newly-formed potassium mineral, but significant amounts of original argon can be trapped mechanically in the lattice structure of certain silicate minerals, and this cannot be distinguished from radiogenic argon. Biotite, muscovite and hornblende, however, rarely contain significant original argon, and these minerals virtually always give reliable minimum K–Ar ages. The chief drawback to K–Ar dating methods is that argon gas diffuses easily from mineral crystals when they are heated to relatively low temperatures or subjected to geological stress for extended periods of time. Thus, K–Ar ages for extrusive igneous rocks or rocks that have never been deeply buried may coincide with the actual age. For rocks that have been deeply buried, K–Ar dates probably indicate the last time the rocks cooled below about 300°C and must therefore be considered only as minimum ages. When K–Ar dating has been applied to rock masses that have been independently dated by U–Pb ratios, the age derived, especially from micas agrees well with those determined from U–Pb radiogenic analysis. This suggests that argon does not escape from the space lattice of the micas, and generally only this type of mineral is now used in K–Ar radiogenic analyses.

The radioactive decay ($^{40}K \rightarrow {}^{40}Ca$) involves emission of a β-particle and ($^{40}K \rightarrow {}^{40}Ar$), K-capture to an excited state followed by emission of β-radiation to the ground state. Although the decay is associated with two partial decay constants λ_β and λ_K, ^{40}K has only one half-life given by:

$$t_{\frac{1}{2}} = \frac{0{\cdot}693}{\lambda_\beta + \lambda_K} \text{ and geological age } t = \frac{1}{\lambda_\beta + \lambda_K} \log_e\left(1 + \frac{\lambda_\beta + \lambda_K}{\lambda_K} \cdot \frac{^{40}Ar}{^{40}K}\right),$$

because the half-life ($t_{\frac{1}{2}}$) is related to the total disappearance of the parent nuclide and is independent of the mechanism by which it decays. For the purpose of geological age determinations the factor for β-emission, which accounts for no more than 2% emission direct to the ground state, is ignored and the specific activity is taken as equal to the rate of electron capture. The natural abundance for ^{40}K is known with some certainty to be $0{\cdot}0119 \pm 0{\cdot}001\%$ and is considered to have no significant effect upon radiogenic age determinations, except in the case of very ancient rocks.

If an igneous rock has not been subject to a later metamorphic event then the K–Ar age of a mineral will be that of the time of emplacement provided that argon has not been lost through diffusion. In the case of young intrusions, some time may lapse before the intrusion has cooled to a point where argon no longer escapes. An emplacement age is one that marks the time at which the diffusion rate does not exceed that of radiogenic argon accumulation. In cases of complete recrystallization, resulting from later metamorphic events, previously accumulated argon is completely lost and the metamorphic event is dated.

In the dating of sediments by the K–Ar method, whole rock samples of sediments are occasionally used, but it is more common to use a separated mineral phase. In each case it is essential that the samples retain radiogenic argon and do not contain fragments of detrital minerals that may contain relict argon. Whole rock samples of metamorphosed sediments such as slates may give acceptable ages provided that these requirements are valid. In addition, apart from true sediments, volcanic ash intercalated between sediments may be used at the time of sedimentation.

During recent years there has been great interest in correlating radioactive age measurements with those of the accepted fossil stratigraphic scale. In general, only approximate ages can be determined by means of intrusive rocks, and an accurate measurement requires the dating of geological formations that contain recognized faunas and floras. Authigenic minerals, such as glauconite, which are formed at the water–sediment interface, which occur with fossil remains and which are widely distributed in both space and time, have proved to be most suitable.

Carbon-14 Method for the Recent Geological past

Naturally-occurring carbon in the atmosphere and in living plants and animals consists of two stable isotopes ^{12}C and ^{13}C with a natural abundance of 98·892 and 1·108%, respectively, whilst a third radioactive isotope ^{14}C is formed by the reaction of cosmic ray particles (thermal neutrons) with ^{14}N in the upper atmosphere. The abundance of ^{14}C in modern wood is 0·000000000107% and it has a half-life of 5730 years which is so low that it is not generally measurable in organic material older than about 40,000 years. Because of the short half-life no existing isotope can be primordial, but instead it is continually being created in the upper atmosphere about 10 miles above the surface of the earth as a by-product of cosmic-ray bombardment. This continually new created ^{14}C is quickly incorporated into carbon dioxide, and thus enters into the terrestrial carbon cycle and is distributed throughout the various carbon phases in a homogeneous manner. It has been estimated that the time for complete mixing of the isotopes is generally less than the half-life.

The age of a carbon sample is related to the calculated radioactive decay by equation (2) ($N = N_0 e^{-\lambda t}$), where N_0 is the activity of ^{14}C present in the sample at the time at which the sample was removed from the natural carbon cycle. For a living organism this would date the time at which it died.

N is the activity of ^{14}C at time t
λ is the decay coefficient

^{14}C isotopes eventually decay to ^{14}N. However, the age of carbon-bearing material is not determined from the normal parent–daughter ratio, but from the ratio of ^{14}C to all other carbon in the sample. ^{14}C, at least until the arrival of man-made thermonuclear explosions, was in equilibrium in the atmosphere; the rate of production was equal to the rate of decay. A plant which removes CO_2 from the atmosphere receives a proportional share of ^{14}C. When the plant dies, it ceases to absorb CO_2, and with time the proportion of ^{14}C progressively decreases. The ^{14}C-method therefore depends on the assumptions:

(1) that the rate of ^{14}C production in the upper atmosphere is nearly constant;
(2) that the rate of assimilation of ^{14}C into living organisms is rapid relative to the rate of decay. These assumptions appear to be valid.

Radiocarbon dating can only be used for the last brief portion of geological time, but has become a most important radiometric method since a great deal of geological activity has occurred within this short time span. It was not until the full potential of ^{14}C-dating was applied that the full extent of these activities were fully appreciated.

Radiocarbon methods have been extensively applied to archaeology and anthropology. Radiocarbon dating helped to disprove the famous Piltdown skull and jaw "hoax" by showing that the skull gave an age of 620 $\pm$ 100 years, whilst the jaw was only 500 $\pm$ 100 years old. More recent applications of ^{14}C dating include the rate of sedimentation of ocean sediments, late Pleistocene geology and climate and also the terrestrial age of meteorites. Other major geological events such as the retreat of the last continental ice sheets, accompanying climatic changes, changes in ocean circulation, the postglacial rise in sea-level and the rise in human civilization have all been studied using radiocarbon dating techniques.

Cosmic Ray Exposure Ages

The so-called cosmic rays are mainly high speed free atomic nuclei. The primary particles consist of 89% hydrogen nuclei (protons), 10% helium nuclei (α-particles) and 1% of other nuclei with atomic numbers greater than 3. They include element nuclei up to and including tin. They originate in part in the sun, but mostly they appear to come from beyond the solar system.

The high energy protons are able to transform elements in many ways. Thus iron may be converted into chlorine:

$$^{56}Fe + {}^{1}H \rightarrow {}^{36}Cl + {}^{3}H + 2\,{}^{4}He + {}^{3}He + 3\,{}^{1}H + 4 \text{ neutrons}$$

The protons cannot penetrate much beyond about 1 m of rock, so that most planetary material will be protected from attack. However small meteorites in space are clearly highly vulnerable. Measurements of quantities of new and frequently rare nuclides which have been produced by cosmic ray activity should give a measure of the time from which exposure first occurred. If the meteorite had been part of a larger (cosmic ray attack free) planetary body then the cosmic ray exposure age should indicate the date when the planet (or asteroid) disintegrated. Consider two nuclides, one radioactive and one stable, e.g. Argon ^{38}Ar (stable) and ^{39}Ar (radioactive $t_{\frac{1}{2}} = 325$ years). Two reactions may be considered:

(1) Formation of ^{39}Ar by cosmic ray attack
(2) Decay of ^{39}Ar.

After a few half-lives and assuming that the half-life is small compared with the ages considered, then a steady state will be reached where the ^{39}Ar is decaying as fast as it forms. This will give a maximum steady state concentration of ^{39}Ar, which cannot be exceeded, whereas ^{38}Ar will accumulate. The rate of decay of the ^{39}Ar in a particular meteorite will therefore ultimately equal its rate of formation. At the same time relative rates of production of ^{38}Ar and ^{39}Ar may be determined in model experiments in targets exposed to cyclotron beams. Finally from a measure of the actual amount of ^{38}Ar which has accumulated in the specimen the cosmic ray exposure age can be calculated. The results are interesting and indicate low ages between about $5–20 \times 10^6$ years for bronzite and hypersthene chondrites compared with about 7×10^8 years for iron meteorites. One could assume that the results suggested that the material from which the specimens came broke up about 5–20 million years ago in the one case and about 700 million years in the other, the implication being the existence of at least two planetary bodies or alternatively several large asteroids or both. On the other hand, the differing degrees of toughness of the various meteorites materials would suggest that some were more readily broken than others. The results clearly must be interpreted with care, but they go some way to supporting a larger body origin of the meteorites.

SUGGESTED FURTHER READING Chapter 2

Books

Eicher, D. L. (1968). "Geologic Time". Prentice-Hall, New York.

Faul, H. (1966). "Ages of Rocks, Planets and Stars". McGraw-Hill, New York.

Hamilton, E. I. (1965). "Applied Geochronology". Academic Press, London.

Hamilton, E. I., and Farquhar, R. M. (Eds) (1968). "Radiometric Dating for Geologists". Wiley-Interscience, New York.

Kirkaldy, J. F. (1971). "Geological Time". Contemporary Science Paperbacks.

Rankama, K. (1954). "Isotope Geology". Pergamon Press, Oxford.

Rankama, K. (1963). "Progress in Isotope Geology". Interscience Publishers, New York.

Zeuner, F. E. (1958). "Dating the Past". 4th edition. Methuen, London.

Articles

Allsopp, H. L., Ulrych, T. J. and Nicolaysen, L. O. (1968). Dating Some Significant Events in the History of the Swaziland System by the Rb–Sr Isochron Method. *Can. J. Earth Sci.* **5,** 605–619.

Black, L. P., Gale, N. H., Moorbath, S., Pankhurst, R. J. and McGregor, G. (1971). Isotope Dating of Very Early Precambrian Amphibolite Facies Gneisses from the Godthaab District, West Greenland. *Earth planet. Sci. Lett* **12,** 245–259.

Bradley, W. H. (1964). Aquatic Fungi from Green River Formation of Wyoming. *Am. J. Sci.* **262,** 413–416.

Clifford, T. N. (1969). Radiometric Dating and Pre-Silvrian Geology of Africa. *In* "Radiometric Dating for Geologists" (Eds E. I. Hamilton and R. M. Farquhar), 299–416. Wiley-Interscience, New York.

Gale, N. H., Grasty, R. L. and Meadows, A. J. (1966). A Potassium–Argon Age for the Barwell Meteorite. *Nature, Lond.* **210,** 620.

Hurley, P. M., Pinson, W. H., Nagy, B. and Teska, T. M. (1972). Ancient Age of the Middle Marker Horizon, Onverwacht Group, Swaziland Sequence, South Africa. *Earth planet. Sci. Letters* **14,** 360–366.

Holmes, A. (1959). A Revised Geological Time-scale. *Trans. Edinburgh geol. Soc.* **17,** III, 183–216.

Moorbath, S. (1971). Measuring Geological Time. *In* "Understanding the Earth" (Eds I. G. Gass *et al.*). Open University Set Book. Artemis Press, Horsham, Sussex.

Tilton, G. R. and Steiger, R. H. (1965). Lead Isotopes and the Age of the Earth. *Science, N.Y.* **150,** 1805–1808.

Ulrych, T. J., Burger, A. and Nicholaysen, L. O. (1967). Least Radiogenic Terrestrial Leads. *Earth planet. Sci. Letters* **2,** 179.

3

The Constituents of the Universe

INTRODUCTION

Imagine a globule of liquid suspended in space. When shaken it breaks up into several smaller globules. When shaken more vigorously the number of globules increases as their size decreases, moreover there are relationships between the amount of energy imparted and the number and size of the smaller globules produced for a given liquid. When the shaking stops the globules will tend to aggregate once more to form larger globules so the whole process is in a sense more or less reversible. This is about as close as one can perhaps hope to get to a simple model of both the Universe and especially of the elementary particles of which it is composed.

The search for the ultimate indivisible constituents of matter has engaged man's attention since the early Greek civilizations and with the discovery of the electron (Stoney, 1881) and the proton (Rutherford, 1920) the search appeared to be largely at an end. However, since that time the number of known particles now exceeds 200 and new ones are being discovered with monotonous regularity. It is beginning to look as if one can have almost as many particles as one wishes, by shaking our hypothetical liquid more and more vigorously, although with a real liquid one would expect some sort of limiting figure. The hypothetical liquid has a name—the luminiferous ether—the medium through which electromagnetic radiation was once thought and required to travel. However, all attempts to measure specific properties of the ether have so far failed. At the same time it has been found possible to a limited degree to dispense with the ether concept and to explain or describe many natural phenomena and elementary particle relationships using various mathematical systems, especially the quantum mechanics and group theory. It must be realized, however, that there is still no evidence either for or against the existence of an ether and in many ways the mathematical treatments merely provide a somewhat roundabout way of avoiding the necessity for such a medium. Thus one can measure vibrations and resonances in any medium without necessarily knowing about the medium itself. One can also, for example, describe the forces occasioned by a woman pushing a perambulator down a street in a multitude of terms which will give precise details about the movement of the objects in space, although they may tell you nothing at all about either the woman or the perambulator.

An ether is therefore certainly a handy thing to have if you wish to obtain some sort of mental picture (albeit maybe illusory) of the physical nature of the Universe and its constituents, something rarely provided by mathematicians. It is important of course to be careful not to apply too rigid a definition to the medium but to allow it to take up such properties (or non-properties) as are required by the observed facts. With these points in mind the various elementary or fundamental particles are seen to be alternative condensed and rarefied portions of ether, but such rarefactions (holes in the ether) and concomitant condensed regions are only produced in discrete sized units according to the quantum theory and the production of one particle is immediately accompanied by the simultaneous production of the other (anti-particle) at some other part (not necessarily adjacent) of the ether. When a particle and an anti-particle meet, the condensed region falls rapidly (exponentially) into the "hole" which may be stepped into different levels. The result is one or other unstable association (e.g. a neutron) or at the bottom of the hole a ripple of energy. Alternatively the process may be reversible and the unstable particle (neutron) or, with increasing difficulty, a ripple of energy may under the right conditions result in the formation of two particles. A photon for instance can yield an electron and a positron, a neutron can yield a proton and an electron, and so on. Gravity is readily seen to be the weak attraction of one piece of "warped" ether (region of much condensed plus rarefied ether) for another placed some distance away, whereas magnetism results from such warped ether (e.g. electrons) in movement relative to other warped ether. On earth increasing depth should therefore lead to an increase in magnetic force since there will be concomitant increase in the "warpedness" of the ether, which must of course be moving (see also Chapter 9), and this leads almost inevitably to a mobile (presumably liquid) core. Stronger interactions only occur when the warped ether regions come close together. An analogy would be a comparison with a ball rolling down a hill to a cliff edge.

It is of particular importance to know why certain particles, e.g. electrons and protons and their derived neutrons, appear (at least in our observable Universe) to be more abundant than others (positrons or anti-protons) in spite of their apparently having no special importance in the particle hierarchy. Our ether concept readily allows us to "explain" even this. We simply have to say that it is a property of the ether to produce preferentially certain sized holes and related condensed regions in much the same way as our globule of liquid when shaken will tend to produce certain preferential sizes of smaller globules according to its various properties, such as for instance in this case its molecular structure and surface energy. The most important property of all space, however, is its propensity to favour the formation of the remarkable complex of materials which we know as the living system.

ELEMENTARY OR FUNDAMENTAL PARTICLES

The known so-called elementary particles, the vast majority of which are produced artificially or found as constituents of cosmic radiation, appear to fall into two main groups characterized essentially by their masses and by their particular type of interaction reactions, which may be weak (e.g. decay reactions), strong (e.g. nucleon collision reactions), electromagnetic and gravitational. The strengths of these interactions differ enormously. Thus if the strong interaction is taken as unity the ratio of electromagnetic interaction: weak interaction: gravitational interaction is $10^{-2} : 10^{-14} : 10^{-39}$. Also the strong interactions, although so much more powerful than the others are restricted to a very short distance (10^{-13} cm $\equiv$ 1 fermi), after which it falls off exponentially. This distance is roughly the size of the atomic nucleus and it is of course the strong interaction which provides the force to bind elementary particles into a nucleus. The electromagnetic interaction has a longer range and the rate of fall-off follows the inverse square law, so that at large distances this becomes all important as is the case in atomic physics. Weak interactions are short-range forces, and only observable because certain transitions are forbidden by the laws of strong interactions. The sort of weak interactions includes the decay reactions of elementary particles with lifetimes varying from 10^{-10} to 10^{3} sec. The gravitational interaction is, however, also a long-range force but much weaker than the electromagnetic force and in micro situations virtually unobservable, whereas in the macro situation the cancellation of charges which is the norm allows detection of

Table 3.01. Some known meson particles

Particle[1]	Mass (MeV)	Mean Life (sec)	Width (MeV)	Principal Decays Partial mode	% Fraction
π: $\pi^{\pm}$	139·58	$2{\cdot}604 \times 10^{-8}$		$\mu\nu$	100
				$e\nu$, $\mu\nu\gamma$ etc.	approx. 10^{-4}
π^0	134·97	$0{\cdot}89 \times 10^{-16}$		$\gamma\gamma$	98·83
				γe^+e^-	1·17
				$\gamma\gamma\gamma$, $e^+e^-e^+e^-$	approx. 10^{-5}
K: $K^{\pm}$	493·82	$1{\cdot}234 \times 10^{-8}$		$\mu\nu$	63·47
				$\pi\pi^0$	20·84
				$\pi\pi^-\pi^+$	5·54
				etc.	
K^0, $\bar{K}^0$	497·76	K^0_S $0{\cdot}862 \times 10^{-10}$		$\pi^+\pi^-$	68·4
				$\pi^0\pi^0$	31·6
		K^0_L $5{\cdot}29 \times 10^{-8}$		$\pi^0\pi^0\pi^0$	25·5
				$\pi^+\pi^-\pi^0$	12·1
				$\pi\mu\nu$	27·3
				$\pi e\nu$	35·2
η	548·8		$2{\cdot}3 \times 10^{-3}$	$\gamma\gamma$, $\pi^0\gamma\gamma$, $3\pi^0$ [2]	71·0
				$\pi^+\pi^-\pi^0$, $\pi^+\pi^-\gamma$ [3]	29·0
				etc.	
ρ: $\rho^{\pm}$, ρ^0	775–770		100–150	$\pi\pi$	approx. 100
				4π	approx. 1
K^*: $K^{*\pm}$, K^{*0}, $\bar{K}^{*0}$	891·4		49·7	$K\pi$	approx. 100
				$K\pi\pi$	<0·2
ω	783·4		12·2	$\pi^+\pi^-\pi^0$	approx. 90
				$\pi^0\gamma$	9·3
				$\pi^+\pi^-\gamma$, $\pi^+\pi^-$ etc.	
ϕ	1019·5		3·4	K^+K^-	47·3
				$K^0_LK^0_S$	38·9
				$\pi^+\pi^-\pi^0$	13·8
				e^+e^-, $\mu^+\mu^-$ etc.	

[1] Many other meson particles are known including axial-vector and 1 = 2 particles.
[2] Neutral decays.
[3] Charged decays.

gravitational effects. An unexcited atom, for example, with the same number of electrons and protons appears from a distance to be neutral and not to interact with an electromagnetic field.

The elementary particles have been placed in two broad groups, namely: (1) The *Hadrons* and (2) the *Leptons* and the *Photon.* The had-

Table 3.02. Some known baryon particles

Particle (or resonance)	Mass (MeV)	Mean Life (sec)	Principle Decays Partial mode	% Fraction
N p	938·3	stable	—	—
N n	939·6	$1{\cdot}0 \times 10^{3}$	$pe^-\bar{\nu}$	100
			$p\pi^-$	65·3
Δ	1115·6	$2{\cdot}52 \times 10^{-10}$	$n\pi^0$	34·7
			$pe\bar{\nu}$, $p\mu\bar{\nu}$	
			$p\pi^0$	52·8
Σ $\Sigma^\pm$	1189·4	$0{\cdot}8 \times 10^{-10}$	$n\pi^+$	47·2
			$n\pi^-$	100
Σ Σ^-	1197·2	$1{\cdot}6 \times 10^{-10}$	$ne\bar{\nu}$, $n\mu\bar{\nu}$ etc.	
Σ Σ^0	1192·3	$<1{\cdot}0 \times 10^{-14}$	$\Delta\gamma$	100
			Δe^+e^-	
			$\Delta\pi^0$	
Ξ Ξ^0	1314·7	$3{\cdot}03 \times 10^{-10}$	$p\pi^-$, $pe^-\bar{\nu}$ etc.	100
			$\Delta\pi^-$	100
Ξ Ξ^-	1321·2	$1{\cdot}66 \times 10^{-10}$	$\Delta e^-\bar{\nu}$, $ne^-\bar{\nu}$, $n\pi^-$ etc.	
N*	1236·0	120	$N\pi$	100
			$\Delta\pi$	91
Y_1^*	1382·2	36	$\Sigma\pi$	9
Ξ	1528·9	7·3	$\Xi\pi$	100
	1533·8			
			$\Xi^0_{\pi^-}$	8 known events
Ω^-	1672·4	Mean Life $= 1{\cdot}3 \times 10^{-10}$	$\Xi^-_{\pi^0}$	3 known events
			ΔK^-	13 known events

rons are the most numerous and have *strong* interactions (often called resonances). They are the heavier particles and include the protons and neutrons, the lightest being the π-meson (pion). They may be further divided into two groups: (a) the bosons which include mesons with integral spin (Table 3.01) and (b) the baryons which are also fermions with half odd integral spin (Table 3.02). The leptons (Table 3.03) which are all fermions are light particles with weak interactions and include the electron, the positron, the muon and neutrino. The photon is a boson. The terms boson and fermion derive from the use of either Bose–Einstein or Fermi–Dirac statistical procedures for their examination. The various interactions (collisions and decay processes) are governed by certain rules of behaviour, the conservation laws. Thus the law of conservation of baryons states that although a baryon may change into another baryon in an interaction the total number before and after must be the same. Each baryon has a baryon number (B)

Table 3.03. Photon and Lepton particles

Particle[1]		Mass (MeV)	Mean Life (sec)	Decays partial mode	% fraction
γ	J = 1; P = −1; C = −1	0	stable		
ν (νe, $\nu \eta$)	$J = \frac{1}{2}$	0	stable		
e^-	$J = \frac{1}{2}$	0·511006 ±0·000002	stable		
μ^-	$J = \frac{1}{2}$	105·659 ±0·002	$2{\cdot}1983 \times 10^{-6}$ $\pm 0{\cdot}0008 \times 10^{-6}$	$e\nu\bar{\nu}$	100
				$e\gamma\gamma$	$1{\cdot}6 \times 10^{-5}$
				3e	$1{\cdot}3 \times 10^{-7}$
				$e\gamma$	$6{\cdot}0 \times 10^{-9}$

[1] Antiparticles $\bar{\nu}$, e^+ and μ^+ have identical masses, spins, lifetimes to ν, e^- and μ^-.

either +1 or −1 and non-baryons have $B = O$. A particle and an anti-particle will have baryon numbers of opposite size. By convention the proton is assigned a baryon number + 1 and other baryon number assignments are deduced from interactions in which the particle and a proton appear; e.g. consider the reactions:

(a) $\pi^- + p \rightarrow \Sigma^- + K^+$ Either Σ^- or K^+ must be a baryon
(b) $\Sigma^- \rightarrow \pi^- + n$ but not both. Σ^- is a baryon

Hence K^+ cannot be a baryon and has $B = O$. Consider also:

$$\begin{array}{lcccc} & K^- + & p \rightarrow & \Delta^0 + & \pi^0 \\ \text{baryon } n^0. & 0 & +1 & (?) & 0 \end{array}$$

Therefore Δ^0 must have a baryon number +1. Also the law states that since the proton is the lightest baryon it cannot decay and must therefore be completely stable with an infinite lifetime. Measurements of the proton's lifetime by Reines (1954) showed that it was greater than 10^{21} years.

The strongly-interacting particles appear to form particle-antiparticle pairs and, when the charges differ, to fit into certain natural multiplet groups according to the SU 3 (Special Unitary Group 3) theory of Gell-Man and Ne'eman (1962) whereby they divide into groups of 1, 8 and 10 particles each and in each group the particles have properties in common. Thus they may all spin with the same angular momentum, or they may all be either baryons or all mesons. If members of a particular group are plotted on a graph (Fig. 3.01)

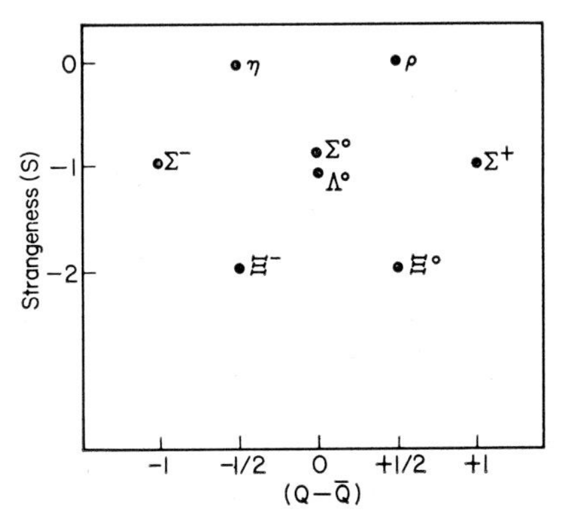

Fig. 3.01. Arrangement of some known baryon particles in a particular grouping of eight particles. The neutron (η) and the proton (p), despite their abundance in the world, do not have any privileged position. The above figure shows that Σ, Ξ and Λ particles are just as basic, essential and important as the neutron and proton.

which plots the quantity $(Q - \bar{Q})$ against a new and necessary quantum number termed the "*strangeness*" (S) which is conserved in *strong* interactions, a pattern results. In the case of eight particles this is in the form of a hexagonal array (Fig. 3.02) and for ten particles a triangular array (Fig. 3.03). The quantity $(Q - \bar{Q})$ refers to groups of like particles (with the same *strangeness*) and is derived as in the following examples:

(1) There are two particles (protons and neutrons) which make up the group of nucleons. Q is the charge on the *individual* particle and $\bar{Q}$ the *average* charge on a nucleon. The proton has charge $+1$ and the neutron charge 0, hence for the nucleons $(\bar{Q}) = \frac{1}{2}(1 + 0) = +\frac{1}{2}$. Therefore for the neutron:

$$(Q - \bar{Q}) = (0 - \tfrac{1}{2}) = -\tfrac{1}{2}$$

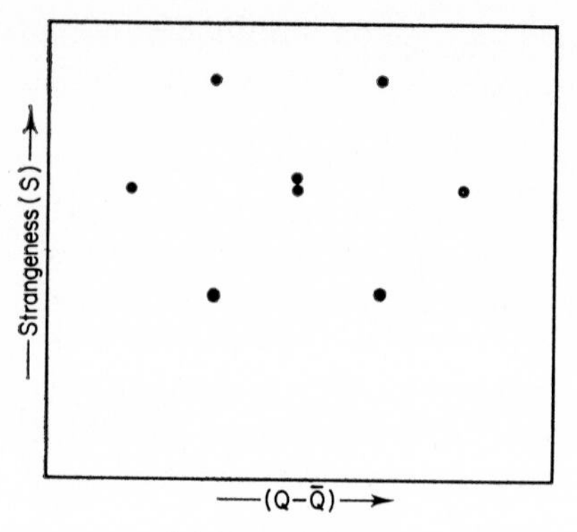

Fig. 3.02. Hexagonal arrangement of eight particles according to the SU 3 scheme groupings.

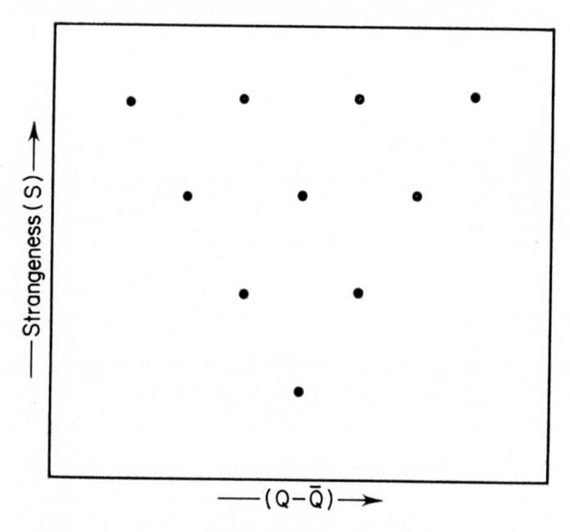

Fig. 3.03. Triangular arrangement of ten particles according to the SU 3 scheme groupings.

and for the proton:

$$(Q - \bar{Q}) = (1 - \tfrac{1}{2}) = +\tfrac{1}{2}$$

(2) There are three sigma particles, namely Σ^+, Σ^0, Σ^- with charges $+1$, 0 and -1 respectively.

Hence for this group $\bar{Q} = 1/3\ (+1\ -0\ -1) = 0/3 = 0$

Therefore

$$(Q - \bar{Q})_{\Sigma^+} = (+1 - 0) = +1$$
$$(Q - \bar{Q})_{\Sigma^0} = (0 - 0) = 0$$
$$(Q - \bar{Q})_{\Sigma^-} = (-1 - 0) = -1$$

The *strangeness* is to some extent an arbitrary figure (new additive quantum number) assigned to elementary particles and has no counterpart in the macroscopic world and no dimensions. It is not apparently universal but confined to particles produced by strong interactions as distinct from those produced by weak interactions such as decay processes. The law of conservation of *strangeness* states that *strangeness* is conserved in strong interactions. The concept arose from the discovery that certain particles are only produced in association with others; for example, it is not possible to produce a single sigma (Σ) particle. It is always accompanied by a K^+ (or K^0) meson but never by a K^- meson. The K^+ meson is allotted a *strangeness* $+1$ unit and the Σ^- -1 unit. The sign is arbitrary as is the precise number—the important thing is that it be the same for both particles but of opposite sign. Pions (π-mesons), neutrons and protons have zero *strangeness*. For example, consider the reactions:

(a) π^- (pion) $+$ p (proton) $\rightarrow \Sigma^- + K^+$ reaction
strangeness $0 \quad + \quad 0 \qquad (-1 \quad +1)$ occurs

and *strangeness* is conserved

(b) $\pi^+ + p \rightarrow \Sigma^+ + K^+$ reaction
$(0 + 0) \quad (?) \quad +1$ occurs

strangeness of Σ^+ must be -1

(c) $K^- + p \rightarrow \Sigma^+ + \pi^-$ reaction
$(?) + 0 \quad -1 + 0$ occurs

strangeness of K^- must be -1

(d) $\pi^- + p \rightarrow \Sigma^+ + K^-$
$(0 + 0) \; (-1 + -1)$

In this case *strangeness* is *not* conserved ($0 \neq -2$). The reaction has *never* been found to occur.

The SU 3 theory forecast in 1962 that a particle—omega minus (Ω^-)—was missing from an array of like particles and in 1964 the particle was discovered, so confirming the predictive powers of the theory in much the same way as the periodic classification of the elements by Mendeleef led to the description of new, and at the time undiscovered, elements. The missing particle had properties almost identical with those forecast by the theory. The SU 3 theory also suggests that there might be three basic particles called *quarks* which would possess unusual properties. In particular they would have fractional charges and be very heavy. Thus three massive quarks would bind together to form a proton, the resulting loss in mass coming from the enormous binding energy conversion which would then occur. In a similar manner the other particles could be described. However all attempts to date to detect such quark particles have failed and many physicists now believe that they may not exist.

It must be remembered at all times that the concept of an elementary particle as presently defined merely reflects the results of mathematical treatment of certain phenomena along specific lines. Whether a given phenomenon is or is not a particle but rather some resonance, as of tones on a violin, is a matter of opinion and indeed it is possible that the concept sought so long, namely the ultimate particle, does not exist.

ORIGIN OF THE ELEMENTS

Although there is a good deal of confusion about the precise manner in which the atomic nucleus is built up from smaller units, it would

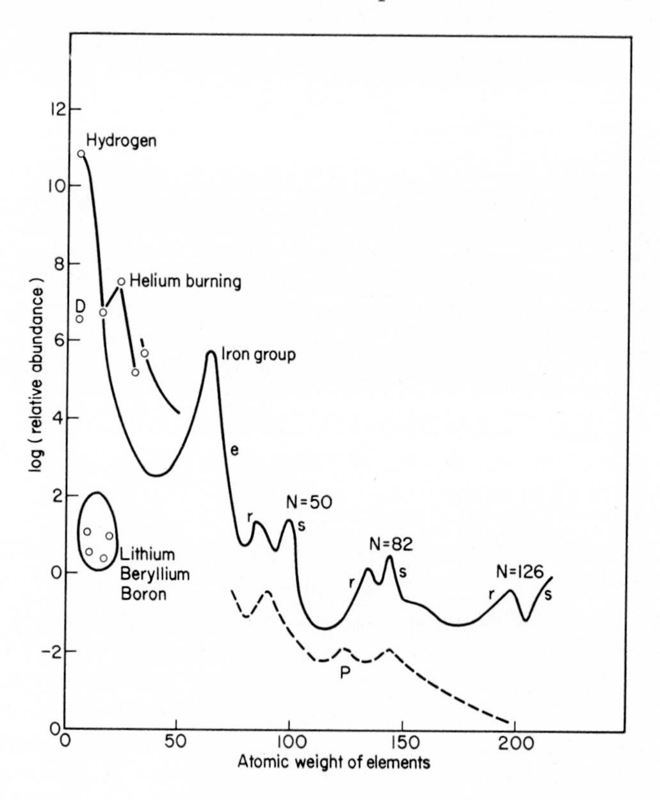

Fig. 3.04. Schematic curve of the relative abundance of the elements in the universe.

perhaps be fair to say that one of the major achievements of science during the last century has been the discovery and confirmation of Prout's thesis that elements are all made up essentially of hydrogen atoms (protons, neutrons, and electrons). Our knowledge of the nature of the elements and the compounds they form is very largely confined to studies of earth matter. However, other information is available from extraterrestrial material such as meteorites, lunar rocks and also from the knowledge brought by spectroscopic studies of the various radiations —light, radio, X-ray, etc.—which emanate from outer space. Certainly all this data has confirmed that the elements are universal and restricted to about 100 of which the heavier ones tend to be unstable (radioactive) and to break down into smaller more stable units. The relative abundance of the elements is of special interest and clearly any theory of planetary or for that matter universe formation must account for the particular observed distribution of elements. Hydrogen appears to be the most abundant element (76% of universe matter), followed by

Table 3.04. Relative abundance of major chemical elements in the human body and in the earth's crust

Human Body % wt	Element	Atomic Weight	Earth's Crust % wt
10·0	Hydrogen	1·008	0·88
18·0	Carbon	12·011	0·087
3·0	Nitrogen	14·007	0·03
65·0	Oxygen	15·999	49·4
0·109	Sodium	22·990	2·64
0·036	Magnesium	24·312	1·94
—	Silicon	28·086	25·5
1·16	Phosphorus	30·974	0·12
0·196	Sulphur	32·064	—
0·20	Potassium	39·102	2·41
2·01	Calcium	40·08	3·39

helium (23% by weight), and all the others account for less than 1% by weight of the total, and the abundance in general decreases with increase in atomic weight (Fig. 3.04). The various estimates of atomic abundance are derived from rock and meteorite analyses and examination of radiation. The figures must, however, be treated with care since the non-radiating parts of the universe are largely unknown and even the composition of our own earth is only known down to a few miles in depth. Indeed the only standard composite of elements which occurs in nature in a reasonably standard form and unquestionably has philosophical importance and which is capable of accurate measurement is the living system. If one assumes that the total function of the universe is ultimately to allow the conversion of all matter into the living system (and this would be a reasonable philosophy), then the elemental composition of the living system assumes a special importance. Apart from elements such as carbon and silicon, the earth's skin composition (Table 3.04) is not unlike that of the human body (Chapter 4). One could intriguingly balance the equation by assuming that the earth's core contained a large amount of carbon in the free form or as, say, carbon dioxide or some other molecular complex.

There seems to be little doubt that elements are continuously being formed in the hot interior of stars from hydrogen and the same processes may also have operated to produce them in *de novo* synthesis if in fact such synthesis ever occurred. The method by which the elements are

Proton Interaction Processes

$^{1}H \rightarrow {}^{4}He \rightarrow {}^{12}C \rightarrow {}^{13}C \rightarrow {}^{14}N \rightarrow {}^{15}N \rightarrow {}^{12}C$ (Carbon-nitrogen cycle)

$^{16}O \rightarrow {}^{17}O$

$^{20}Ne \rightarrow {}^{21}Ne \rightarrow {}^{22}Ne \rightarrow {}^{23}Na$

α—Particle Interaction Processes

$^{4}He \rightarrow {}^{12}C \rightarrow {}^{16}O \rightarrow {}^{20}Ne \rightarrow {}^{24}Mg \rightarrow {}^{56}Fe \rightarrow$ Higher elements

$^{14}N \rightarrow {}^{18}O$

$^{15}N \rightarrow {}^{19}F$

Neutron Capture Processes

$^{13}C \rightarrow {}^{16}O$

$^{21}Ne \rightarrow {}^{24}Mg$

$^{23}Na \rightarrow {}^{32}S$

Fig. 3.05. Synthesis of elements in stars.

formed within stars was conceived by Hans A. Bethe as two major processes: the proton–proton fusion and the carbon–nitrogen cycle (Fig. 3.05). In the first of these processes hydrogen may first fuse (at 2×10^{7} degrees K) to form helium (the major source of internal energy in the star), and this may fuse with itself (2×10^{8} degrees K) to form the highly unstable nucleus ^{8}Be ($t_{\frac{1}{2}} = 3 \times 10^{-16}$ sec) which although not found in nature may be produced transiently in laboratory experiments and has a good ability to capture further helium to produce an excited ^{12}C. Stars (carbon stars) whose emitted light contains abnormally strong absorption bands of carbon molecules are readily observed. In some there is evidence for the presence of substantial amounts of ^{13}C as well as ^{12}C. By similar so-called helium-burning reactions ^{12}C may absorb more He to produce nuclei whose atomic masses are multiples of four, including $^{16}0$, ^{20}Ne, ^{24}Mg. As the He is used up the star contracts and hence heats up again, so producing further fusion reactions (5×10^8 degrees K) between C, O, etc., to produce elements such as ^{28}Si. The process goes on until the most stable elements of the iron group are formed (^{56}Fe) in which the balance of nuclear and electrostatic forces in the nucleus are most stable. Nuclear reactions involving the iron group elements must absorb energy instead of releasing it, hence these nuclei cannot serve as fuel or allow the fusion processes to continue. It has been suggested by Hoyle that the apparent relatively large abundance of iron group of elements in the Universe is due to this, and when the star perhaps ultimately explodes, large amounts of iron-containing material are flung into space. Elements larger than iron are then presumably produced by slow (s) or fast (r) neutron capture reactions or by proton capture. In this sense two main types of stars are recognized, first generation or primeval stars starting from hydrogen and helium and second or later generation stars which contain elements up to the iron group which have been merged into interstellar hydrogen–helium matrix to produce the new star. In the second-generation star ^{12}C may capture hydrogen with conversion into elements such as ^{13}C, ^{14}N, ^{15}N, etc. ^{15}N ultimately adds a proton and breaks down into ^{12}C and He. This is the carbon–nitrogen cycle assumed to be the major energy source in this type of star with an internal temperature of more than 15 million degrees. ^{16}O may capture H to produce ^{17}O, and ^{20}Ne produces ^{21}Ne. The three isotopes ^{17}O, ^{21}Ne, and ^{13}C react with He to produce unstable nuclei which liberate neutrons as has been shown in laboratory experiments. Here then is the source of neutrons to allow extension (by neutron capture) beyond ^{56}Fe into the higher elements. There is good evidence now that heavier elements are produced in the stars and indeed they have also been shown to be formed during nuclear explosions. Slow neutron capture occurs by processes such as

$$^{13}\text{C} + {}^{4}\text{He} \rightarrow {}^{16}\text{O} + {}^{1}\text{n}$$
$$^{56}\text{Fe} + {}^{1}\text{n} \rightarrow {}^{57}\text{Fe}$$
$$^{57}\text{Fe} + {}^{1}\text{n} \rightarrow {}^{58}\text{Fe} \rightarrow \text{Heavy elements} \rightarrow {}^{209}\text{Bi}$$

in which addition of each neutron adds one mass unit to the nucleus. The nucleus may of course undergo β-decay to the next higher element so long as another neutron has not been absorbed in the interim period. This slow (s) process terminates in ^{209}Bi since neutron capture beyond this leads to elements in which α-decay occurs more rapidly than the neutron capture. Clearly one would expect a natural transgression of elements formed according to the ease with which they may capture neutrons and one would expect that those elements least able to capture neutrons would occur maximally. This seems to be so since the elements Sr and Ba with low neutron-capturing propensity occur in quantity in certain stars (barium stars). A further proof of neutron capture in stars (so-called *S*-stars) comes from the presence therein of the element Technetium which has no stable isotopes and does not occur on earth. It has a half-life of 2×10^5 years and is produced in the same chain process as Sr and Ba.

Since α-decay of unstable elements may result in loss of certain nuclei it is necessary to invoke another type of neutron capture—the so-called rapid or *r*-process—in which rapid capture occurs over a few seconds as distinct from the 10^5 year span required for the slow capture process. Evidence for this fast type of neutron capture has come from observations of hydrogen bomb explosions in which neutrons released were captured by the metallic parts of the bomb to produce transuranic elements up to and including Californium (^{254}Cf). Clearly, of course, although such processes are specifically required for the production of the very heavy elements, there is no reason why they should not also operate to produce at least a fraction of the lighter elements.

In addition to the *s*- and *r*-processes a third group of elements beyond Ni requires a further *p*-process which involves proton capture by neutron-rich nuclei and a final *x*-process is required to explain the degree of local abundance of certain light elements such as deuterium, Li, Be and B which at temperatures ruling in the interior of stars would be expected to be rapidly destroyed.

These various nuclear reactions combined produce a valuable hypothesis of stellar nucleogenesis, but there is little doubt that many modifications to the theory will have to be made before it is anywhere near fully acceptable, and the many anomalies explained.

THE UNIVERSE CONSTITUENTS

All studies of the various portions of the known Universe have combined to suggest that the basic constituent elements and elementary particles are universal. The universe is then seen to possess a fairly well-defined shape and structure. Essentially it appears to consist of a very large number of bodies termed galaxies separated from each other by quite vast distances. Between the galaxies there appears to be little or nothing except a high vacuum which contains minute amounts of atoms and molecules such as perhaps hydrogen and dust. Each galaxy consists of large numbers of stars of varying types of which our own sun is apparently an average sort of sample and presumably many stars have planetary systems such as our own solar system and these in turn may have sub-planetary satellite (moon) systems. The latter of course may include not only the large obvious bodies but also large numbers of lumps of rock (asteroids) which may range in size from millimetres to miles in diameter. The galaxies also contain gas-dust

Fig. 3.06. "Horsehead" Nebula in Orion south of Zeta Orions. IC.434. Barnard 33. Photographed in red light. 200-in. Hale. Photograph PAL 8 is published by kind permission of the Royal Astronomical Society, London and is a photograph from the Hale Observatories.

clouds which sometimes form morphological units (nebulae) (Figs. 3.06, 3.07) or may be more diffuse. All the galaxies appear to be moving away from each other at great speeds which with increasing distance approach the speed of light. Evidence for this comes from the red-shift in the spectrum of light (Fig. 3.08) emitted by the galaxies, and by extrapolating backwards in time it is possible to arrive at an age of the Universe, assuming that the extrapolation can be made in that way and that the whole Universe had a specific beginning in time.

These conclusions resulted from the researches of E. P. Hubble (1889–

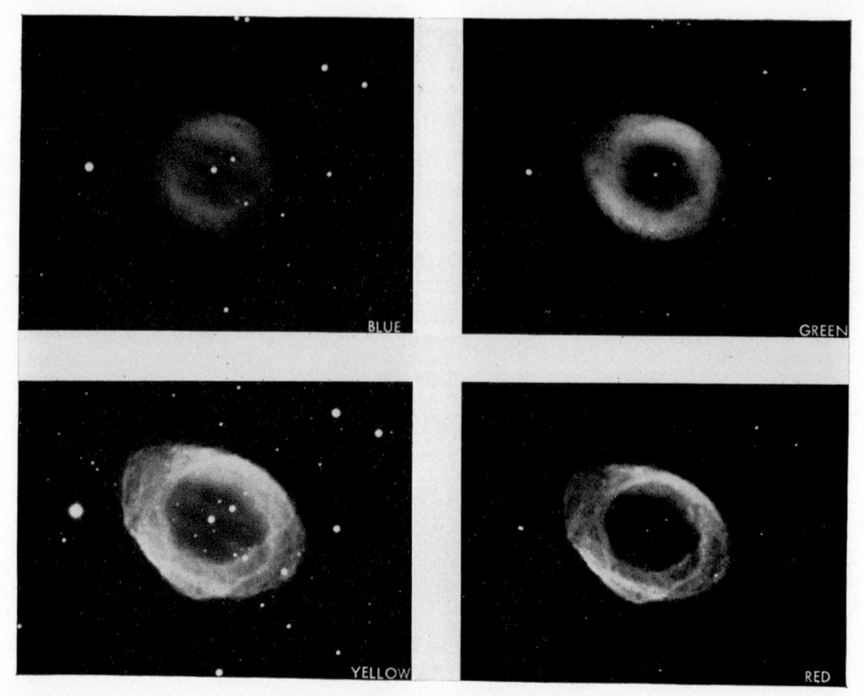

Fig. 3.07. Planetary nebula ("Ring") in Lyra photographed in blue, green, yellow and red light. 200-in. Hale. Photograph (PAL 66) from the Mount Wilson and Palomar Observatories and published with permission of the Royal Astronomical Society, London.

1953) with the Mount Palomar telescope and have given rise to the so-called "Hubble's Law" namely that the recessional velocity (v) of a stellar system is proportional to its distance.

The recessional velocity may be expressed by:

$$v = cz\ (1 + z^2)^{-\frac{1}{2}}$$

where c = velocity of light (186,000 miles per sec)

$z = \frac{d\lambda}{\lambda}$ where $d\lambda$ is the amount by which any spectral line of wavelength λ has shifted.

A plot of the recessional velocities of known galaxies against their distances gives a straight line so Hubble's Law may be expressed:

$$v = HD$$

where H is Hubble's constant (10^{-10} year $^{-1}$) or $1/H = 10^{10}$ years, and D is the distance.

Assuming the linearity of the above plot the time $D/v = 1/H$ for any galaxy to travel to its present position is therefore independent of D and this suggests that all galaxies were at a single point some 10^{10} years ago. There are of course errors in the red shift measurements and

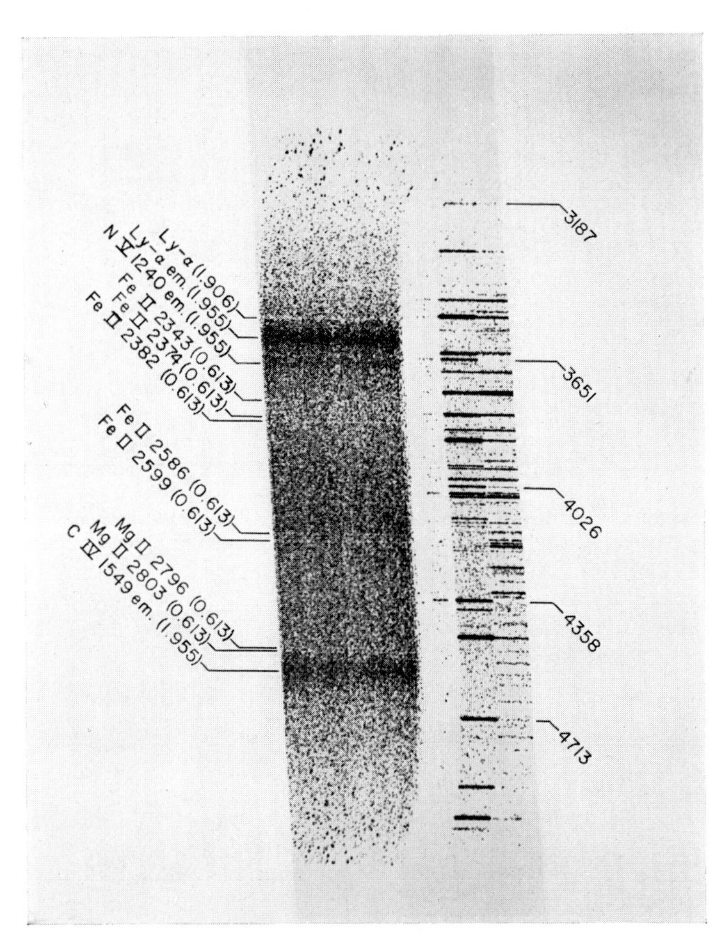

Fig. 3.08. "Red shifts". Quasar PHL 938 showing emission and absorption lines with different red shifts. Values in parentheses. Kitt Peak. Photograph (RAS 700) published by kind permission of the Royal Astronomical Society, London.

these are outlined in the Fig. 3.09 which indicates a starting point for the Universe somewhere between $6 - 13 \times 10^9$ years ago.

A lot of cosmological theory rests on these interpretations of the red shifts and on Hubble's Law. The passage of time has generally led to increasing acceptance of the "law" and its consequences, but there are

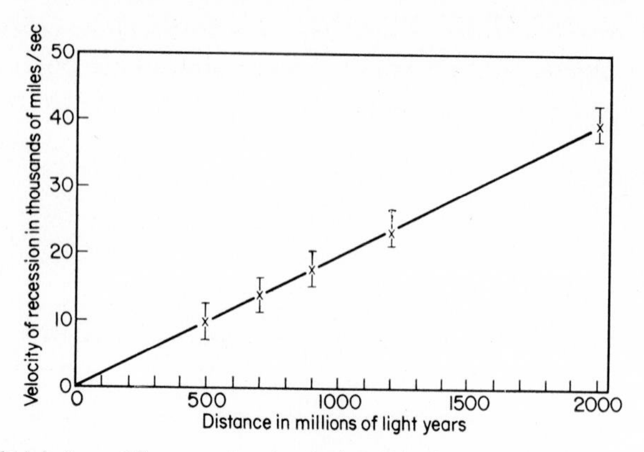

Fig. 3.09. Hubble's Law. The recessional velocity (V) of a galaxy at distance (D) from its origin:

$$V = \text{H}D \quad \text{and} \quad 1/\text{H} = D/V$$

where H is Hubble's constant. Observations (shown on figure) show that galaxies at:

10 million light years recede at 190 miles per second.
100 million light years recede at 1900 miles per second

$$V = 19d$$

where d = distance in million light years receded at $19d$ miles per second.

$$\text{Also } D = 10^6 d$$
$$= 10^6 \times 365 \times 24 \times 3600 \times 186{,}000 \; (d \text{ miles})$$

$$1/\text{H} = \frac{10^6 \times 365 \times 24 \times 3600 \times 186{,}000}{19 \times d} = \frac{D}{\text{V}}$$

$$= 9{\cdot}7 \times 10^9 \text{ years}$$

1/H is the time taken by a galaxy to travel its present distance D at velocity V and is independent of D. Thus the age of the galaxy (i.e. time to its present position) $= 9{\cdot}7 \times 10^9$ years. Possible errors in calculations using Hubble's Law can occur in the measurements of the *velocity of recession* (thousands of miles per second: error = a) and in *the distance* (in millions of light years: error = b).

Errors in Hubble's Law: $1/\text{H} \times 9{\cdot}7 \times 10^9 \pm (\text{a} + \text{b})$ years.

Percentage error	Possible error in age minimum	maximum
1	9·506	9·894
2	9·312	10·088
3	9.118	10·282
4	8·924	10·476
5	8·730	10·670
10	7·760	11·640
20	5·820	13·580

a few nagging problems, most important of which perhaps is the existence of some galaxies which are apparently as much as 25×10^9 years old.

Fig. 3.10. A spiral galaxy. Great Spiral Nebula in Andromeda. Messier 31. Satellite nebulae NGC 205 and 221 are also shown. Photograph (PAL 150) from the Hale Observatories is published with permission of the Royal Astronomical Society, London.

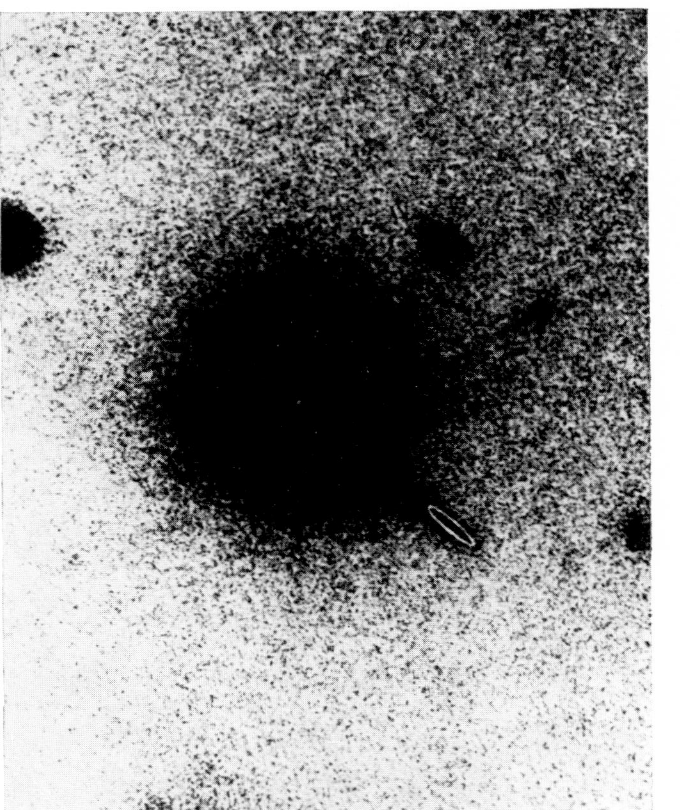

Fig. 3.11. Quasar 3C273. The white dot in the centre of the object represents one component of the quasar, whilst the white oval over the jet represents a second component of the source. Photograph RAS 682 is published with permission of the Royal Astronomical Society London.

THE GALAXIES

The galaxies may be classified into seven major groups:

(1) Spiral galaxies (Fig. 3.10): they are weak radio sources and have a spiral or irregular spiral structure, examples include M82 M31 (M = Messier).
(2) Elliptical galaxies (E) of generally ovoid appearance.
(3) D galaxies. They have an ovoid nucleus surrounded by a large extended envelope.
(4) Intermediate galaxies between E and D (termed DE or ED according to the dominant species).
(5) Dumb-bell galaxies (db). They contain two "nuclei" in a common envelope.
(6) N galaxies. These contain brilliant star-like nuclei in which most of the luminosity of the system resides. They are analogous to quasars but less powerful radio emitters.
(7) The so-called quasi-stellar or quasars—QSRs (Fig. 3.11). These are the most powerful sources of radio-luminosity, ranging from 2×10^{44} to 2×10^{45} ergs per sec compared with a spiral galaxy at about 10^{39} ergs per sec. The quasars are thought to be possible precursors of galaxies and therefore to represent primeval material from which the Universe arose.

ORIGIN OF THE UNIVERSE AND THE GALAXIES

There are two main theories which suggest how the Universe and the galaxies arose. They are the "big bang" theory and the "continuous creation" or "steady state" theory.

Big Bang Theory

This is undoubtedly the most widely held theory to account for the formation of galaxies and indeed of the whole Universe and almost all physical observations made by astronomers and radio-astronomers have tended to confirm the theory's basic concepts. The assumption is that somewhere about 10×10^9 years ago the Universe consisted of a single piece of matter, a sort of giant primeval atom concentrated at a single point in space somewhat analogous to our liquid globule model mentioned at the beginning of this chapter. This concentrated matter then exploded, thus flinging out either hydrogen atoms and related nuclei which subsequently became organized into discrete units (galaxies) and/or lumps of specifically condensed matter which could be regarded as *de novo* galaxies. These lumps of matter or more tenuous gas have ever since continued to flee away from each other at ever-increasing speed. Presumably, since we can only observe those objects which travel at speeds less than that of light, then as the speed approaches that of light the objects will disappear and ultimately the whole galactic Universe will disappear, presumably leaving behind only our own galaxy—a peculiarly egocentric viewpoint of the Universe. Certainly observations seem to confirm these arguments since at greater distances the concentration of galaxies apparently increases as one would expect. In addition there is evidence (Penzias and Wilson, 1965) that the whole Universe is emitting radiation at radio wavelengths like a very cold black body at about 3°K. Such radiation had been predicted by Gamow and Dicke as a natural consequence of the primeval big bang. Isotopic radiation (equally intense from all directions) has been observed at wavelengths of 0·86, 1·58, 3·2, 7·35, and 20·7 cm. Moreover absorption lines of cyanogen (CN:CN) in the spectra of the stars Zeta Ophiuchi and Zeta Persei show that a significant proportion of the cyanogen is at a level of excitedness to be expected if it were bathed with radiation of wavelength 2·6 cm. The strength of this radiation equally is compatible with background radiation of a black body at 2–3°K. The super condensed primeval atom is seen to result from an earlier contraction of a rarefied Universe and the assumption is that at a certain degree of rarefaction our own present Universe will begin to contract and ultimately be yet again reborn in another giant explosion.

This particular theory, whilst fitting what few facts there are, has nevertheless had an uneasy reception largely because it involves tightly scheduled starts and finishes in a Universe which somehow is felt to be endless and infinite by many.

Continuous Creation or Steady State Theory

This particular theory one suspects was largely designed to overcome psychological rather than mathematical objections to the big bang theory and indeed has been criticized on this basis as a sort of search for perfection in nature. It suggests that matter (in the form of hydrogen) is always being created from nothing to counteract the dilution of material which occurs by the drift away of the receding galaxies, so that the concentration of galaxies in our observable Universe always remains constant. The amount of matter required to be produced is very small indeed—about 1 atom of hydrogen per litre per 10^{12} years—although the rate of creation could obviously be much greater in specific pockets. However it must not be created in too non-random a fashion or it might not do the job intended, i.e. to renew lost galactic material. In addition the newly created matter will have the same velocity as its immediate neighbours so that the red shift effect will not be contravened. A theory of this sort might seem at first sight to be absurd. However, one can get an idea as to how matter can be created from "nothing" by invoking our very useful "ether" concept. Ether we will say has the property of being "nothing" in the sense that it cannot be measured, but warped (condensed and rarefied) ether becomes something in that it can be measured. Thus we say that ether has the property that allows it to take up specific discrete, not continuous, small condensed forms (elementary particle), but its total properties are such that when such a condensed form appears a rarefied equivalent piece of ether (elementary anti-particle) must also appear. So from apparently nothing we obtain two particles. The mathematician however claims to produce the same sort of result without invoking an ether and in addition to provide quantitative data to support his theory.

Formation of the Galaxies and of Stars

Galaxies are generally presumed to form from a primeval hydrogen gas cloud and it is of great importance for an understanding of the Universe to know why it should consist of essentially similar sized large bodies separated by high *vacuo* and not as a mere finely-dispersed cloud of gas and dust. If one assumes that a big bang creation produced roughly similar sized lumps of aggregated matter then there is no problem. In the case of the tenuous gas extending over light years, however, it is necessary to assume that accretion of particles occurs by mutual gravitational and magnetic attractions to form an enormous but increasingly unstable mass. This in turn begins to break up by processes say of condensation and rarefaction into smaller clouds. This process can continue with formation of more and more sub-clouds from each parent until a sub-cloud structure is reached in which the kinetic energy is translated into rotatory motion which will tend to increase with increasing shrinkage. The type and degree of initial rotation will determine the shape of the galaxy, i.e. whether flat disc-shaped or thicker and more ovoid in structure when gravitational and rotational forces are finally balanced. As the cloud shrinks the liberated energy will cause the temperature to rise. At lower temperatures (say 10,000–25,000°C) little or no energy is radiated as heat if the cloud is of a size much greater than that of ten billion average suns. The cloud cannot therefore shrink further, since if it tried to do that, the energy content would simply expand it to its initial size. Consequently the cloud must fragment into smaller clouds and when these reach the size of 10 billion suns (the average galactic mass) half the gravitational energy is released and radiated as heat. The cloud continues to shrink until its size is about 100,000 light years in diameter, when it becomes too dense to undergo further shrinkage and so fragments again into sub-clouds; these are the embryo Population II stars (see later).

Mechanisms such as this can at best be pale shadows of reality but they give some idea of the type of possibilities available. The resulting galaxies vary in size from our own (100,000 light years diameter) to small ones such as the Andromeda (4000 light years across) (Fig. 3.10), the average size being about 12,000 light years. These variations nevertheless are relatively small ones. The numbers of contained stars however appear to vary more, thus the large galaxies may have up to 200 billion stars and the dwarf ones only about 300 million with the average about ten billion stars.

The Stars and Their Origins

Stars can be broadly classified into two types: Population I and Population II types. The stars at the centre of the galaxies are known as Population II stars and are produced from primeval hydrogen and consequently are relatively poor in metals, whereas Population I stars which occur at the outer edges and in the spiral arms of the galaxies have been formed from dust ejected from Population II stars after explosion (nova and supernova) (Fig. 3.12), together with a hydrogen–helium matrix, and consequently are relatively rich in higher elements including metals. Our own sun is such a Population I star. Indeed the galaxies in addition to containing stars also contain large quantities of gas and dust and this material is constantly being converted into Population I stars. Evidence for quite recent formation of stars (1·5 million years ago) has come from observations of certain star associations (e.g. Zeta Persei in Perseus) which must have had a common centre some $1{\cdot}5 \times 10^6$ years ago, and another example occurs in Orion with a common centre $2{\cdot}8 \times 10^6$ years ago.

The Population I star is considered, then, to form by condensation of dust and gas by gravitational attraction to give initially relatively dark globules. Such "Bok's globules" can be observed especially when viewed against a luminous background as in the Rosette nebula (Fig. 3.13). Further evidence to show that stars are constantly being born

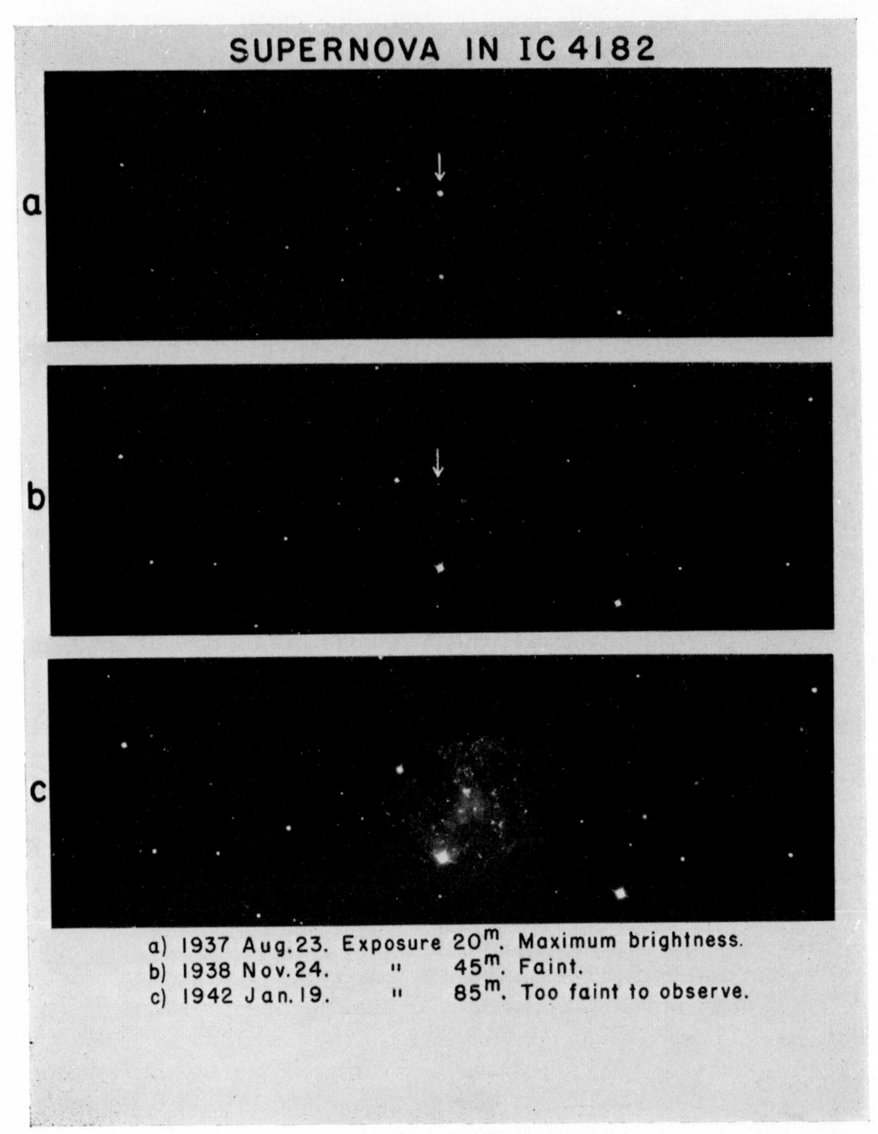

SUPERNOVA IN IC 4182
a
b
c
a) 1937 Aug.23. Exposure 20m. Maximum brightness.
b) 1938 Nov.24. " 45m. Faint.
c) 1942 Jan.19. " 85m. Too faint to observe.

Fig. 3.12. An exploding supernova. Supernova in IC 4168. Three views, photographed in 1937, 1938 and 1942 showing the various stages of explosion. Photograph PAL 111 from the Mount Wilson and Palomar Observatories is published with permission of the Royal Astronomical Society, London.

by processes of this type has come from examination of other celestial bodies including the so-called Herbig–Haro objects, e.g. the T. Tauri star observed in a dark cloud near the Orion nebula. These so-called

T-association stars which have peculiar spectra and contain large amounts of lithium appear to be stars in the making and probably still undergoing gravitational contraction.

A plot of star magnitude and spectral class (Hertzsprung–Russell—

Fig. 3.13. A rosette nebula. NGC 2237. Nebula in Monoceros. 48-in. Schmidt. Photograph PAL 151 from the Mount Wilson and Palomar Observatories is published with permission of the Royal Astronomical Society, London.

H.R.—diagram) of many stars (Fig. 3.14) shows that stars fall into specific classes. These include white dwarfs—extremely dense stars in the last stages of their life. They are said to be degenerate. The recently discovered quasars (enormous energy emitters) or neutron stars are even more densely packed (several tons per cubic inch) very distant, and may represent matter as it was early in the life of the Universe. Then we have a broad band of the main sequence stars which include

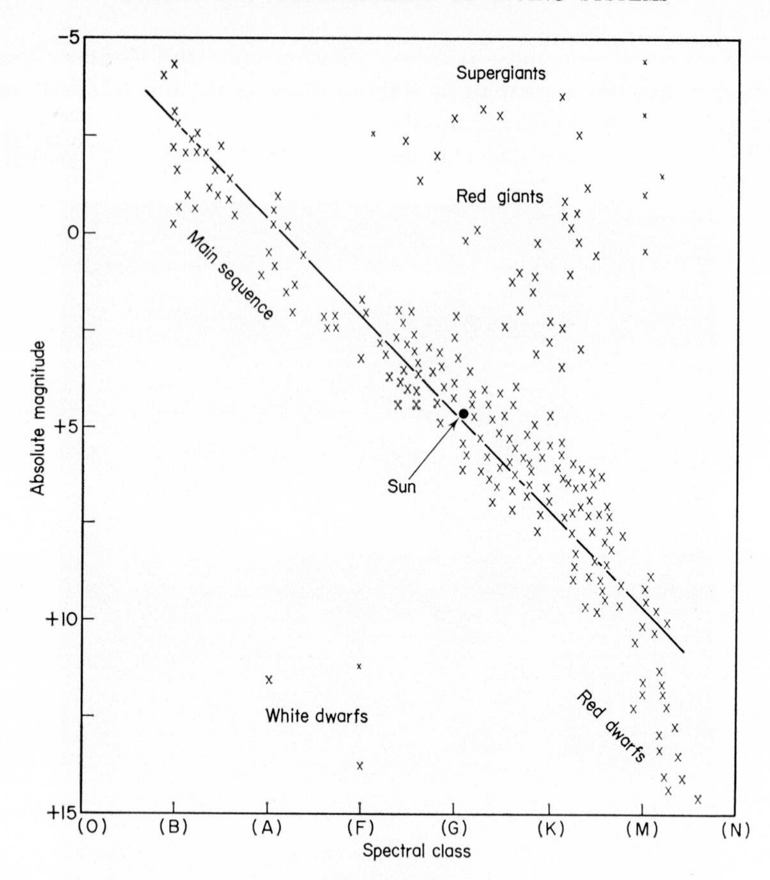

Fig. 3.14. Star types.

our own sun. These in turn, as they grow older, cool down and enlarge as they radiate redder light to become red giant stars. These ultimately explode in a supernova, scattering their material into space to provide new fodder for the production of new Population I stars, but they leave behind a small core which becomes the white dwarf, or neutron star produced by some implosion process. These finally give rise to dark dead stars.

SUGGESTED FURTHER READING Chapter 3

Books

Cameron, A. G. W. (1966). "Space Science". Blackie, London.

Chand, R. (Ed.) (1970). "Symmetries and Quark Models". Gordon and Breach, London.

Hoyle, F. (1959). "Nature of the Universe". Pelican Books, London.
Hoyle, F. (1966). "Galaxies, Nuclei and Quasars". Heineman Press, London.
Martin, A. D. and Spearman, T. D. (1970). "Elementary Particle Theory". North Holland, Amsterdam.
McKay, H. A. C. (1971). "Principles of Radiochemistry". Butterworths, London.
Prentki, J. and Steinberger, J. (Eds) (1968). Proc. 14th Intern. Conf. on High Energy Physics, Vienna.
Rankama, K. (1963). "Progress in Isotope Geology". Interscience Publishers, New York.
Singh, J. (1970). "Modern Cosmology". Penguin, Harmondsworth.
Thiele, R. (1958). "And There was Light—the Discovery of the Universe". Deutsch, London.
Urey, H. C. (1952). "The Planets". University of Chicago Press.

Articles

Fowler, William A. (1956). The Origin of the Elements. *Sci. Am.* September 1956.
Fowler, William A. (1957). Nucleosynthesis in Big and Little Bangs. *In* "High Energy Physics and Nuclear Structure" (Ed. A. Gideon), 203–225. Wiley.
Fowler, W. A., Greenstein, J. L. and Hoyle, F. (1962). Nucleosynthesis During the Early History of the Solar System. *Geophysics* **6,** 148–220.
Ringwood, A. E. (1966). Chemical Evolution of the Terrestrial Planets. *Geochim. cosmochim. Acta* **30,** 41–104.
Rutherford, E. R. (1912). *Philos. Mag.* **24,** 461.
Stoney, G. J. (1891). *Scientific Transactions of the Royal Dublin Soc.* **4,** 11th Series.

4

The Earth, its Origin, Atmosphere, Structure and Development

INTRODUCTION

The planet Earth is an ellipsoid, but so close does it approximate to a sphere that it is sometimes called an oblate spheroid, in particular a spheroid with a slight bulge at the equator. The average polar radius is 3950 miles and the average equatorial radius 3964 miles. It has a mass $5{\cdot}978 \times 10^{24}$ kg, volume $1{\cdot}08 \times 10^{21}$ m^3 and an average density 5515 kg m^{-3} ($5{\cdot}515$ g cm^{-3}). Since the average density of the surface rocks is about 2900 kg m^{-3} some of the material at least which is present in the interior of the earth must have a higher than average density, although since only the very thin outer skin has been observed directly these figures are in fact of much less use than many would have us believe.

The earth rotates about an axis and orbits round the sun. It does so however in a complex fashion which involves precessional motion of the axis which is at an angle of 23·5° to the plane of the ecliptic (the plane of the apparent solar orbit). In its motion it traces out a conical figure in space but the precession of the axis takes about 26,000 years to complete one precessional revolution. In addition, other generally minor movements are imposed on the precession and result from various factors which include the effect of the gravitational pulls of the sun and the moon, and the oblate shape of the earth itself in which the equatorial bulge may take up different relative positions at different times. Other types of minor movement which occur on earth include the Chandler wobble which results in the anticlockwise movement of a point near the north pole in a 43 ft diameter circle over a 14 months' period. This movement may result from changes in the distribution of mass throughout the earth because of various earth movements.

COMPOSITION OF THE EARTH

Information about the nature and composition of the earth's interior which is not of course accessible (the deepest drill is only about 5·5 miles) has come largely from seismographic observations. This method involves the observation with delicate instruments (seismometers) of two distinct types of waves which can pass through the earth and which result from disturbance whether of a natural or induced kind. The two types of waves are the so-called primary or P-waves which are compression waves and move at a velocity of about 5–8·7 miles s^{-1}. and secondary or S-waves which are shear or longitudinal waves and move at about half the velocity of the P-waves. The terms primary and secondary relate to the order in which the waves arrive at a measuring source from a particular disturbance. The velocity of the waves are not normally required to be known but merely the time interval between their arrival at the measuring device. The velocity of a longitudinal or P-wave is given by

$$u = \left(\frac{\psi}{\rho}\right)^{\frac{1}{2}}$$

where ρ = the density of the medium and ψ is the axial modulus (stress/strain). The velocity of the S-wave is given by

$$w = \left(\frac{\mu}{\rho}\right)^{\frac{1}{2}}$$

where μ is the rigidity modulus (stress/strain—the strain is different in this case). Now liquids or gases cannot conduct S-waves and μ and hence $w = 0$ for such media. For waves passing through the earth

$$\mu \fallingdotseq 2w \text{ and } \frac{\psi}{\mu} \fallingdotseq 4.$$

It is clear therefore that seismology offers the possibility of obtaining information about the physical state of the earth's inner materials.

THE INTERNAL STRUCTURE OF THE EARTH

Such evidence as there is about the nature and composition of the inner parts of the earth (Figs. 4.01 and 4.02) has come almost exclusively from seismic studies, using either natural (earthquakes, etc.) or man-made explosions and making simultaneous observations of the arrivals of the S- and P-waves at different parts of the earth's surface. The most outstanding observation is that both S- and P-wave velocities increase

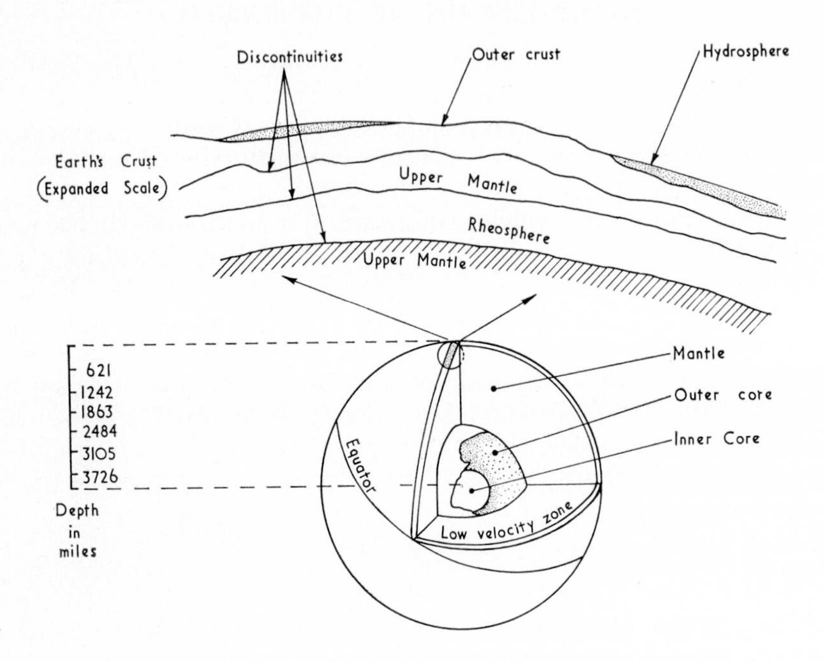

Fig. 4.01. Cut-away section of the earth showing its main subdivisions.

with depth although not in a regular fashion. The P-wave velocity from about 4·0 miles s^{-1} at the surface to about 8·7 miles s^{-1} at the earth's centre, whereas the velocity of the S-waves varies from about 2·5 to 4·4 miles s^{-1}. At certain depths, however, the rate of velocity increase varies and may even decrease (Fig. 4.02). These discontinuities in the velocities have led to concepts about related discontinuities in the innermost parts of the earth. In particular the earth is seen to consist of an

inner core	about	840 miles radius
outer core	about	1310 miles thick
lower mantle	about	1180 miles thick
upper mantle	about	590 miles thick
crust		3–40 miles thick

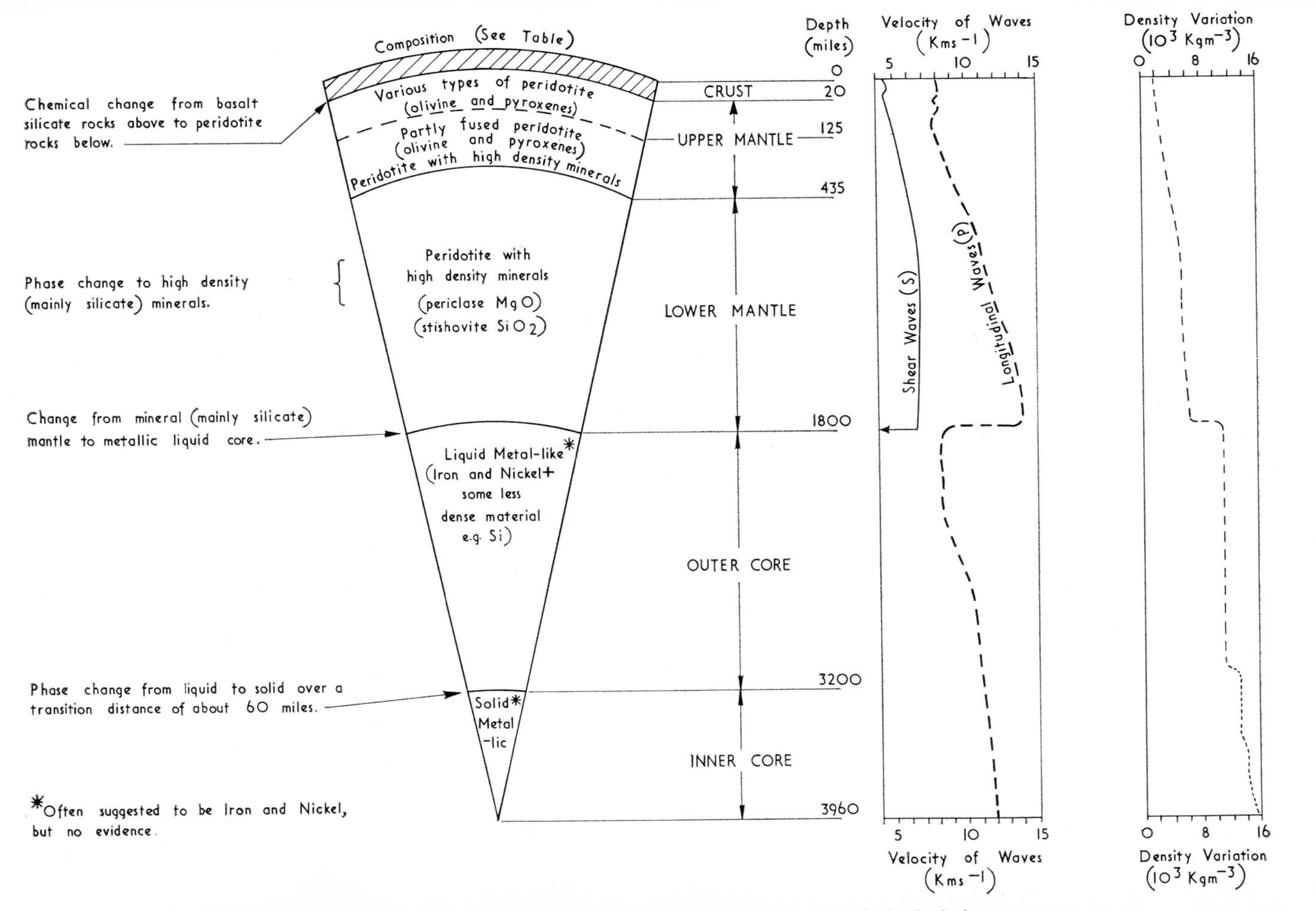

Fig. 4.02. Schematic cross-section of earth showing major chemical and physical changes.

The Core

At a depth of about 1800 miles there is a major discontinuity as shown by marked changes in the behaviour of the S- and P-waves. Thus the velocity of the P-waves drops sharply from 8·5 to 5·0 miles s^{-1}, whereas the S-waves are suddenly brought to a stop (Figs 4.02 and 4.03).

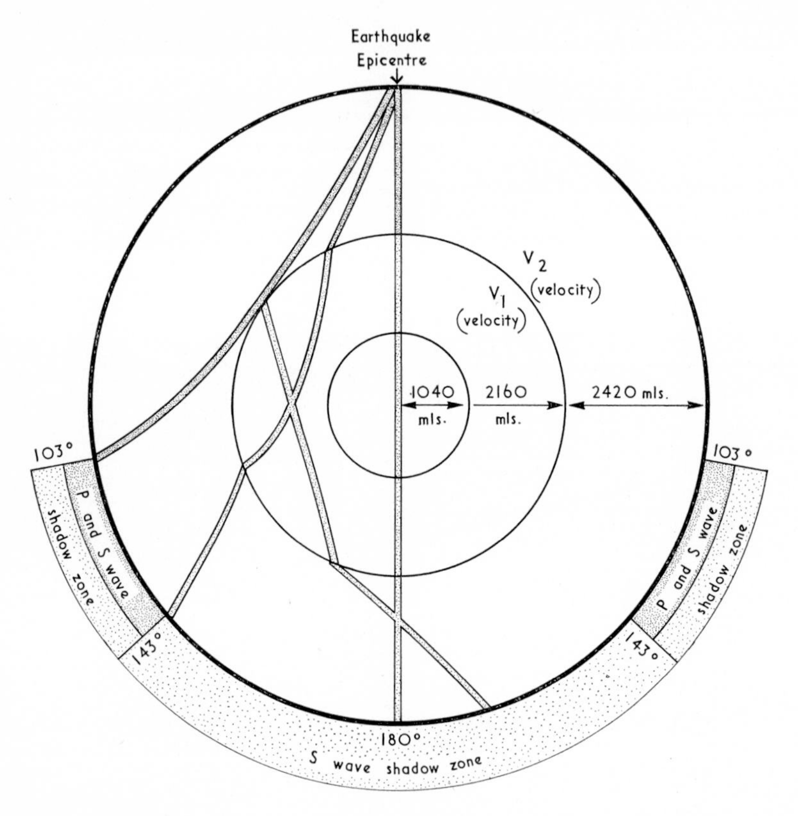

Fig. 4.03. Structure of the earth using earthquake wave studies. P = longitudinal waves, S = shear waves.

The conclusion drawn from these results is that at the core mantle boundary we see a transition from a solid (mantle) to a liquid (core) since S-waves cannot pass through liquids, because their resistance to shear is essentially zero. At the same time it would seem reasonable that all earth outside the core is solid since both S- and P-waves pass through it. In addition, however, P-waves are found to pass through the innermost part of the core (inner core) at slightly increased velocity and this suggests that the inner core is solid. It will be appreciated that the terms solid and liquid as used in this context must be seen in terms of the enormous pressures which exist, at the depths of the earth's central region and that little or nothing is known about the behaviour of matter under such conditions.

Further evidence for the solid nature of the inner core has come (Dziewonski and Gilbert, 1971) from an analysis of particular kinds of vibrations or "normal modes" of the earth which are excited by earthquakes, with an appreciable part (20%) of their energy in the inner core and whose frequency therefore depends on the core properties. The results were found to be compatible only with a solid inner core. A consequence of the liquid outer core structure is that internal refraction will occur at the core–mantle phase boundary (Fig. 4.03).

It is found that no direct S-waves emerge at distances from the epicentre corresponding to angles greater than 103°. Between 103° and 143° no direct waves emerge whether S or P. The particular angles mentioned result if one assumes the presence of phase discontinuities at the particular depths mentioned above in the core.

Nature of the Core Materials

There have been many speculations about the nature of the material which makes up the inner and outer core materials. Most of the assumptions have been based on average density computations and comparison with densities of model materials under atmospheric pressures and temperatures, whereas of course at the enormous depths of 3730 miles the nature of matter will undoubtedly be much different from what it is under normal atmospheric conditions. Some speculations have assumed that the earth's general structure ought to reflect the average mineral structure of the various meteorites and this has led to a fairly widespread belief that the core is made of iron or iron nickel. The main evidence for this is that it should be conducting so as to provide a self-exciting dynamo. Such properties however are not exclusive to metals (see later). One could equally assume that the core was made of carbon dioxide or other carbon compounds and a philosophical reason for this was mentioned in Chapter 3. However, the fact is that we simply do not know but can only make inspired guesses about the composition of the earth's core materials.

The Mantle

The mantle forms the major (80% by volume) part of the earth. At depths of about 30–150 miles there occurs a low velocity layer in which the P- and S-wave velocities drop by about 4% before returning to their initial velocities. It is assumed that the material in this low velocity layer is largely composed of the mineral peridotite (mixture of magnesium and iron silicates) and occurs because under certain conditions of temperature and hydration the mineral will melt at least in part (5% would account for the results). It is also suggested that part of the rock above the low velocity layer be called the *lithosphere*. The low velocity layer is called the *asthenosphere* and will be a very important region since it is one where major movements between the inner and outer parts of the earth will occur. The mantle below the low velocity layer is called the *mesosphere*. The deepest parts of the mantle down to the core mantle boundary show density increases of from 3·3 to 5·6 $\times$ 10^3 kg m^{-3}, pressures change from 9 to 1400 kilobars (*ca* 1·4 $\times$ 10^6 atmospheres); the deep mantle may also be composed of magnesium iron silicates of the peridotite or eclogite (a type of basalt with densely packed atomic structure requiring pressure for formation) types associated with heavier minerals.

The Crust and Upper Mantle

A Jugoslavian seismologist Mohorovicic showed that a discontinuity exists about 62 miles below the earth's surface. He noticed that two

Table 4.01. The structure of the earth

Section	% Mass of Earth	Major Chemicals	Physical Properties
Core	32·4	Iron-Nickel Alloy (postulated)	Central portion probably solid Outer part probably liquid
Mantle	67·2	Silicate materials at high pressures	Solid
Crust	0·4	Silicate materials at more normal pressures	Solid
		Organic matter (0 ⟶ approx 2%)	Mainly insoluble matter
Hydrosphere	0·024	Salt and fresh water, also ice and snow	Liquid Solid
Biosphere		Water	Liquid
		Organic matter	Solid and liquid
Atmosphere	0·00009	Nitrogen, oxygen, carbon dioxide, inert gases	Gas
		Water	Vapour
		Traces of other matter	Mainly solid

sets of S- and P-wave deflections (Fig. 4.03) occurred on seismographs when an earthquake was recorded within a radius of 500 miles and with a focus within 25 miles of the surface. The first set travelled directly whereas the second set of waves reached the recording device by refraction from the discontinuity. The discontinuity is known as the Mohorovicic discontinuity or Moho. It divides the earth into two main parts—the upper is called the crust and the lower is called the mantle at least to the core-mantle boundary (Table 4.01).

The Crust

The thin layer of rocks which constitute the earth's crust may be up to about 12 miles thick and separated from the mantle by the Moho discontinuity. The crust varies in thickness considerably. Thus under

Table 4.02. Main chemical elements present in the earth's crust

Element	Weight %	Atoms %	Atomic Radius (Å)	Volume %
Oxygen	46·60	62·55	1·40	93·77
Silicon	27·72	21·22	0·42	0·86
Aluminium	8·13	6·47	0·51	0·47
Iron	5·00	1·92	0·74	0·43
Calcium	3·63	1·94	0·99	1·03
Sodium	2·83	2·64	0·97	1·32
Potassium	2·59	1·42	1·33	1·83
Magnesium	2·09	1·84	0·66	0·29
	98·59	100·00		100·00

Figure from V. M. Goldschmidt (1954). "Geochemistry", Chapter 2.

The earth's crust consists almost entirely of oxygen compounds, especially silicates of iron, magnesium, calcium, sodium, potassium and aluminium.

The crust consists of more than 62% oxygen atoms, and when the corresponding volume of different atoms (or ions) is considered, oxygen makes up more than 93% of the total volume occupied by the elements.

Goldschmidt has remarked that the *lithosphere* could quite well be called the *oxysphere*.

the major mountain chains it will usually have maximum thickness, whereas under the ocean basins the crust is thin perhaps as thin as about 3 miles. Over the flatter parts of the continents the crust depth averages 12–18 miles. Within the continental crust there is yet another discontinuity, the so-called Conrad discontinuity which divides the crust into upper and lower parts. The thin ocean crust may represent the lower part of the continental crust. The upper part of the crust is termed the sialic layer, the term being derived from the predominance of the two elements Si and Al in the upper crystal rocks which are dominated by granite and related materials. The lower part of the

Table 4.03. Major minerals present in the earth's crust

Mineral	% Weight
SiO_2	60·18
Al_2O_3	15·61
CaO	5·17
Na_2O	3·91
FeO	3·88
MgO	3·56
K_2O	3·19
Fe_2O_3	3·14
TiO_2	1·06
P_2O_5	0·30
	100·0

Results from Clarke, F. W. and Washington, H. S. (1924) U.S. Geol. Survey, Profess. Paper 127.

Compositional data determined on 5159 samples and results calculated to a percentage value with the elimination of amounts of water and minor constituents.

Table 4.04. The mass of the earth's crust

Oceanic crust	$0{\cdot}48 \times 10^{25}$ g
Continental shelves	$0{\cdot}48 \times 10^{25}$ g
Young folded belts	$0{\cdot}53 \times 10^{25}$ g
Shield areas	$1{\cdot}08 \times 10^{25}$ g
Total mass of crust	$2{\cdot}61 \times 10^{25}$ g

crust is less well known, but assuming that the oceanic crusts are part of this layer the major rock material is basalt. This is the so-called Sima layer (from Si and Ma (magnesium!)) and its minerals differ from granite in possessing a high magnesium content.

The crust is very heterogeneous and there is much variation in the composition of rocks from region to region (Tables 4.02, 4.03, 4.04).

The younger marginal platforms adjacent to a continent consist mainly of sedimentary deposits derived from continued erosion of the continental surface and subsequent transport (normally by water) to the coast and deposition in shallow water on the continental shelf. These platforms often form beds up to several miles in thickness. Alternatively the oldest known rocks of the Precambrian Shields generally consist of igneous rocks such as granite and highly metamorphosed gneiss rocks. There are vast differences in the chemical composition within very narrow areas, and the largest compositional differences appear to occur

Table 4.05. Chemical composition of some common minerals

Mineral	Chemical Composition
Anhydrite	$CaSO_4$
Biotite	$K(Mg,Fe)_3[(AlSi)O_{10}](OH)_2$
Calcite	$CaCO_3$
Chlorite	$(Mg,Fe)_5\ Al(AlSi_3)O_{10}(OH)_9$
Chromite	$FeCr_2O_4$
Diopside	$Ca(Mg,Fe)[Si_2O_6]$
Dolomite	$CaMg(CO_3)_2$
Enstatite	$(Mg,Fe)SiO_3$
Glauconite	$K(Fe,Al)_2(Si,Al)_4O_{10}(OH)_2$
Gypsum	$CaSO_4,2H_2O$
Haematite	α-Fe_2O_3
Hornblende	$(Mg,Fe,Ca,Na,Al)[Si_6Al_2O_{22}](OH,F)$
Hypersthene	$(Mg,Fe)\ SiO_3$
Ilmenite	$FeTiO_3$
Lepidolite	$K(Li,Al)_3(Si,Al)_4O_{10}(OH,F)_2$
Magnetite	Fe_3O_4 or $MgFe_2O_4$
Microcline	$KAlSi_3O_8$
Muscovite	$K_2Al_4[Si_6Al_2O_{20}](OH,F)_4$
Oligoclase	$[NaAlSi_3O_8]\ [CaAl_2Si_2O_8]$
Olivine	$(Mg,Fe)_2[SiO_4]$
Orthoclase	$KAlSi_3O_8$
Pigeonite	$(Mg,Fe,Ca)(Mg,Fe)[Si_2O_6]$
Plagioclase	$Na[AlSi_3O_8]$–$Ca[Al_2Si_2O_8]$
Pyrite	FeS_2
Pyroxene	$(Fe,Mg)SiO_3$
Quartz	SiO_2
Sanidine	$KAlSi_3O_8$
Serpentine	$Mg_3[Si_2O_5](OH)_4$
Talc	$Mg_3(Si_4O_{10})(OH)_2$
Troilite	FeS
Uraninite	UO_2
Zircon	$ZrSiO_4$

among sedimentary rocks. During sedimentation it appears that sorting and separating of the chemical components often takes place. Weathering of surface rocks produces soil-like materials which contain high concentrations of the cations calcium, magnesium, sodium and potassium which because of their relatively high solubility are transported in solution mainly into the oceans and seas.

In the marine environment calcium is deposited as almost pure calcium carbonate (limestone, calcite ($CaCO_3$)), whilst under special conditions evaporites of dolomite ($CaMg(CO_3)_2$), anhydrite ($CaSO_4$), gypsum ($CaSO_4,2H_2O$) and even sometimes salt (NaCl) may be deposited. The more insoluble components, such as silica and alumina, produced by the decomposition of rocks are carried away as aqueous suspensions and emulsions and deposited as sandstones and clays. Sandstones are amongst the most common of all rocks and consist mainly of more or less rounded grains of quartz. Although silica (quartz) is generally considered insoluble in water, there still exist large amounts of soluble silicates produced during the weathering of silicate minerals. This soluble silica is eventually re-precipitated and contributes to the formation of muds. The silica and alumina insoluble phases may be physically separated (the rates of sedimentation of different particle sizes differ) to form clays. There are many different kinds of minerals in the outer crust and the more common rock-types occur in different proportions in different rocks. Many estimates have been made of the average composition of the outer crust, but the wide variation in composition of the rocks and large compositional changes from one area to another have given widely varying figures. Recent calculations based on the relative volumes and composition of the various types of sedimentary, igneous and metamorphic rocks (Table 4.05) suggest that the overall composition of the outer crust may be considered intermediate between granite and basalt.

THE EARTH'S MAGNETIC FIELD

The liquid nature of the earth's outer core has been invoked to explain the origin, sustenance and variability of the earth's magnetic field. It is assumed or required (1) that the fluid core is conducting (hence the popularity of an iron or iron–nickel constitution), (2) that there exists a small non-uniform magnetic field initially and (3) that the fluid material of the core is in motion, (4) that the earth's rotation is relatively (e.g. as compared with some other planets) rapid. The earth is then seen to function as a so-called self-exciting dynamo in which a current is induced in the moving fluid core as it rotates through the initial magnetic field. This induced current produces a magnetic field and the magnetic field produces a further current and so on. So long as we have sufficient energy to keep the core fluid moving, we can continue to produce a magnetic field. In other words, the earth behaves like a dynamo, but one which is self-exciting. Its function depends on the movements of massive bodies relative to one another and can similarly be seen as movements of heavily warped "ether" (Chapter 3). The concept is of course theoretical but has been shown to be feasible and model dynamos of this type have been made.

The concept tends to add weight to a metal composition for the core, but we must remember that at the pressures and temperatures existing at the centre of the earth many non-metal materials may possess properties of the desired metallic type.

The geomagnetic field is at least $2{\cdot}6 \times 10^9$ years old since rocks of that age have been found to contain original magnetization. Older rocks have yet to be examined.

ORIGIN OF PLANETARY BODIES

Most theories of planetary origin have special reference to our solar system (Table 4.06) and see it as a special case of the almost universal

Table 4.06. Planets and asteroids in the solar system

Planet	Actual distance[1] (in astronomical units) from the sun	Titius-Bode's Law[2]	Mass[3]	Density[4]
Mercury	0·39	0·4	0·05	4·5
Venus	0·72	0·7	0·8	4·8
Earth	1·0	1·0	1·0	5·5
Mars	1·52	1·6	0·1	3·9
Asteroids	2·805	2·8	undetermined	undetermined
Jupiter	5·2	5·2	318·0	1·3
Saturn	9·5	10·0	95·0	0·7
Uranus	19·1	19·6	14·5	1·5
Neptune	30·0	38·8	17·2	2·4
Pluto	39·4	77·2	0·9	4·0(?)

[1] Based on unit distance for the sun–earth.
[2] Titius-Bode's Law $r^n = 0{\cdot}4 + 0{\cdot}3(2)^{n-1}$, where r^n is the distance of the nth planet from the sun.
[3] Based on unit mass for the earth.
[4] in g/cm^3.

process of binary star formation. Thus almost all stars appear to form in twins, triplets, quintuplets or multiplets by the type of sub-cloud formation mentioned earlier. It is suggested by Kuiper and others that whether a nebular disc condenses into planets or stars will depend on the density of the disc material, only a small density range allowing planet formation. Any theory of origin for the solar system has to account for the various specific characteristics of the system, namely:

(1) The planets revolve around the sun in the same direction in circular orbits which lie almost in a single plane. The planets also revolve around their axes in the same direction as the orbital revolutions. The planetary orbits lie largely in the equatorial plane of the sun.
(2) The distance of the planets from the sun follows the Titius–Bode "law", namely: $r_n = 0{\cdot}4 + 0{\cdot}3\ (2)^{n-1}$ where r_n is the distance of the *n*th planet from the sun using the earth's distance as unity. However the outer planets Neptune and Pluto fail to follow the law.
(3) The rotation of the system is concentrated almost wholly in the planets and their satellites, whereas the mass of the system (99%) is concentrated at the centre in the sun with only 2% of the angular momentum.
(4) The four inner planets have low masses but high densities (about 4000 kg m^{-3}), whereas the four outer planets have high masses and low densities (about 1000 kg m^{-3}) (Table 4.06), with Pluto (density *ca* 4000 kg m^{-3}) an exception. Presumably a corollary to this would be that the inner planets contain a greater proportion of the heavier elements.
(5) As far as planet earth is concerned we must explain the formation of the particular and characteristic abundance of elements and molecules which make up its crust and atmosphere.

One of the earliest and best-known theories of planetary formation is the nebular theory due to Laplace (1796) and with various modifications it still remains one of the more attractive hypotheses. The theory suggests that the solar system originated from a rotating gaseous–dust mass (nebula) which spun off gaseous rings. The material of the rings condensed further into planets and the central portion remained and after suitable contraction became the sun. A major difficulty in this theory is concerned with the angular momentum of the planets and of the sun. It appears that the nebula could not in fact have spun sufficiently fast to fling out rings. Thus if one adds the current measured angular momenta of the planets to that of the sun the result would increase the rate of rotation of the sun by about 50 times to about 1 revolution per half day, instead of the actual observed 1 revolution per 25 days. However, this increase in rotation would only result in an increase of the centrifugal force at the sun's equator by about 5% of the gravity force and would be insufficient to overcome the force of gravity and to throw off rings.

These general conclusions led to a temporary abandonment of the Laplace theory (until 1943) and its replacement by alternative theories, which served to explain the enhanced angular momentum of the planets relative to the sun by invoking a collision or near collision type mechanism involving one or more large bodies which would closely approach our nebular mass and cause portions to be torn away, to fall ultimately into appropriate circular orbits with momenta enhanced by the process. Such mechanisms assume what must be a very rare occurrence, namely the close approach of two stars which are normally separated by such enormous distances, and would, if true, indicate that planet formation was a very rare occurrence indeed. However, examination (de Kamp) of perturbations in nearby stars suggests the presence of planetary systems other than our own, indicating that planetary systems may be quite common.

An alternative analogous mechanism proposed by Schmidt suggests that our primeval sun produced the planetary system by slow movement (calculations suggest not more than 0·5 mile s^{-1} compared with the sun's present speed relative to the stars of 12 miles s^{-1}) through vast extended gas–dust clouds of which there are certainly plenty available in our galaxy. In so doing it would gather the material to it in the form of rings. However, many believe that in fact, under these circumstances, it would simply enlarge to form ultimately a giant star. To overcome this objection Schmidt has had to postulate that, in addition, a close (but not as close as in the collision hypotheses) approach to some other large body is required. The process then leads to accretion of material at relatively low temperatures in agreement with the conclusions of Urey that the crustal abundance of various light elements such as boron or mercury would suggest that the primordial earth's surface temperature would not have been very high, perhaps not more than 1–200 degrees.

However, since 1943 the original objections to the Laplace theory have been largely overcome, and it is now possible to account for the differences in angular momentum of the sun and planets. There are numerous modern variations on the Laplace type mechanism postulated by Alfven, Weizsacker, Kuiper, Hoyle and others. They mostly suggest that the primordial sun had a moderate magnetic field unlike the very weak field possessed at the moment. The theories differ in the amount of field strength required. Thus Alfven requires an enormous field (*ca* 300,000 gauss) compared with the sun's probable very weak field of 1 or 2 gauss, and the problem here is to explain the subsequent loss of field. Hoyle, on the other hand, only requires a small field (1 gauss). The magnetic fields are required to slow down the central portion of the protosun and speed up the outermost edges.

In Hoyle's theory it is assumed that the original sun was rotating more rapidly than at present so as to develop a disc of planetary material. The lighter elements, hydrogen, etc., are then seen to accelerate and separate out at the edge of the disc, taking with them finer dust particles and leaving the heavier elements and larger particles nearer the central sun. This accounts for the dense nature of the inner planets including earth and the less dense nature of the outer planets, although Pluto seems to be an exception. The theory would also go some way to eliminating the likelihood of any reducing (H-containing) atmosphere on the primeval earth which has been frequently suggested but for which there is no geochemical evidence. Some of the other mechanisms achieve elemental separation by invoking ionization of materials which will differ according to the particular element and play a part in the resultant separation in electromagnetic fields. However, the amount of field required to produce sufficient ionization is generally thought to be too large. None of the current theories is entirely satisfactory and the theorists are handicapped by a lack of data.

THE PRIMEVAL EARTH'S ATMOSPHERE

Perhaps the most commonly held view about the formation of the earth's primitive atmosphere dates back to the original concepts of Urey, Kuiper and others. As we have seen in these theories, the earth is seen to form by the accretion of gas and dust. The chemistry of such material is generally assumed to reflect the chemistry of present dust particles in the solar system (meteorites) and this gives rise to the concept of an earth rich in iron (core) with much silicate, aluminium, etc., type elements (Table 4.07). The total mass of catalogued falls of iron meteorites, however, contribute only about 10% of the total mass of recorded meteorites; this conflicts with the large iron content ascribed to earth materials if this is to be based on meteorite analyses and ratios of falls to finds of iron to stoney meteorites. The theory also suggests that the accretion process was slow, that it occurred to give eventually an earth with a low initial temperature and with a gaseous reducing atmosphere, although it is considered (Urey) that the closeness of the primeval or proto-earth and for that matter the other three inner proto-planets (from which Mercury, Venus and Mars were evolved) to the sun would result in loss of volatile hydrogen and helium and concentration of the heavier elements. This would account for the much greater density of the four inner planets. The low average density of the outer planets is of course due in large measure to their enormous gaseous atmospheres. The reducing atmosphere, it is believed, would be

Table 4.07. Some estimates of the main chemical elements present in the earth

Element	% Weight
Iron	34·63
Oxygen	29·53
Silicon	15·20
Magnesium	12·70
Nickel	12·39
Sulphur	1·93
Calcium	1·13
Aluminium	1·09
Sodium	0·57
Chromium	0·26
Manganese	0·22
Cobalt	0·13
Phosphorus	0·10
Potassium	0·07
Titanium	0·05
Other elements	>0·01
	100·00

Results from Mason, B. (1966) "Principles of Geochemistry", Chapter 3. Wiley, New York.

ideal for the abiogenic synthesis of organic compounds and their subsequent elaboration into living systems by further abiogenic processes (see Chapter 8).

However in recent years there has been a change of thinking about the origin of the earth and its atmosphere. On the one hand we have seen earlier that there is evidence which suggests that the core of the earth was formed rapidly and this would suggest a very hot primitive world. More extensive dating has underlined the increasingly old nature of the oldest rocks on earth (now about 3980 million years) and there is accordingly a concomitant increasing requirement for a relatively early core formation. On the other hand, the amounts of rare gases in the current earth's atmosphere are very low; thus, as Abelson has pointed out, the amount of Ne in the current earth's atmosphere is only 10^{-10} that of its cosmic abundance whereas Ar, Kr and Xe are virtually absent. Since these gases cannot readily be removed by chemical reactions it may be argued that they were never able to accumulate (partly because of high temperature earth formation, processes) and hence much smaller gases such as hydrogen, methane, etc., would certainly not be able to accumulate under whatever conditions were extant at the time.

It has generally been suggested that the accretion processes occurred rather slowly over a period of about 10^8 years so that energy could be lost equally slowly and the earth formed in a cool unmolten condition. The formation and segregation of the core would occur by heating, the required energy being derived from radioactive decay of long-lived elements. This frequently promulgated theory would suggest that differentiation of the core and presumably other layers in the body of the earth would have probably occurred much later than the primary accretion process, and it has been suggested that the process might still be going on today.

However, recent work (Ringwood and co-workers, 1971) suggests that during core formation the Pb/U ratio of the metal phase would be much higher than that of the silicate phase and a large amount of the total Pb in the earth would enter the core. The age of the earth is about $4{\cdot}6 \times 10^9$ years and this age is based on the observation that the isotopic composition of modern Pb falls on the meteoritic isochron which records a differentiation of lead relative to U in meteorites at this time. Both the meteor and ore body ages assume that the Pb/U ratio of the upper mantle crust had been established at the time of earth formation and then remained constant. If the core was formed subsequently, then the Pb/U ratio of the mantle should alter since a molten iron matrix would carry much Pb but little U into the core. Since this does not appear to have happened, it has been concluded that the core was formed very shortly after formation of the earth. In support of this it was shown in model experiments that the distribution of lead between relevant metal and silicate systems favoured the metal phase. Various models used included $Fe_{89}Si_{11}$, $Fe_{83}S_{15}C_2$ and $Fe_{70}S_{30}$ which were chosen according to various hypotheses about the core composition, i.e. whether dominant in Si, and with or without sulphur and carbon. The silicate phase was modified basalt doped with various amounts of Pb.

Outgassing as a Source of the Primitive Atmosphere

There is good geological evidence that gases and other volatile components have been constantly added to the atmosphere of the earth for at least $3{\cdot}9 \times 10^9$ years by various degassing processes, notably volcanism and orogeny. Moreover the composition of volcanic gases today is of a type (Table 4.08) which resembles a major part of

Table 4.08. Volcanic gases

Steam	Major Amount
Carbon dioxide	
Nitrogen	
Sulphur dioxide	
Hydrogen	
Carbon monoxide	
Sulphur	
Chlorine	
Hydrogen sulphide	
Hydrochloric acid and other acids	
Volatile chlorides of iron, potassium and other metals	Trace Amounts

Decreasing amounts present ↓

the current atmospheric constituents. Many now believe that at time t_0 there was little or no atmosphere present on earth and the current atmosphere resulted firstly from degassing of the earth's interior and later and additionally of course from the cumulative effects of increasing amounts of biological material, leading especially to oxygen formation by photosynthesis, or allied processes. Growing evidence in favour of this concept of the earth at time t_0 includes the likelihood of an initially hot earth with its requirement of an early core formation, the very large abundance of ^{40}Ar in the current atmosphere relative to ^{36}Ar and ^{38}Ar. The ^{40}Ar can be accounted for (Turekian, 1964) by the radioactive decay of ^{40}K in the lithosphere. Other similar rare gas abundance anomalies have been mentioned earlier.

Fanale (1971) has outlined how these various pieces of evidence suggest that the earth at the time of accretion underwent a period of catastrophic and thorough melting, leading (1) to concentration of most of the U, Th, K and Pb in the outermost parts, (2) rapid core formation, (3) almost complete expulsion of inert gases such as Ar, Ne and Kr from its interior. Substances such as water, carbon dioxide and the like would be less efficiently expelled he concludes. These concepts certainly would be in full accord with the expectations of chemistry in this area (see later).

In particular during the accretion process we have dust particles and associated gases. Now under the high vacuum and low temperatures of outer space we would expect a substantial perhaps almost complete absorption of gases and other volatile materials on the rock surfaces. The compounds which will be most strongly absorbed will be those compounds most able to undergo, in general, hydrogen bonding and polarization processes (Fig. 4.04). They will include especially water, hydrogen sulphide, carbon dioxide, ammonia, alcohols and similar molecules.

(A) van der Waals adsorption processes:

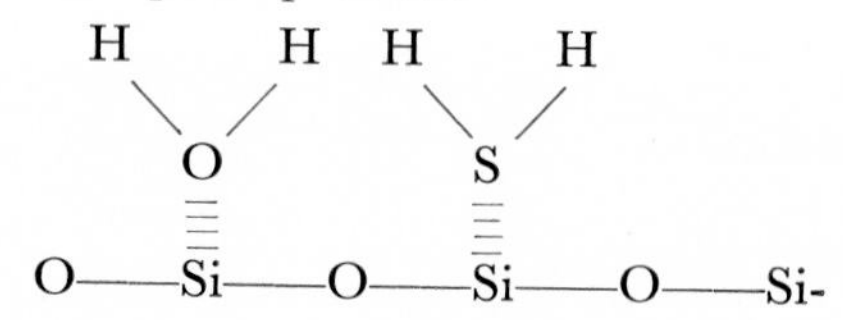

(B) Chemisorption processes involving covalent and ionic bond formation:

O——Si——O——Si——O——Si- + $H_2$0 → O——SiOH HO-Si——O-

Also:

(a) withH_2S

→ O——Si-SH HO-Si——O—

(b) with ammonia

→O——Si-O^- NH_4^+ H_2N-Si——O-

(c) with $CO_2 + H_2O$

→ O——Si-OH ^-O_2C-O-Si——O-

Fig. 4.04

Compounds such as this will be strongly bound to rock surfaces, whereas less polarizable substances including nitrogen, oxygen, hydrogen, methane and especially the rare gases will be but weakly held. Incidentally the accretion process itself should favour those particles with maximal polar adsorbed materials since the act of adsorption of such compounds will in itself tend to polarize the rocky particles and make them more likely to adhere. Following accretion we will have a primitive planet with virtually no free atmosphere. The primitive planet is hot and within it the core and mantle structures are rapidly formed and the magnetic field soon established.

Meanwhile the atmosphere will begin to form by degassing processes. In the initial stages these will be catastrophic and lead quickly to an atmosphere containing water, hydrogen sulphide, ammonia, carbon dioxide and similar compounds in equilibrium with large reservoirs of magma with little free elemental gases or weakly adsorbed material such as alkanes. The degassing processes continue in ever-decreasing intensity to the present day. At the same time modifications to the primeval atmosphere will occur by chemical interactions with the environment and especially by the effect of biological material, which was probably introduced to the planet at a very early stage and would begin to survive and multiply as soon as the surface had cooled to a reasonable temperature. This was probably about 4×10^9 years ago and roughly corresponds with the appearance of the first life-bearing sediments (Chapters 9 and 10) which approach this age and the presence of thoroughly worked adjacent basement rocks of about the same age.

During the magmatization processes most of the U, Th and K in the earth will have been concentrated in the outer parts and isotopic studies suggest that this process of elemental concentration occurred during the first 100 million years of the earth's history.

SUMMARY OF THE PHYSICAL AND PHYSICO-CHEMICAL CHANGES IN THE EARTH FROM TIME t_0

t_0

This is about $4{\cdot}56 \times 10^9$ years ago. The figures are based both on meteorite age data and on age determinations of earth rocks. The earth was formed by accretion of dust particles with a preponderance of those particles which will have been rendered "sticky" by adsorption of polar molecules such as water, hydrogen sulphide, ammonia, carbon dioxide and similar materials which will confer polarity to the rock particles. Saturation of adsorption sites on the rock particles with the readily adsorbed polar materials will of course leave little room for the weakly adsorbed neutral materials such as the simple elemental gases nitrogen, oxygen, hydrogen (less so) and especially the inert gases. This would go some way to explaining the paucity of inert gases in the current atmosphere, whereas the higher concentrations of nitrogen and oxygen today must be explained by subsequent chemical and biological processes.

$t_0 \rightarrow t_{1\times10^8}$ years

The earth's core and mantle structures are formed and the magnetic field established. The primitive atmosphere is formed by degassing processes and will consist largely of polar molecules. The most polar molecules will be last to be degassed and their concentration will accordingly increase with time. The atmosphere will contain but small amounts of neutral molecules (nitrogen) and very little indeed of the rare gases, reflecting their weak adsorption characteristics. However, since neutral molecules will be first to be desorbed their concentration should be maximal in the initial stages.

$t_{1\times10^8} \rightarrow t_{0\cdot58\times10^9}$ years

In the first part of this period the earth passed through a violent stage involving catastrophic magmatization processes at high temperatures. It slowly cooled down, until at the end of the period it had reached temperatures approaching those sufficiently moderate to support primitive living systems. In addition the atmosphere would have undergone various chemical changes. These would lead to an increase in the percentage of the more highly polarized molecules such as water, carbon dioxide and hydrogen sulphide by continuing degassing processes but, in addition, formation of many other substances by various chemical processes would undoubtedly occur. Such compounds may include alkanes, much free sulphur, elemental nitrogen, etc. Hydrogen sulphide could be of special importance as a ready source of hydrogen in these reactions.

e.g. possibilities include:

$$CO + 3H_2S \rightarrow CH_4 + H_2O + 3S$$
$$CO_2 + 4H_2S \rightarrow CH_4 + 2H_2O + 4S$$
$$2NH_3 + 3S \rightarrow N_2 + 3H_2S$$

$t_{0\cdot58\times10^9} \rightarrow t_{0\cdot75\times10^9}$ years

During this relatively brief period it seems reasonably certain that living systems first arose on earth but whether pro- or eu-karyotic or both is uncertain. The atmosphere would be essentially neutral with large quantities of carbon dioxide and other carbon compounds derived therefrom, but little oxygen. The time available for any abiogenic synthesis let alone chemical evolution is very small (see Chapters 8 and 12).

$t_{0\cdot75\times10^9} \rightarrow t_{3\cdot6\times10^9}$ years

Throughout this enormous span of time the simple primitive organisms extant on the evolving earth underwent little or no evolution. At the end of the period some degree of sophistication of the micro-organisms was becoming evident (worm-like creatures and other multicellular complex organisms occurred). The atmosphere, however, underwent slow but inexorable changes resulting from a complex of degassing, chemical reactions and especially biological chemical reactions. It is the latter reactions which led to the formation of the continually increasing amounts of oxygen in the atmosphere, as a result of photosynthetic processes.

$t_{3\cdot6\times10^9} \rightarrow t_{4\cdot0\times10^9}$ years

A period in which further limited biological evolution occurred. Towards the latter part of this period the ratio of oxygen/carbon dioxide in the atmosphere apparently reached a figure which favoured extensive mutation and evolution of the simple micro-organisms. It is a period about which we should know much more than we do.

$t_{4\cdot0\times10^9} \rightarrow$ to date ($t_{4\cdot56\times10^9}$) years

An enormous sharp increase in evolutionary processes occurred during this period (the Cambrian) with formation of all the highly differentiated and integrated metazoan living systems including man, with which we are familiar today.

SUGGESTED FURTHER READING — Chapter 4

Books

Gass, I. G., Smith, P. J. and Wilson, R. C. L. (1971). "Understanding the Earth". Open University Set Books. Artemis Press, Horsham, Sussex.
Holmes, A. (1965). "Principles of Physical Geology". Nelson, London.
Mason, B. (1966). "Principles of Geochemistry". Wiley, New York.
Milne, J. (1939). "Earthquakes and Other Earth Movements". Routledge and Kegan Paul, London.
Phinney, R. A. (1968). "The History of the Earth's Crust". Princeton University Press.
Strangway, D. W. (1970). "History of the Earth's Magnetic Field". McGraw Hill, New York.

Urey, H. C. (1952). "The Planets". University of Chicago Press.
Marsden, B. G. and Cameron, A. G. W. (Eds) (1966). "The Earth–Moon System". Plenum Press, New York.
Munk, W. H. and MacDonald, G. J. F. (1960). "The Rotation of the Earth". Cambridge University Press.

Articles

Chinnery, M. (1971). The Chandler Wobble. *In* Understanding the Earth (Eds I. G. Glass *et al.*), Open University Set Book. Artemis Press, Horsham, Sussex.
Moore, R. C. (1970). Stability of the Earth's Crust. *Geol. Soc. Am. Bull.* **81,** 1285–1323.
Urey, H. C. (1962). Evidence Regarding the Origin of the Earth, *Geochim. cosmochim. Acta* **26,** 1–14.

5

The Chemicals of Life

The simple cell contains many types of chemicals both organic and inorganic, ranging from simple substances like common salt or ammonia to complex molecules with enormous molecular weights. The latter substances are the central and universal units of all living matter and contain in their macromolecular structures the complex variety of function required for the smooth self-contained operation of the cell. The compounds which are of particular and dominant importance in this respect are the nucleic acids, the proteins and the polysaccharides, their respective monomeric precursors, the nucleotides, α-amino-acids and monosaccharides, and members of that large and somewhat ill-defined group of substances—the lipids.

CARBOHYDRATES

Monosaccharides

Monosaccharides are a group of compounds which may be given the general formula $(CH_2O)_n$, where n is 3 or more. They are essentially polyhydroxy aldehydes (aldoses) or ketones (ketoses) or more commonly structural (cyclic) isomers (see below) and the most common may be formally (and chemically) derived from the simplest members of each family, namely from D-glyceraldehyde (aldoses) (Fig. 5.01) and dihydroxyacetone (ketoses) (Fig. 5.02).

They are termed trioses, tetroses, pentoses, hexoses, etc. according to the number of carbon atoms present in the molecular stem and aldoses or ketoses according to their possession of an aldehyde or ketone group respectively. Glucose is therefore an aldohexose, ribose an aldopentose and fructose a ketohexose. They form groups of diastereoisomers, the number of isomers being given by 2^n, where n is the number of different asymmetric carbon atoms. Thus in the aldehexoses with

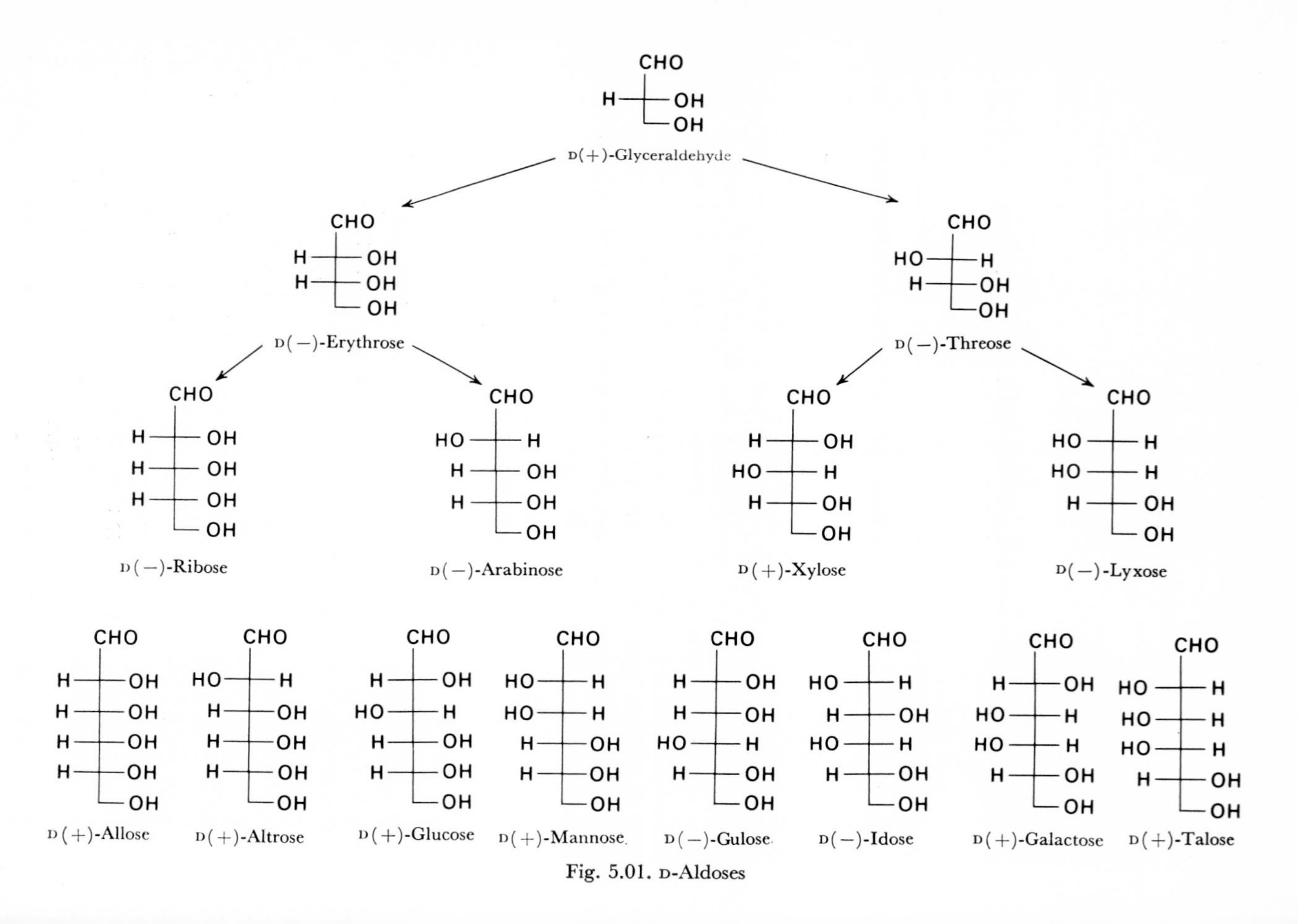

Fig. 5.01. D-Aldoses

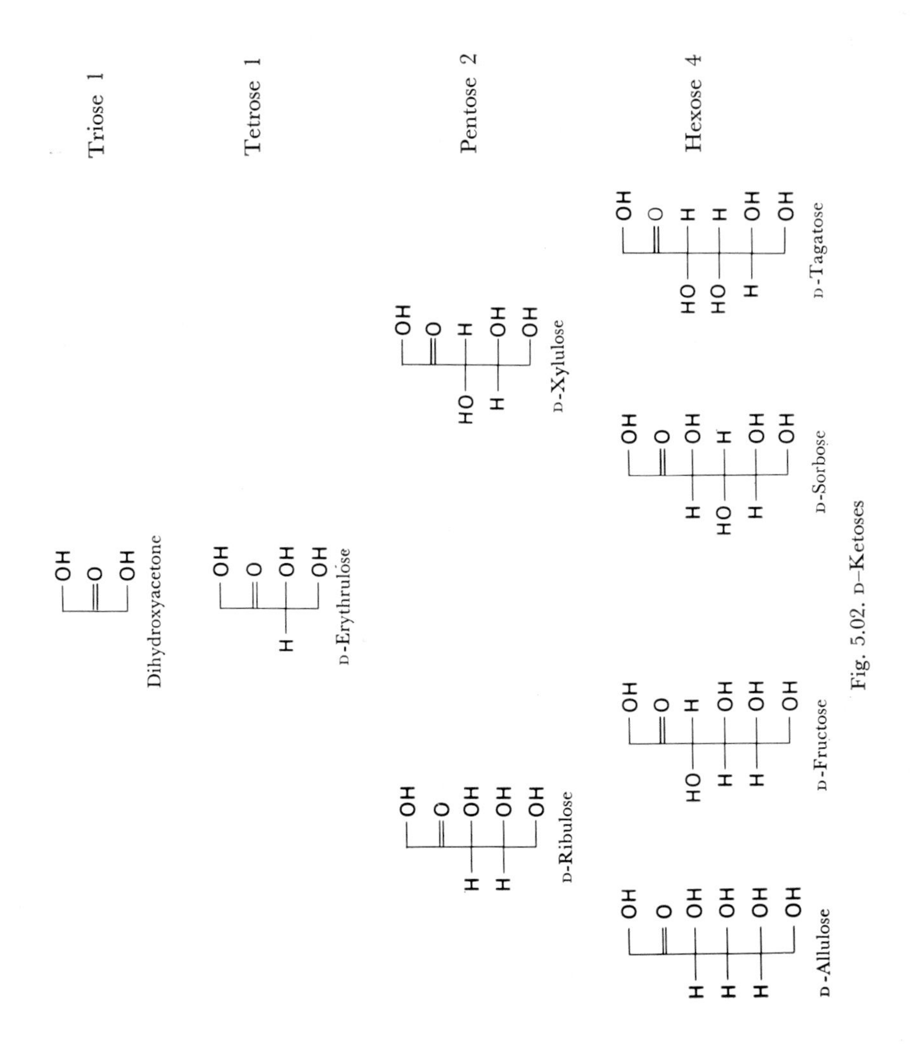

Fig. 5.02. D-Ketoses

5 chiral centres there are $2^5 = 32$ isomers, or 16 DL pairs. These may be further resolved into 8 pairs each of the α- and β- forms of the D- and L- isomers. Each of the 8 are given individual names (Fig. 5.01). The α- and β- nomenclature refers to the anomeric configuration at C1 which because of its hemiacetal nature is especially labile (hence the special label) and readily undergoes equilibration reactions in aqueous solution. Thus a solution of α-D-glucose ($\alpha_D^{20} = +112{\cdot}2°$) in water rapidly reverts to a mixture of α- and β-D-glucose ($\alpha_D^{20} = 52{\cdot}7°$). Similarly the β- form ($\alpha_D^{20} = 18{\cdot}7°$) gives the same equilibrium mixture

(Fig. 5.03). The formation of α- and β- anomeric forms of the sugars requires that they exist in cyclic hemiacetal forms (Fig. 5.04) which may be 5-(furanose) or 6-(pyranose) membered ring systems. Generally the pyranose form is the most common and preparation of furanose

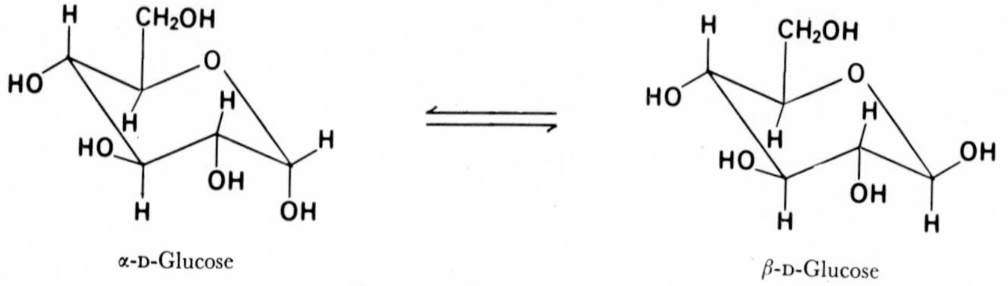

α-D-Glucose β-D-Glucose

Fig. 5.03. Mutrarotation of α-D ⇌ β-D-glucose

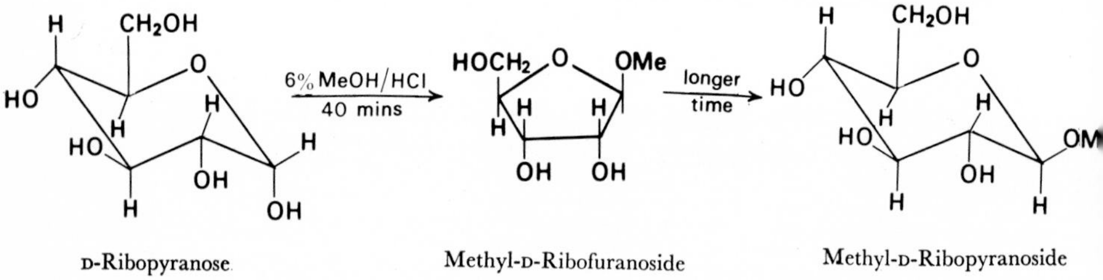

6% MeOH/HCl 40 mins; longer time

D-Ribopyranose Methyl-D-Ribofuranoside Methyl-D-Ribopyranoside

Fig. 5.04. Formation of Furanose derivative of D-ribose

derivatives needs carefully defined conditions. It must be remembered, however, that in solution glucose, for example, will exist as a mixture of chemical entities which include a small amount (0·06%) of the open chain aldehyde (Fig. 5.05).

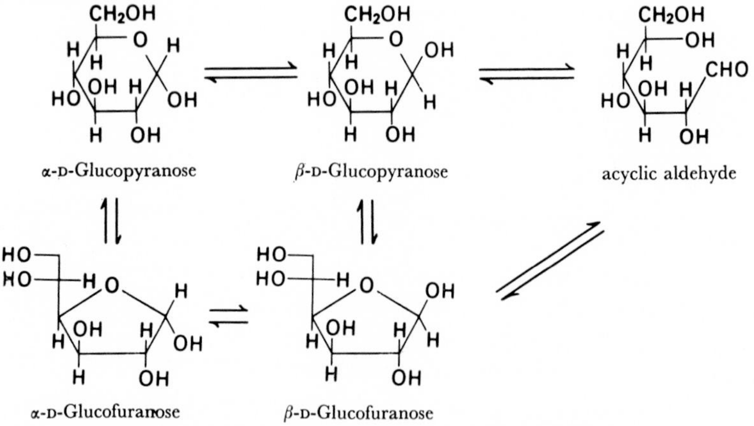

α-D-Glucopyranose β-D-Glucopyranose acyclic aldehyde

α-D-Glucofuranose β-D-Glucofuranose

Fig 5.05. Types of equilibrium in aqueous solutions of D-glucose

Polysaccharides

The nucleic acids and the proteins form a dominant central partnership in the processes of cell growth and metabolism, whereas the major role of polysaccharides appears to be of a dual nature, concerned on the one hand with storage of sugar monomer units and on the other with formation of structural or skeletal units in membrane or cell wall formation.

Polysaccharides are condensation polymers of monosaccharides in which the glycosidic C-1-hydroxyl of one of the units is linked in an ether bond to a non-glycosidic hydroxyl of the next.

The simple polysaccharides (especially starches and celluloses) differ from both proteins and nucleic acids in that the polymer chain rarely contains exocyclic substituents and consequently its function is limited; also materials like cellulose will largely differ among themselves only in chain length, and starches in both chain length and in the mode of connection between the inter-glycosidic units. Other less well-known polysaccharides, however, including those concerned with wall structures of both lower organisms and of vertebrates, etc., have protruding exocyclic groups and in this sense are more closely allied in structure to proteins or nucleic acids.

Starch and Cellulose

The two quantitatively most important naturally-occurring polysaccharides are starch and cellulose (Fig. 5.06). Both are condensation

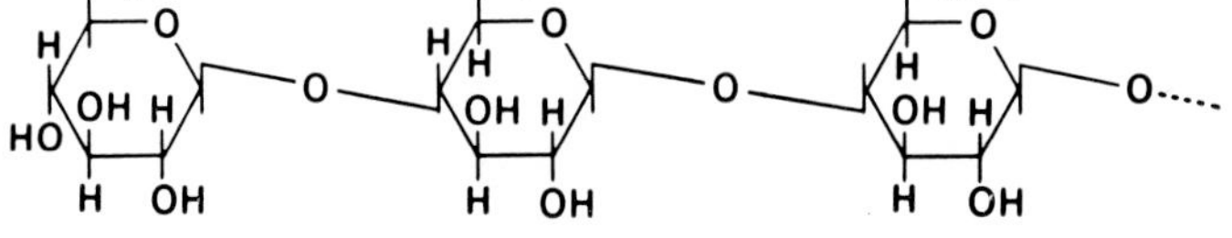

Fig. 5.06. Cellulose

polymers of D-glucose, but their physical and chemical properties are markedly different. Thus whereas starch is soluble, readily hydrolysed and easily digested, cellulose is insoluble, hydrolysed with difficulty and indigestible. The reasons for these differences are to be found in the stereochemical configuration of the molecules. Cellulose, which makes up the major material of the extra-cellular fibrous and skeletal material of plants and trees and 50% of the biosphere organic matter, consists of a long chain of D-glucose units joined by a β-glucoside link of one unit to the C_4 hydroxyl of another, the configuration of the simple disaccharide, cellobiose, and in a typical specimen of cotton there may be approximately 3000 glucose units per average molecule. The particular configuration results in polymer molecules which readily align in an orderly sense, forming bundles of parallel chains or fibrils and combined with the strong electro-static forces which exist between adjacent chains serve to give cellulose its strength and fibrous qualities. Cellulose of higher fibre quality consists of larger chains and fibrillar bundles. *Starch*, on the other hand, is a mixture of two polysaccharides, amylose and amylopectin, and is essentially a source of food, generally occurring as intracellular granules. The amounts vary, but most starches contain about 20% amylose and 80% amylopectin (Fig. 5.07). The components

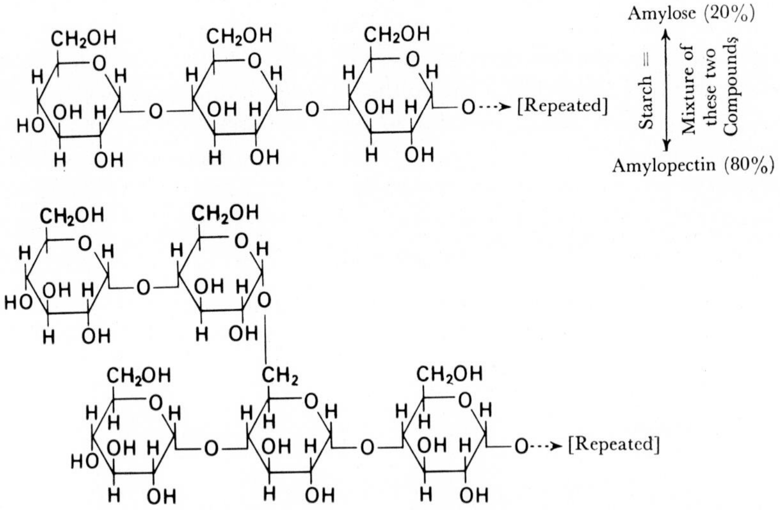

5.07. Polysaccharides present in starch

may be separated by swelling starch granules in warm water when amylose enters solution as a helically coiled hydrated micelle, leaving amylopectin as an insoluble residue. Amylose is a linear polyglucoside containing, like cellulose, 1,4-linked glucose units but joined by α-glucosidic linkages with the disaccharide maltose configuration. It is the amylose part of starch which gives the characteristic blue colour with iodine—a result of inclusion of the element in the chain; amylopectin gives a red to violet colour. Amylopectin also contains 1,4-linked glucose units with α-glucoside linkages but is also branched so that each branch contains about 20–25 glucose units attached by 1,6-linkages to a backbone of α-1,4-linked (amylose) units as illustrated (Fig. 5.07).

Amylose is hydrolysed by the enzyme α-amylase [α-(1 → 4) glucan 4-glucano hydrolase] to a mixture of glucose and maltose. The enzyme occurs in pancreatic juice and saliva and participates in starch digestion. Starch is also hydrolysed by β-amylase, a constituent of malt, and this cleaves maltose units from the polymer chain from the non-reducing end. Intermediate polymers are formed and called dextrins. α- and β-Amylases also attack the 1–4 linkages in amylopectin but are unable to attack the 1–6 branching link, so that the product is a highly branched core polysaccharide. The core may be further degraded by other enzymes [α (1 → 6) glucosidases] to give full hydrolysis to glucose and maltose.

Glycogen

Glycogen or animal starch is a food reserve storage D-glucose polysaccharide which occurs in animal tissues, especially muscle and liver. It is similar to amylopectin, but is generally more highly branched with 1–6 linkages at intervals of about 10 glucose units.

Dextrans

Dextrans function as storage polysaccharides in yeast and fungi and are also branched polymers of D-glucose in which the main linkage, however, is 1–6 and the branching varies from 1–2, 1–3 and 1–4.

Nitrogenous Polysaccharides

D-Glucose polymers (especially starch and celluloses) are the most ubiquitous of polysaccharides but there are many similar polymers of

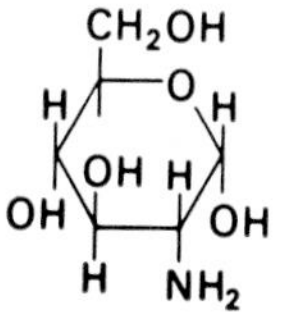

CH_2OH H O H OH H OH OH H NH_2

D-glucosamine

5.08. Glucosamine

considerable importance. An important group contains amino deoxy D-glucose units, especially D-glucosamine (Fig. 5.08) in which the amino group is invariably acetylated. Chitin (Fig. 5.09) for example is a homopolymer of N-acetylglucosamine (NAG) and forms an

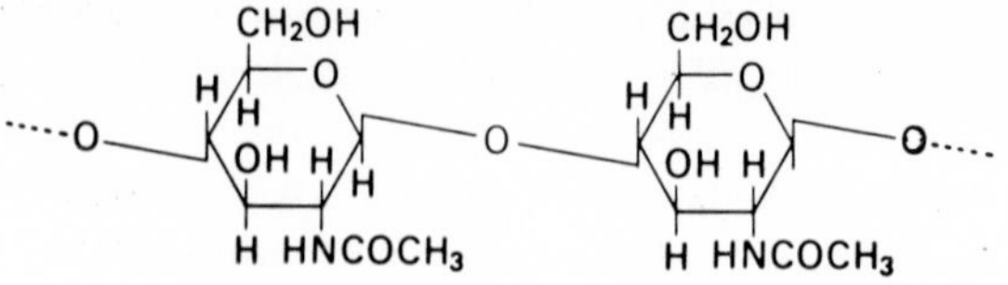

5.09. Chitin (polymer of acetylglucosamine)

important structural unit of the exo-skeleton of insects and crustacea. It has the same (1–4) linkage as in cellulose and like the latter is highly insoluble and relatively inert. Many bacterial cell walls contain a polysaccharide associated with peptides, i.e. peptidoglycan or murein (*murus* = wall). In this material the recurring unit is the muropeptide (Fig. 5.10) which consists of a disaccharide formed from N-acetyl-glucosamine and N-acetylmuramic acid (an ether at C3 of NAG with

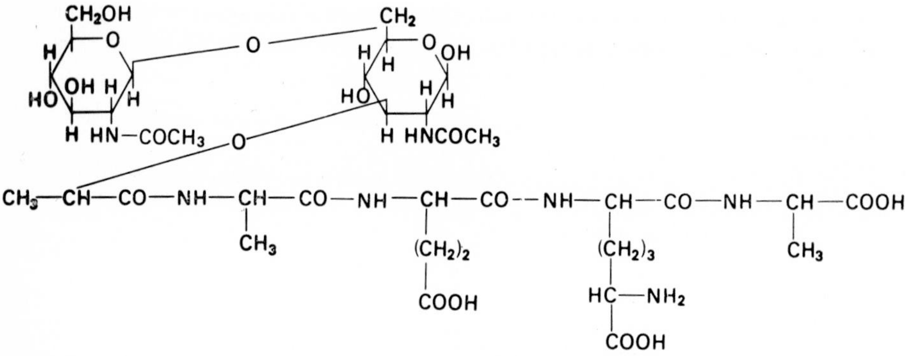

5.10. Muropeptide C_6 from *E. coli*

the α-hydroxyl group of lactic acid) in a β-1-6 linkage. The lactic acid carboxyl group of the acetylmuramic acid acylates a tetrapeptide consisting of L-Ala, D-Ala, D-Glu and L-Lys (or diaminopimelic acid).

In murein itself the long parallel polysaccharide chains of the poly-muropeptide are cross linked by peptides. In *Staphylococcus aureus* for example the linkage peptide consists of pentaglycine between the D-Ala and L-Lys of adjacent chains. Bacterial cell walls also contain accessory components which vary according to whether the bacteria are Gram-positive or Gram-negative (i.e. whether they react positively or not to the Gram stain—crystal violet, iodine and safranine).

In Gram-positive bacteria these are largely polymers interwoven in the murein network structure. They consist of various recurring polysaccharides of glucose, rhamnose, mannose or galactose or their deoxyamino derivatives, of peptides or proteins, and of substances known as teichoic acids (Fig. 5.11). The latter are poly-glycerol or ribitol phosphates some of which possess, on alternating hydroxyl groups, D-Ala and D-glucose or N-acetyl D-glucosamine.

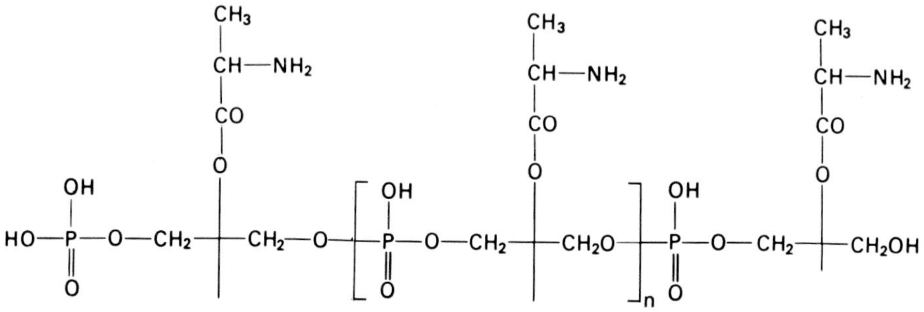

Glycerol teichoic acid from intracellular portion of *Lactobacillus casei*

5.11. Example of teichoic acid

Both the teichoic acids and the other polysaccharide components have antigenic activity. The walls of Gram-negative bacteria (e.g. *E. coli*) have more accessory units directly attached to the murein framework. They include polypeptides, lipopolysaccharides and lipoproteins. These components tend to overlay the murein network and combine to make such bacterial coats more superficially fatty than those

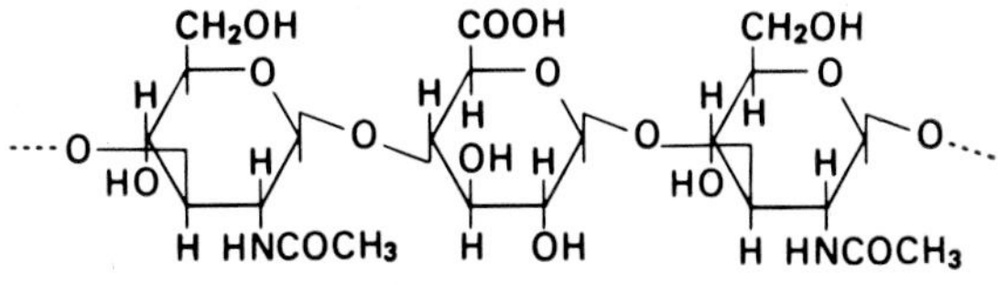

Fig. 5.12 Hyaluronic Acid

of their Gram-positive colleagues, and this undoubtedly accounts for the different reaction to the Gram stain.

In higher organisms there is little precise information about cell coat structures although a certain amount of work has been carried out on red blood cells. The coats apparently contain, in addition to various glycoproteins and lipopolysaccharides of unknown structure, the acidic mucopolysaccharides. The most abundant of these is hyaluronic acid (Fig. 5.12), a polymer of the disaccharide compounded of D-glucuronic acid and NAG in an α-1–3 linkage. Chondroitin is similar but contains N-acetyl-D-galactosamine in place of the NAG residues. Chondroitin occurs as a major component of cartilage, bone, cornea and other connective tissues in vertebrates, in the form of sulphate esters on C4 (Chondroitin A) (Fig. 5.13) or C6 (Chrondroitin C), and forms highly viscous solutions. The material heparin (Fig. 5.14) which occurs in the

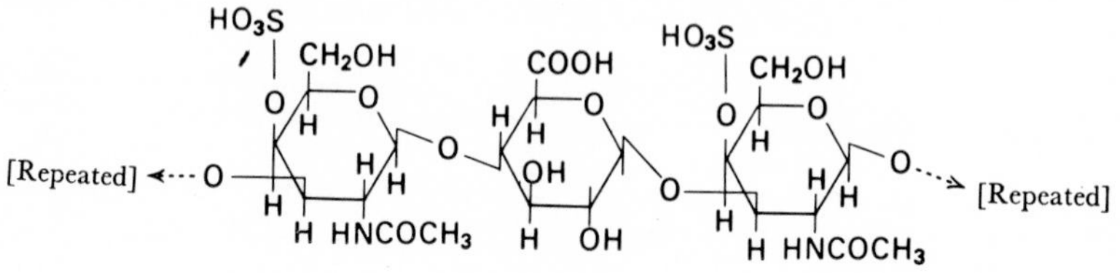

Fig. 5.13. Chondroitin sulphate A (e.g. as in cartilage and bone and the ocular cornea of vertebrata)

Acid mucopolysaccharides

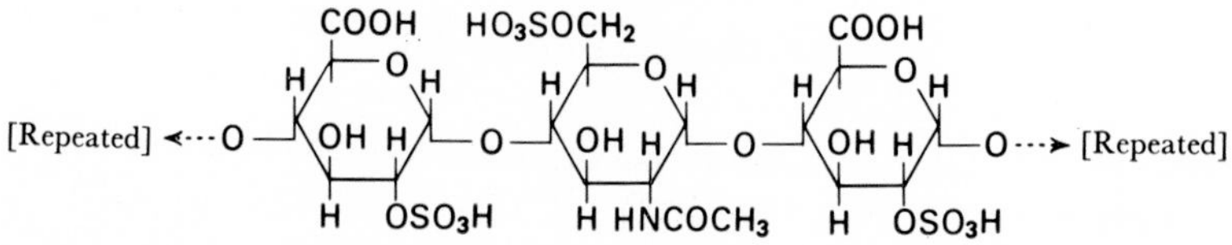

Fig. 5.14. Heparin (e.g. as in the liver, lung and blood of vertebrata; acts as a blood anti-coagulant)

extracellular spaces of lung, liver, etc., is similar, and consists of a 1–4 linked polymer of glucuronic acid with the sulphamic acid and sulphonate (C3) ester of D-glucosamine. It is concerned with the regulation of blood clotting and widely used as an anti-coagulant in the treatment of thromboses.

NUCLEIC ACIDS

Introduction

Nucleic acids are a group of naturally occurring condensation polymers of purine and pyrimidine nucleotides and are derived from two different but closely related sugars, D-ribose and 2-deoxy-D-ribose, so producing ribose nucleic acid (RNA) and deoxyribose nucleic acid (DNA). DNA was first isolated by Friedrich Miescher (1869) from pus cells and later from fish sperm by extraction with aqueous alkalis and precipitation of the acid polymers with mineral acid. The name, originally nuclein, then later nucleic acids (proposed by Altman in 1889), arose from the known association (especially from histochemical studies) of the polymers with cell nuclei. The early preparations were necessarily

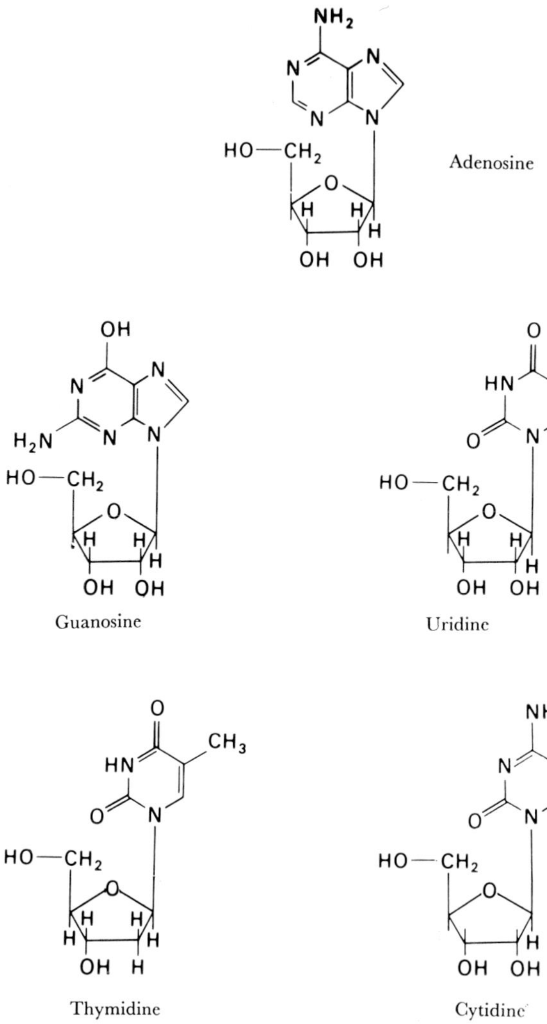

Adenosine

Guanosine

Uridine

Thymidine

Cytidine

Fig. 5.15. Structure of Nucleosides

much degraded by the extraction procedure, but later more refined extraction methods using phenol or salt solutions has given material of high molecular weight and considerable purity.

Hydrolysis of the nucleic acids ultimately gives rise to the unit sugar–base–phosphate: these are termed mono-nucleotides and may be regarded as the basic monomer precursors of the polymers. Further hydrolysis of the nucleotides gives base–sugar structures, e.g. adenosine, guanosine, uridine, cytidine or thymidine, known as nucleosides (Fig. 5.15). More vigorous hydrolysis may finally result in removal of the sugar and formation of the free heterocyclic bases. Early studies with nucleic acids clearly revealed a four-base component structure and

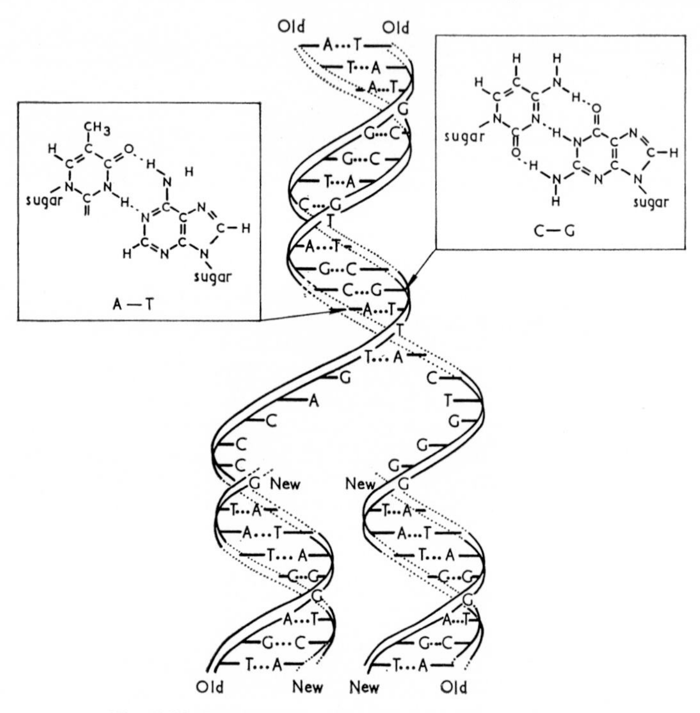

Fig. 5.16. Replication of DNA (after Watson and Crick)

suggested that they might be derived by polymerization of a tetra-nucleotide unit containing the four different bases, but this was subsequently shown to be incorrect.

Nucleic acids from different sources vary widely. Accurate analyses of base compositions in purified nucleic acids ultimately revealed a consistency in the ratios of certain of the bases. In particular the ratios of adenine (A) to thymine (T) or of guanine (G) to cytosine (C) in DNA were always unity. These results, together with knowledge gleaned from X-ray analysis of the α-helical structure of the polynucleotide molecules, led Watson and Crick in 1953 to put forward their theories on nucleic acid structure (Fig. 5.16).

The most outstanding feature of this theory is concerned with an explanation of the constant ratio of A to T and G to C and suggests that a given molecular strand of DNA may be paired with a complementary strand of another molecule such that hydrogen bonding occurs between A and T or G and C in a manner illustrated in Fig. 5.16. This concept of nucleic acid structure has led to a greater understanding of the principles of genetics and in particular to an understanding of the mechanisms of cell division and protein biosynthesis, and to the establishment of specific relationships between different nucleic acids and between nucleic acids and proteins.

Mononucleotides

The mononucleotides are the simple monomeric precursors of the nucleic acids and consist of a phosphorylated sugar (D-ribofuranose or

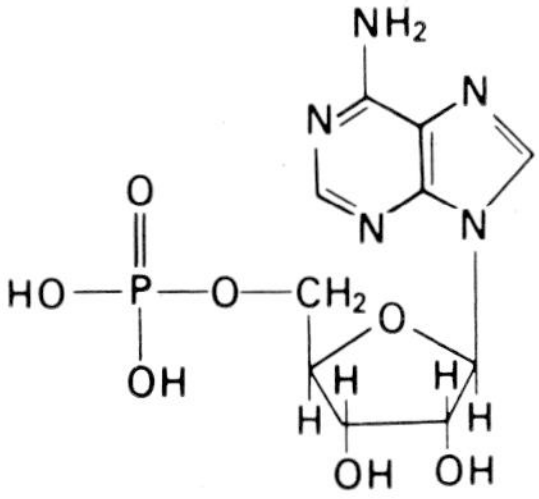

Adenosine 5′-Monophosphate
(AMP)

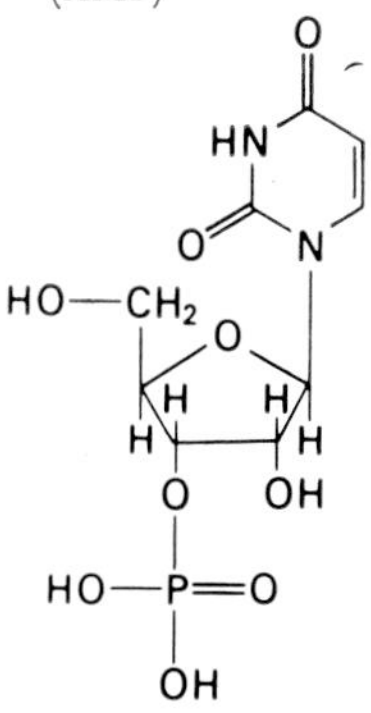

Uridine 3′-Monophosphate
(3′-UMP)

Fig. 5.17. Examples of mononucleotides

2-deoxy-D-ribofuranose according to the nucleic acid) with a heterocyclic base attached at the glycosidic position (C1), (Fig. 5.17), with the β-configuration. In the major nucleic acid macromolecules (see later) there are only four bases present, two of which are purines (adenine and guanine) (Fig. 5.15) and common to both DNA and RNA, and two are pyrimidines, uracil and cytosine (Fig. 5.18) in RNA,

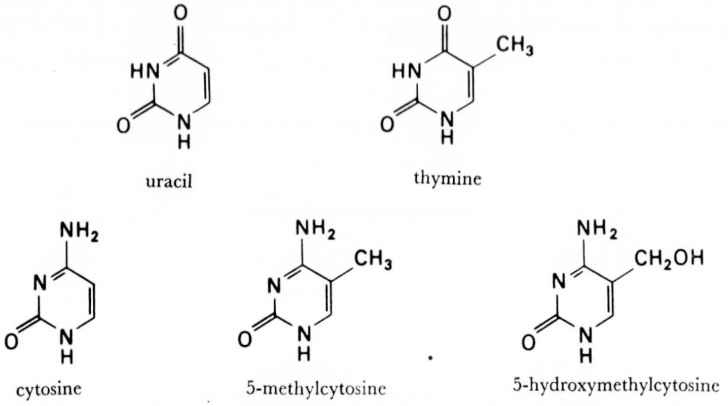

Fig. 5.18. Pyrimidine bases found in nucleic acids

and thymine and cytosine in DNA. There is another type of RNA, the transfer RNA (tRNA), which is concerned with protein biosynthesis (see later) and is relatively small (75–100 mononucleotides units long) and contains additional bases, many of which however are closely related to the simple pyrimidines and purines outlined above. Thus they include thiouracils, methylated (both on N and on sugar OH) adenines and guanines and exceptionally a pseudouridine in which the ribose is linked at the 5-position of the uracil (Fig. 5.19).

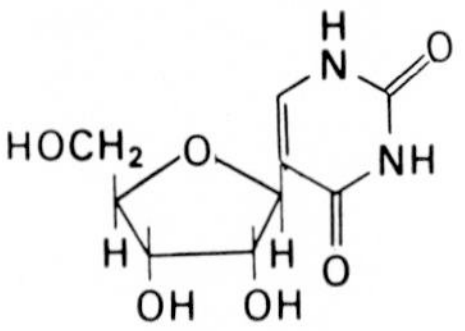

Fig. 5.19. Pseudouridine

In addition to the simple nucleoside monophosphate derivates there also occur the nucleoside polyphosphates, particularly the di- and tri-phosphates (Fig. 5.20) and these are the direct precursors of the nucleic acids to which they may be polymerized by different enzymes

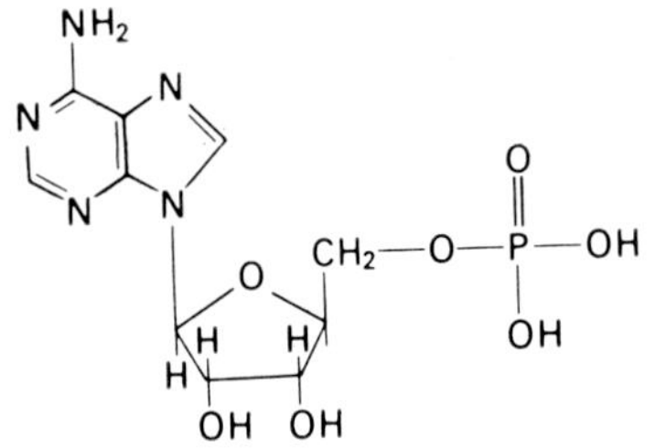

Adenosine Monophosphate
(AMP)

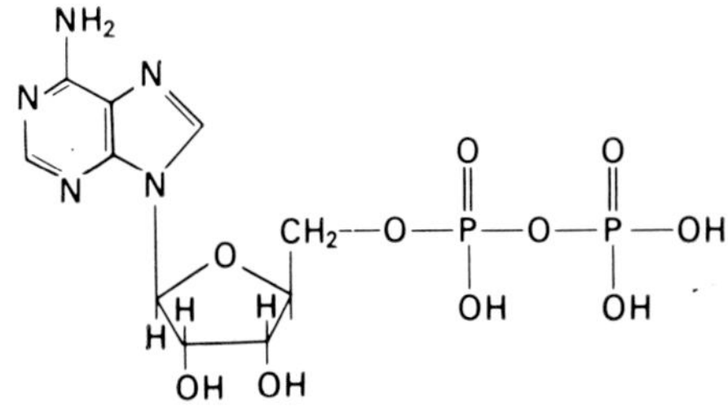

Adenosine Diphosphate
(ADP)

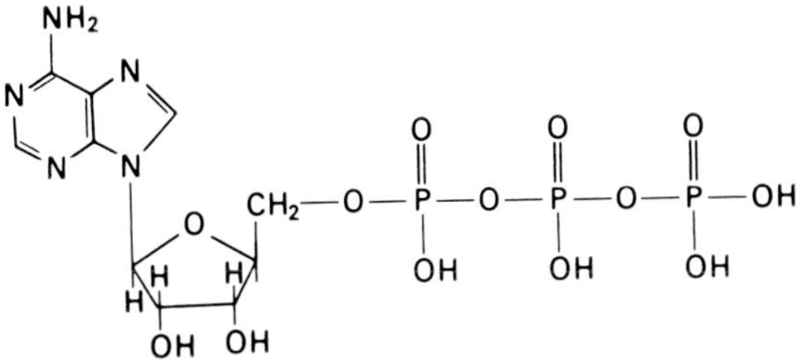

Adenosine Triphosphate
(ATP)

Fig. 5.20

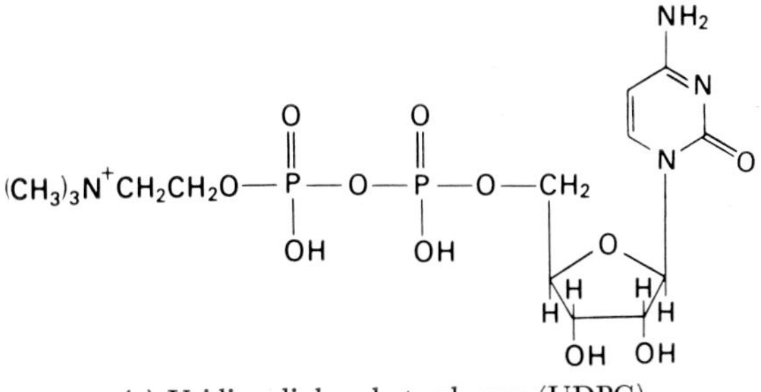

(a) Uridine diphosphate glucose (UDPG)

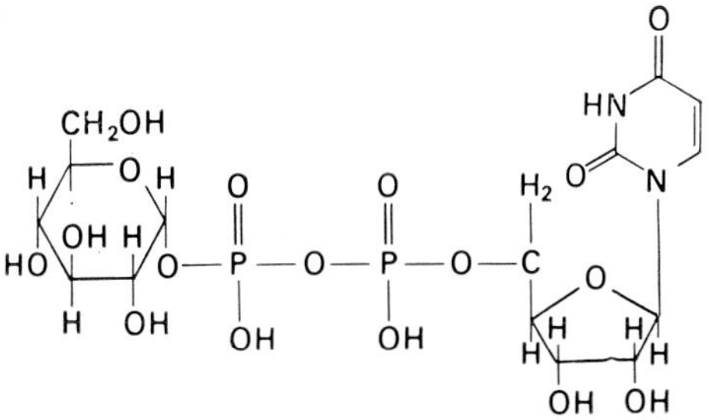

(b) cytidine diphosphate choline

Fig. 5.21

(polymerases). The nucleoside polyphosphates and derivatives also have other functions. Thus adenosine triphosphate (ATP) (Fig. 5.20) is one of the main carriers of chemical energy in the cell and operates by transfer of phosphate groups.

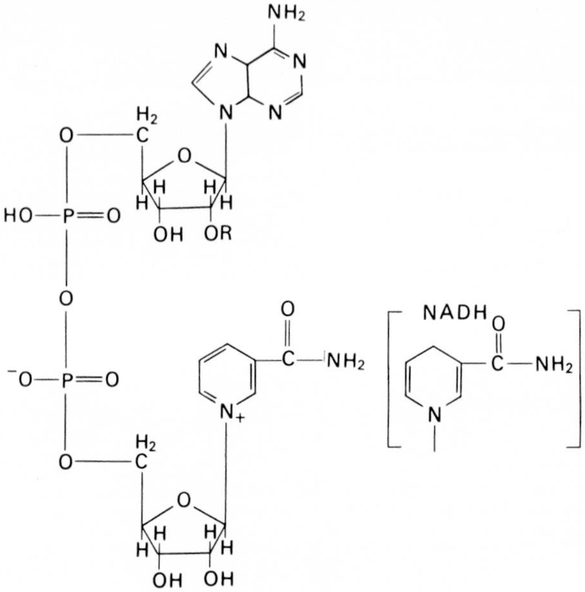

(a) Nicotinamide adenine dinucleotide (NAD) (Coenzyme I where R = H; Coenzyme II where R = $PO(OH)_2$)

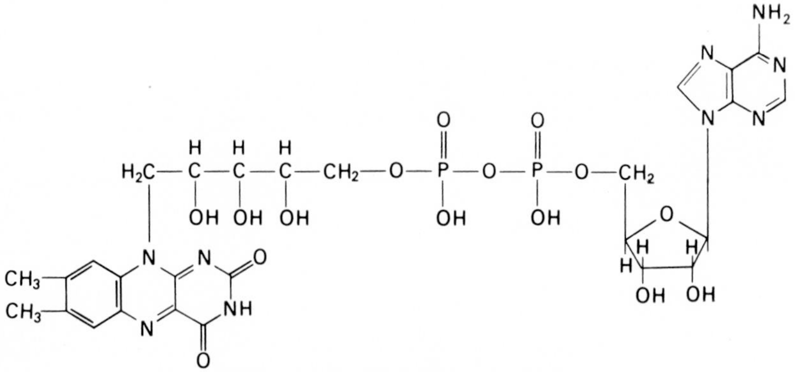

(b) Flavin adenine diphosphate (FAD)

Fig. 5.22

The formed adenosine diphosphate (ADP) or monophosphate (AMP) may be rephosphorylated during respiration and the lost energy regained. ATP is especially concerned and implicated with many biochemical reactions in which intermediate formation of an anhydride-type compound is required as in biochemical acylation reactions.

The nucleoside di- and tri-phosphates also form a part of many molecules which may act as co-enzymes. These include compounds derived from the nucleic acid bases such as uridine diphosphate glucose (UDPG) which serves as a donor of glucose units, or cytidine diphosphate choline (Fig. 5.21), a donor of choline residues in the biosynthesis of certain phosphoglycerides. There are, however, also a number of other related compounds with similar co-enzyme functions which contain heterocyclic or other bases in addition to or in place of pyrimidines or purines. They include: nicotinamide adenine dinucleotide (NAD) (Fig. 5.22) and its 3-phosphate (NADP); important oxidation–reduction coenzymes; flavin adenine dinucleotide (FAD) (Fig. 5.22), also concerned with oxidation reduction; and co-enzyme A (Fig. 5.23) which is involved in fatty acid metabolism.

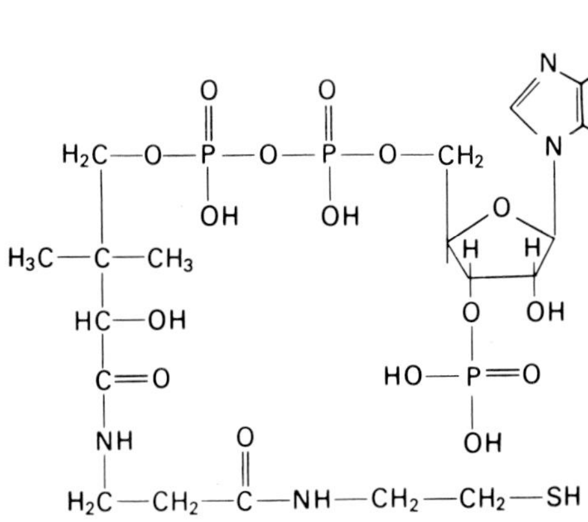

Fig. 5.23. Co-enzyme A

DNA

The polynucleotide structure of DNA (Fig. 5.16) is formally considerably simpler than that of RNA since the 2′-hydroxyl group is absent. The major bases present are adenine, guanine, cytosine and thymine (5-methyluracil) although small amounts of methylated derivatives have been isolated in some polymers especially viral DNA. The mononucleotides termed dAMP, dGMP, dTMP and dCMP are linked by 3′-5′-phosphodiester linkages and in various sequences. The molecular weights of pure DNA specimens are enormous. Thus in primitive prokaryotic cells containing a single chromosome, the DNA may be present as a single molecule with a molecular weight greater than 2×10^9. In more complex eukaryotic cells with several chromosomes and consequently many different types of DNA the molecular weights are also very high indeed. Simple bacterial cells, which contain a single molecule of DNA not generally associated with protein, contrast with the DNA of eukaryotic cells which is very commonly associated with strongly basic proteins, namely histones which are characterized by possession of large amounts of the basic amino-acids arginine and lysine.

In contrast to bacteria in which the DNA which may be as much as 1% of the cell weight and floats free in the cytoplasm, eukaryotic cells normally have the DNA in a more discrete body—the cell nucleus.

RNA

There are three major types of ribose nucleic acid which co-exist in living cells. These are ribosomal RNA (rRNA), messenger RNA (mRNA) and transfer RNA (tRNA). They are all linear single-stranded polynucleotides which contain the normal purine and pyrimidine bases and occasionally modified derivatives (especially in tRNA). They also occur in multiple molecular species. Thus there are at least three rRNAs, several hundreds of mRNA and probably up to about 100 tRNA molecular species. Living cells usually contain roughly 10 times as much RNA as DNA. The distribution of the different RNA molecules in a living cell depends to a large extent on the type of cell involved. Thus in prokaryotic cells (e.g. bacterial cells) which do not contain a nucleus but a single-coiled DNA molecule—double helix—the RNA is found in the cytoplasm. In a more complex cell such as mammalian liver about 11% of total RNA is present in the nucleus largely as mRNA, 15% in the mitochondria (rRNA and tRNA), 50% (largely rRNA) in the ribosomes and 24% (mainly tRNA) in the cytosol. RNA also occurs in all plant viruses and in some bacterial and animal viruses (see also Chapter 7).

Ribosomal RNA (rRNA)

These RNA molecules make up a major part (>50%) of the ribosome weight. In *E. coli* they are obtained in three characteristic forms which are differentiated by their sedimentation coefficients in the ultracentrifuge denoted as the 23S (corresponding to a mol. wt. $1{\cdot}1 \times 10^6$) 16S (mol. wt. $5{\cdot}5 \times 10^5$) and 5S (mol. wt. $3{\cdot}5 \times 10^4$) respectively. The function of rRNA is not clear.

Messenger RNA (mRNA)

Messenger RNA is synthesized in the nucleus of the cell during the transcription process (see later)in such a manner that the sequence of bases in a strand of chromosomal DNA is transferred to an RNA strand which thereby has a complementary sequence. The mRNA then passes to the ribosomes, where a message is transcribed via the tRNA such that a triplet of bases in the RNA (originally DNA complement (codons)) specify the order of amino-acids in a produced protein (see Chapter 7 for full details).

The Sedimentation Coefficient

The sedimentation coefficient (s) is defined by the equation

$$s = \frac{dx/dt}{w^2 x}$$

where x is the distance from the centre of rotation;
w is the angular velocity in radians per sec;
t is the time in sec.

A sedimentation coefficient of 1×10^{-13} is called a Svedberg unit, denoted by S. Thus 5S indicates a sedimentation coefficient of 5×10^{-13}. Increases in the sedimentation coefficient corresponds to an increase in the molecular weight.

AMINO ACIDS AND PROTEINS

Proteins are naturally occurring condensation polymers of L-α-amino acids and of two (proline and hydroxyproline) L-α-imino acids, in which head to tail formation of amide (peptide) bonds has occurred. They are conventionally drawn with the N-terminal amino-acid to the left, and the constituent amino acids which are readily produced by vigorous acid hydrolysis are represented by the abbreviated notation in Fig. 5.24a and b.

$$H_2N\,\underset{R^1}{\underset{|}{CH}}\,CO_2H + H_2N\,\underset{R^2}{\underset{|}{CH}}\,CO_2H + \ldots H_2N\,\underset{R^3}{\underset{|}{CH}}\,CO_2H$$

$$\downarrow$$

$$H_2N\,\underset{R^1}{\underset{|}{CH}}\,CO - NH\,\underset{R^2}{\underset{|}{CH}}\,CO \ldots NH\,\underset{R^3}{\underset{|}{CH}}\,CO_2H$$

General formula for α-amino acids

$$\overset{+}{N}H_3-\overset{COO^-}{\overset{|}{\underset{\text{R group}}{\underset{|}{C}}}}-H$$

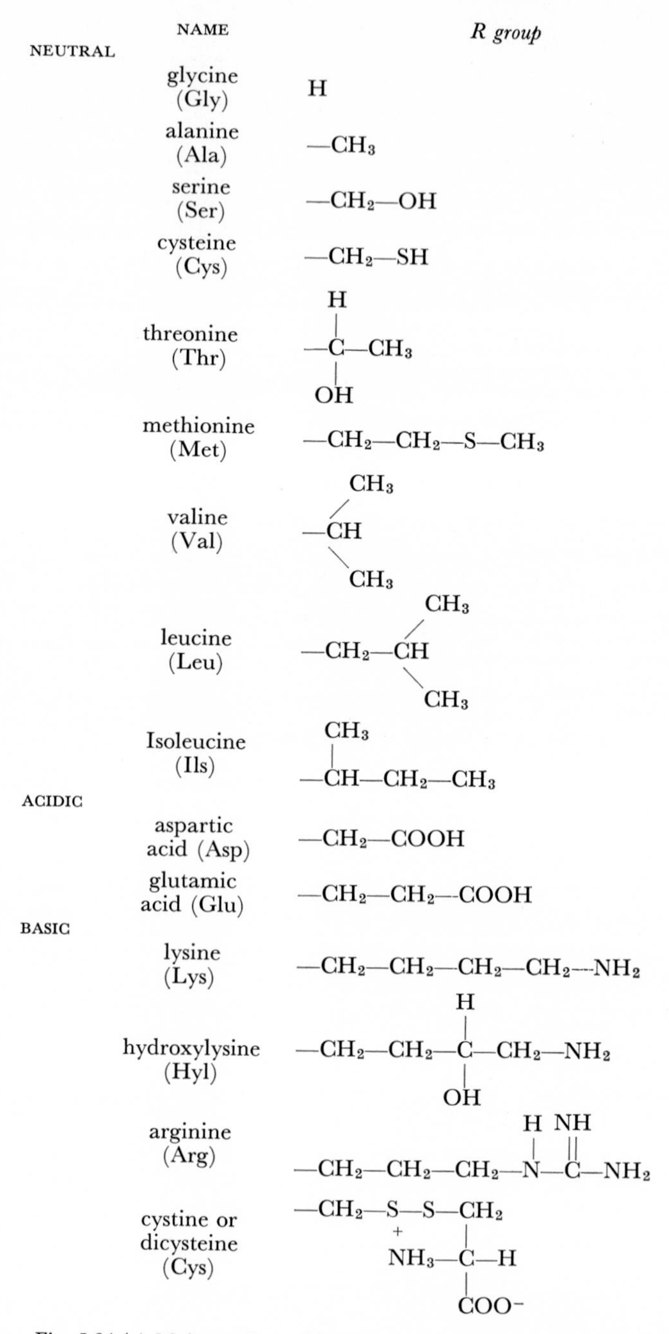

Fig. 5.24.(a) Major amino acids present in proteins and peptides (aliphatic)

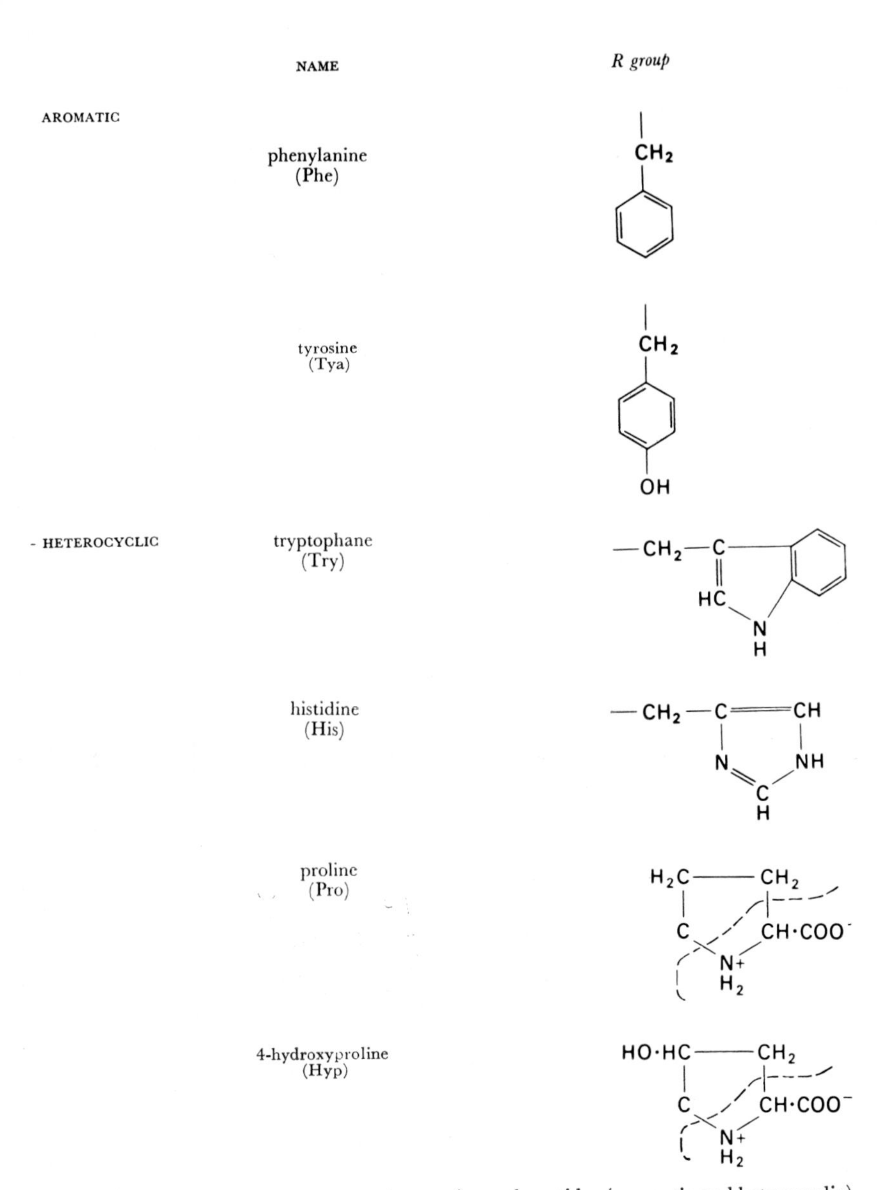

Fig. 5.24. (b) Major amino acids present in proteins and peptides (aromatic and heterocyclic)

Protein amino acids are defined as those substances which have been isolated, preferably by more than one person, from the hydrolysate of a naturally occurring protein of demonstrable purity. The acids should have been characterized by elemental and other analyses and their structure confirmed by synthesis and comparison of the synthetic material with the natural. Such criteria have led to the acceptance of twenty compounds, eighteen of which are α-amino acids and two (proline and hydroxyproline) related analogues (see Fig. 5.24a and b). They are all, with the exception of glycine, optically active and of the L-series, and comparison of natural and synthetic material must include a comparison of stereochemistry.

In addition, typical proteins also invariably contain simple derivatives of some of these amino acids, such as the amides of aspartic acid (asparagine) and glutamic acid (glutamine) and the hydroxy amino acids often occur in acylated or phosphorylated form. The remarkable constancy of occurrence of the twenty or so compounds in protein molecules from all forms of living matter is not accidental and is closely associated with the structure of certain nucleic acids (see later) which genetically control both the type of amino acid present in a given protein, and by a simple code the orders in which the monomeric units are aligned along the polymer chain.

If one examines the side chains associated with the amino acid structures they are seen to provide, for the derived protein, a wide variety of simple functional groups. Indeed a classification of the groups reads something like the chapter headings of an elementary textbook of organic chemistry. Thus there are represented alkyl groups (alanine, valine, etc.), alcoholic hydroxyl groups (serine, threonine, hydroxyproline), and these could clearly be a source, if required, of related aldehyde or ketone groups, thiol (cysteine) and alkylthiol (methionine), carboxylic acid (aspartic and glutamic acids) and derivatives (asparagine, glutamine) aromatic (phenylalanine), phenolic (tyrosine), and amino groups (lysine, hydroxylysine). Heterocyclic groups are represented in tryptophane and histidine, and arginine (a guanidine derivative) completes the series of compounds. The protein which contains a reasonable proportion of the main amino acids has clearly at its command a substantial armoury of reactive functional groups with which to exercise its role, either as a structural unit (e.g. silk, fibroin, muscle) or as an enzyme.

In addition to the twenty or so protein amino acids there has been isolated from various natural sources a large number (about 100) of other α-amino acids. These may occur free, in peptides, or in association with other types of molecules, and include optical antipodes of the

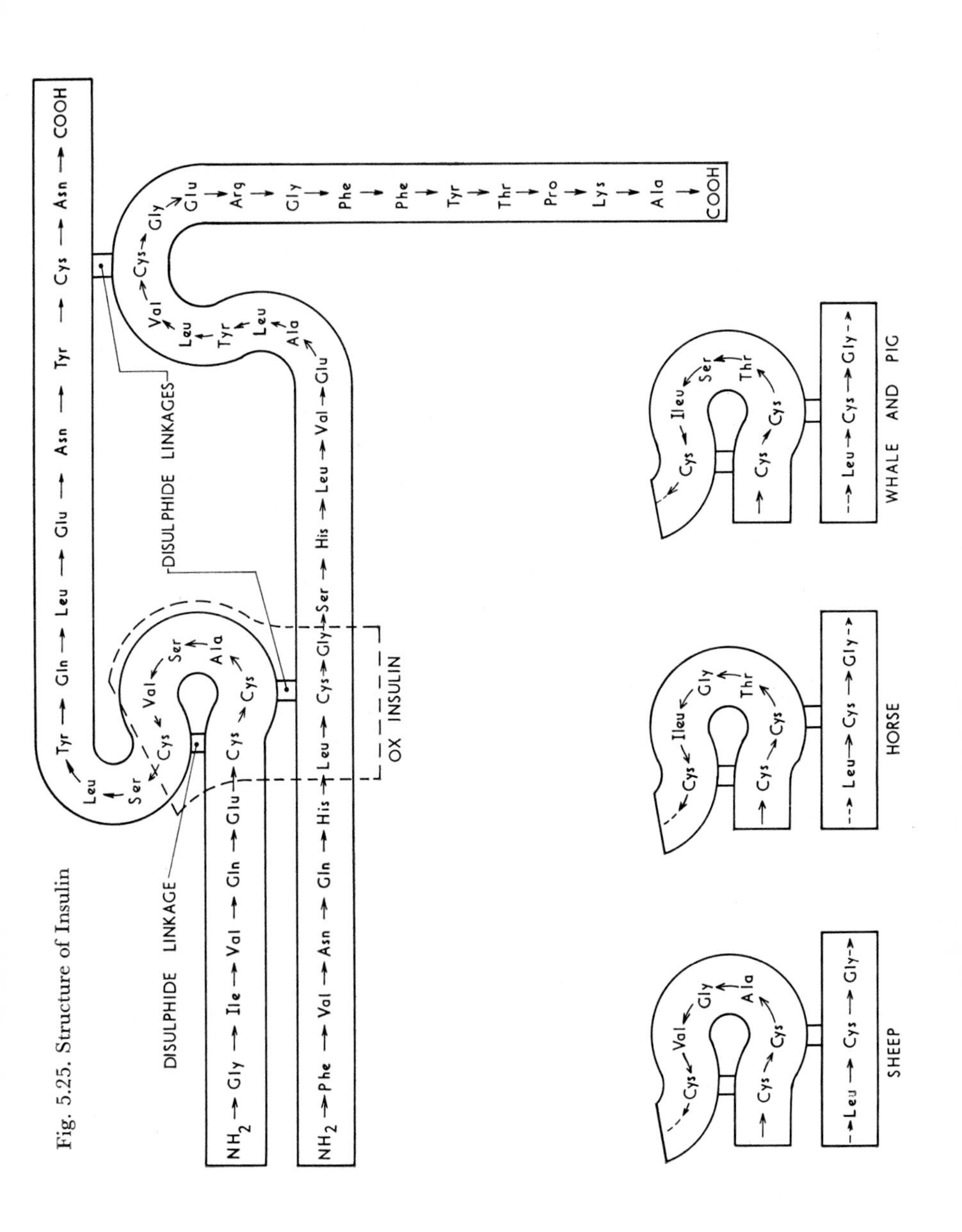

Fig. 5.25. Structure of Insulin

protein amino acids in addition to a wide variety of novel systems. Many of these compounds are of considerable biological importance and include hormones (thyroxine), constituents of antibiotics (penicillamine in penicillin), intermediates in metabolic pathways (citrulline, ornithine) and many other compounds with no known apparent function.

The simple typical protein will thus consist of one or more molecular polyamide backbones from which a wide variety of functional groups will protrude. The physical and chemical properties of a protein will consequently depend not only on its size but also on the nature of the particular amino acids from which it is derived, their relative numbers, and the particular sequence in the molecular chain.

The amide bond linking together the amino acid units is also, for historical reasons, called a peptide bond. The difference between a polypeptide and a protein is arbitrary. A protein would normally have a molecular weight $>$10,000 and be naturally occurring. A polypeptide is a general term and would cover all non-naturally occurring materials and also naturally occurring products with molecular weights $>$10,000. The formal polypeptide chain will be linear, but opportunities exist for cross-linking of two or more chains, or intramolecular cyclization, especially by the disulphide linkages present in cystine. Insulin (Fig. 5.25), the antidiabetic hormone, with 51 amino-acid units and a molecular weight of 6000, consists of two polypeptide chains cross-linked with disulphide bonds, and was the first polypeptide of any great size to have its primary structure determined.

Consideration of the structure of a protein will involve at least three main areas.

1. *The primary structure* outlines the specific sequence of amino acids present in the polymer molecule.

2. *The secondary structure* involves the manner in which the chain is coiled. A particularly favourable configuration is the α-helix in which hydrogen bonds between the amide hydrogen atoms and the carbonyl groups four peptide bonds apart, are found. The hydrogen bonds are almost parallel to the long axis of the helix and the spacing between the turns is about 5·4 Å; the side chains lie outside the helix (Fig. 5.26).

3. *The tertiary structure* details the way in which the coiled chains are folded and hydrated under natural aqueous conditions.

Proteins occur in all types of shapes and sizes and with many different functions. They may act as skeletal units as in skin, hair, muscle, silk, as carriers of oxygen (in conjunction with metals, especially iron) as with haemoglobin, as hormones regulating various metabolic processes (insulin, pituitary hormones) and perhaps most importantly as bio-

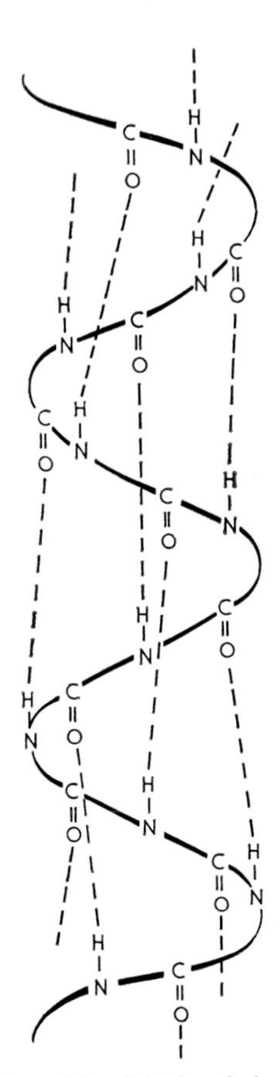

Fig. 5.26. A schematic representation of the right-handed α-helix with L-amino acid residues

logical catalysts or enzymes. The enzymes are a very large group of proteins which catalyse almost all the many and diverse reactions which exist in the living cell in a specific and characteristic manner. The structures of a number of enzymes are now known, as are their relationships with the specific substrates with which they interact. In spite of this, however, the mechanism by which such catalysed reactions operate, especially the ability of enzymes to increase reaction rates enormously, is still little understood. In general, however, their particularly efficient catalytic properties may be ascribed to the ability of the macromolecule to provide a system whereby active functional groups operating in profitable unison (e.g. histidine and serine in many acylases), although far apart in the chain may be very close together in space. This allows intramolecular catalysis to occur and this is always more favourable than intermolecular catalysis.

LIPIDS

Lipids are a vaguely defined group of naturally occurring substances characterized frequently more by their immediate physical properties (water insolubility, extractable by non-polar solvents, etc.) than by their chemical structures. Nevertheless it is possible to group these specific compounds into two reasonably well-defined classes (fats, fatty acids) and (isoprenoid lipids) which result generally from their mode of biosynthesis (Chapter 6).

Fats, fatty acids and derivatives

Fatty (alkan-, alken- and alkyn-oic) acids and various oxy- and oxo-derivatives are of wide occurrence in living systems although they normally occur as esters with various alcohols. Most of the 70 or so different fatty acids which have been isolated possess an even number of carbon atoms and chains 14–22 carbon atoms long. The acids with 16 or 18 carbon atoms, i.e. palmitic $CH_3(CH_2)_{14}CO_2H$ and stearic $CH_3(CH_2)_{16}CO_2H$ acids, are much the most abundant. Although many of the acids are fully saturated, the unsaturated acids normally predominate in those cells which exist at low temperatures, presumably because the unsaturated fat derivatives (glycerol alkanoates) have lower melting points than their saturated analogues and are generally liquid at room temperature. The unsaturated acids generally contain mostly *cis* double bonds at position C_9-C_{10}, e.g. oleic acid—$CH_3(CH_2)_7CH{:}CH(CH_2)_7CO_2H$, and palmitoleic acid—$CH_3(CH_2)_5$–$CH{:}CH(CH_2)_7CO_2H$, and additional unsaturation when present is never in conjugation, e.g. linoleic acid—$CH_3(CH_2)_4CH{:}CHCH_2CH{:}CH(CH_2)_7CO_2H$, and linolenic acid—$CH_3CH_2CH{:}CHCH_2CH{:}CHCH_2CH{:}CH(CH_2)_7CO_2H$.

Fatty acids derived from prokaryotic cells as distinct from eukaryotic cells usually form a smaller spectrum of compounds between C_{12} and C_{18} which rarely contain more than a single double bond but may include unusual derivatives such as lactobacillic acid—$CH_3(CH_2)_5$)-CH-CH-$(CH_2)_9CO_2H$, tuberculostearic acid—$CH_3(CH_2)_7CH(CH_3)$-$(CH_2)_8CO_2H$ and cerebronic acid—$CH_3(CH_2)CH(OH)CO_2H$.

(a) mono-	(b) di-	(c) tri-acyl glycerols
CH_2OCOR	CH_2OCOR	CH_2OCOR
$CHOH$	$CHOCOR'$	$CHOCOR'$
CH_2OH	CH_2OH	CH_2OCOR''

Fig. 5.27

Fats and oils are glyceryl esters of fatty acids in which the hydroxy groups are acylated by 1 (monoglyceride), 2 (diglyceride) or 3 (triglyceride) acyl groups which may be similar or different (Fig. 5.27).

The phosphatidic acids or phosphoglycerols are similar compounds in which one of the acylating groups is phosphoric acid (Fig. 5.28),

$$\begin{array}{l} CH_2OCOR \\ | \\ CHOCOR' \\ | \\ CH_2—O—PO_3^{2-} \end{array}$$

(a) (phosphatidic acid)

e.g.

$$\begin{array}{l} CH_2OCOR \\ | \\ CHOCOR' \\ |\qquad\quad O \\ CH_2O—P—OCH_2CH_2NH_2 \\ \qquad\quad O— \end{array} \qquad \begin{array}{l} CH_2OCOR \\ | \\ CHOCOR' \\ |\qquad\quad O \\ CH_2O—P—OCH_2CH_2\overset{+}{N}(CH_3)_3 \\ \qquad\quad O— \end{array}$$

(b) (phosphatidyl ethanolamine) (c) (phosphatidyl choline)

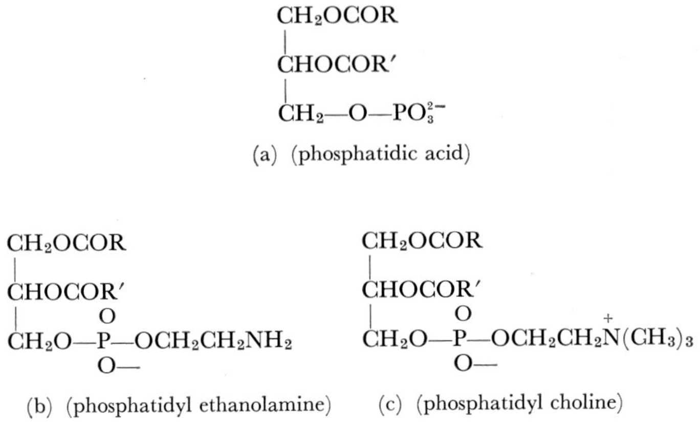

Fig. 5.28

and they are almost entirely found as constituents of cell membranes. Various ester derivatives of phosphatidic acids occur and ether analogues (plasmalogens) also exist, e.g.

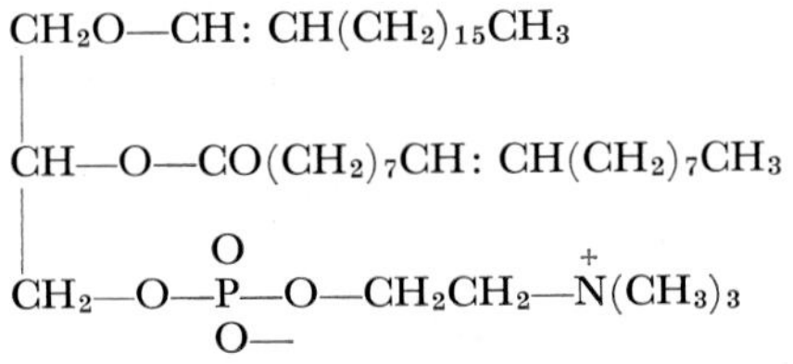

$$\begin{array}{l} CH_2O—CH{:}\ CH(CH_2)_{15}CH_3 \\ | \\ CH—O—CO(CH_2)_7CH{:}\ CH(CH_2)_7CH_3 \\ |\qquad\qquad O \\ CH_2—O—P—O—CH_2CH_2—\overset{+}{N}(CH_3)_3 \\ \qquad\qquad O— \end{array}$$

(phosphatidal choline)

Fig. 5.29

Other fatty acid esters include the waxes which are esters of long-chain fatty acids with long-chain (generally monohydric) alcohols, e.g. Spermaceti from the Sperm whale is cetylpalmitate—$CH_3(CH_2)_{14}COO(CH_2)_{15}CH_3$. They are found as protective coats on leaves and berries of plants, and skin, fur, etc., of animals and may include acyl steroids. Cutins are related cross-linked polyesters derived from hydroxy fatty acids.

A somewhat diverse group of lipids with a formal resemblance to the fats and with functions largely concerned with both plant and animal cell membrane structure include substances, in which the glycerol unit is partly glycosylated (glycolipids) or where the glycerol unit is replaced by a somewhat analogous amino-diol structure (sphinogolipids) and various combinations of both systems. Examples include:

(a) CH_2OCOR
$|$
$CHOCOR'$ (simple glycolipid)
$|$
CH_2O-galactosyl

(b) $HO—CH—CH:CH(CH_2)_{12}CH_3$
$|$
$CH.NHCO(CH_2)_7CH:CH(CH_2)_7CH_3$ (sphingomyelin)
$|$
$CH_2O–P–OCH_2CH_2\overset{+}{N}(CH_3)_3$

(c) $HO—CH—CH:CH(CH_2)_{12}CH_3$
$|$
$CH—NHCO(CH_2)_{22}CH_3$ (lignoceric acid—a cerebroside)
$|$
$CH_2—O—\beta$-galactosyl

Fig. 5.30

Gargliosides are similar to the last compound but generally contain an extended oligosaccharide unit.

Isoprenoid lipids

The second major group of lipids are biochemically produced by an *n*-merization of a C_5 (isoprene) unit derived from mevalonic acid (C_6) and consisting of either isopentenyl pyrophosphate or dimethylallyl pyrophosphate (Fig. 5.31).

(a) $CH_2 {=} C(CH_3)—CH_2—CH_2OP_2$

Isopentenyl pyrophosphate

(b) $(CH_3)_2C:CHCH_2OP_2$

dimethylallyl pyrophosphate

Fig. 5.31

The products formed vary from simple minor cell components (terpenes) produced by di- or tri- or *n*-merization including geraniol —$Me_2C{:}CHCH_2CH_2CMe{:}CHCH_2OH$, farnesol—$Me_2C{:}CHCH_2CH_2$

$CMe{:}CHCH_2CH_2CMe{:}CHCH_2OH$ and cyclic analogues (Fig. 5.32).

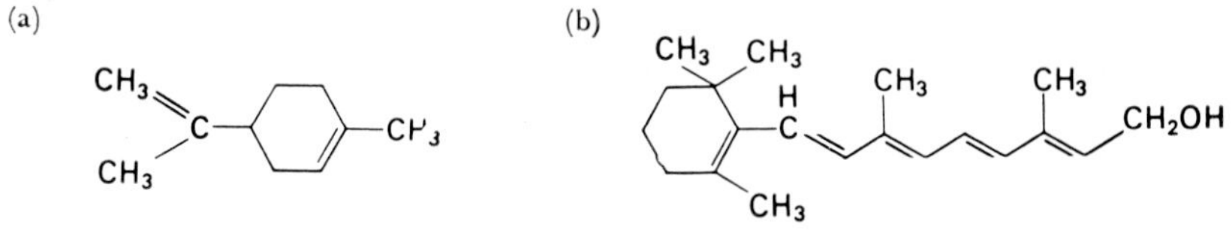

Fig. 5.32. Limonene and vitamin A_1

to more extended structures of wider occurrence such as the carotenoids which also occur as fatty acid esters (Fig. 5.33).

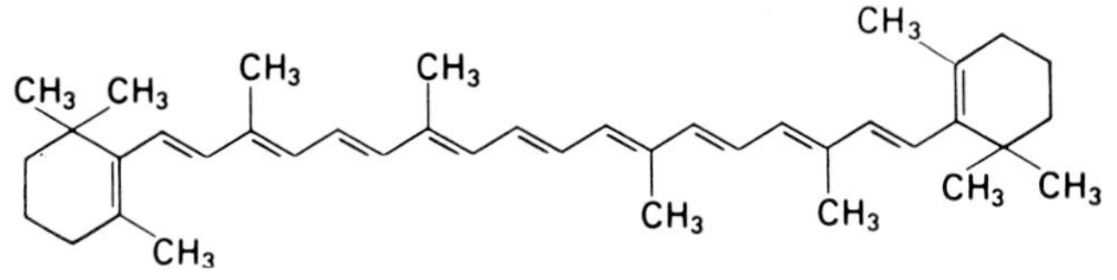

Fig. 5.33. β-carotene

and products formed by cyclization of acyclic substances of this type include the sterols (Fig. 5.34).

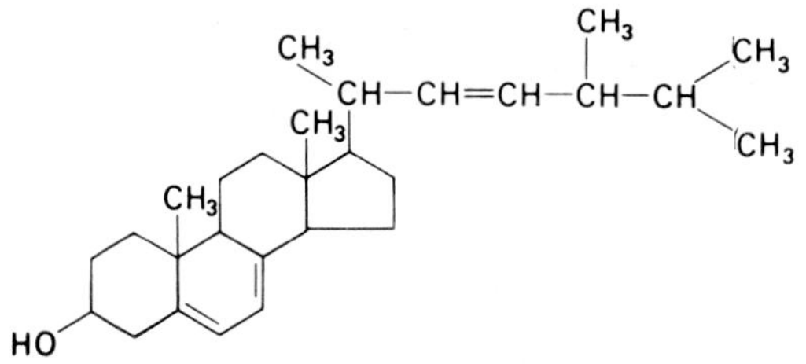

Fig. 5.34. Ergosterol

The functions of the isoprenoid lipids are most obvious and dominant in highly evolved differentiated living systems. Thus vitamins A and D are concerned with vision and bone formation respectively and many steroids are intricately concerned with stabilization and preservation of sexual characteristics and processes in higher living systems. Indeed until recently sterols were not thought to be present in prokaryotic cells, but it is known that they do occur, apparently in very small quantities, in blue-green algae but are of unknown function. Carotenoids and carotenoid esters (e.g. antheraxanthin palmitate—Fig. 5.35) are also concerned with complex processes of vision and possibly photosynthesis; recently, however, it has been found by Brooks and Shaw that they have a major function both in lower organisms and higher plants in providing a source (by oxidative polymerization) of the group of substances known as sporopollenins which make up the resistant part of the walls of various micro- and megaspores of these organisms.

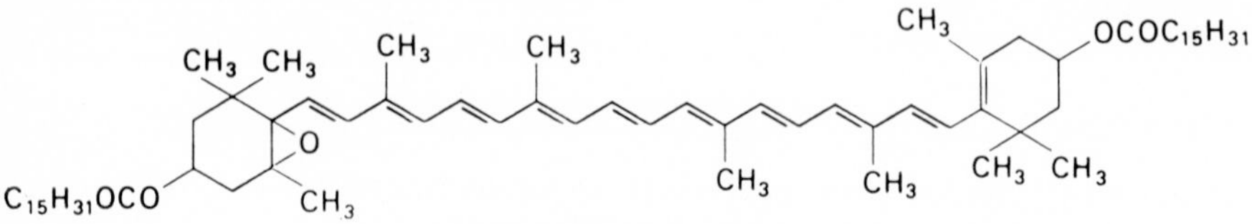

Fig. 5.35. Antheraxanthin dipalmitate

In addition to the specialized case of the carotenoids most of the lipids mentioned are concerned in various ways with cell membrane structure and function, but mostly in eukaryotic cells where about half the membrane may consist of lipid. Apparently the lipids exist in membranes of this type in association with both proteins (lipoproteins) and polysaccharides (lipo-polysaccharides), but little or nothing is known either about the structure of such compounds and numerous hybrids thereof or their function as cell membranes. Indeed the only eukaryotic cell membranes to be studied in any detail are the somewhat specialized erythrocytes which are readily separated from their environment. There are of course numerous theories and models of membrane structure based on the very slim chemical evidence but the field is in such a state of flux and uncertainty at this time that it would perhaps be unwise to describe any of these here.

SUGGESTED FURTHER READING Chapter 5

Books

Bernhard, S. (1968). "The Structure and Function of Enzymes". Benjamin, Philadelphia.

Blackburn, G. M. (1970). "Nucleic Acids". *In Ann. Rep. Chem. Soc.* 489–522.

Calvin, M. and Jorgenson, M. J. (Ed.) (1968). "Bio-organic Chemistry". Freeman, San Francisco.

Chapman, D. (1965). "The Structure of Lipids by Spectroscopy and X-Ray Techniques". Methuen, London.

Davidson, E. A. (1967). "Carbohydrate Chemistry". Holt, Rinehart and Winston, New York.

Davidson, J. N. (Ed.) (1972). "Biochemistry of Nucleic Acids". Wiley, New York.

Dickerson, R. E. and Geis, I (1969). "The Structure and Action of Proteins". Harper and Row, New York.
Ferrier, R. J. and Collins, P. M. (1972). "Monosaccharide Chemistry". Penguin, Harmondsworth.
Fieser, L. F. and Fieser M. (1961). "Organic Chemistry". Reinhold, New York.
Florkin, M. and Mason, H. S. (Ed.). (1962). "Comparative Biochemistry", Volumes I–VII. Academic Press, New York.
Goodwin, T. W. (Ed.). (1965). "Chemistry and Biochemistry of Plant Pigments". Academic Press, London.
Gunstone, F. D. (Ed.) (1968). "Topics in Lipid Chemistry". Logos Press, Toronto.
Karrer, P. and Jucker, E. (1950). "Carotenoids". Elsevier, Amsterdam.
Lehninger, A. L. (1970). "Biochemistry". Worth, New York.
Needham, A. E. (1956). "The Uniqueness of Biological Materials". Pergamon Press, Oxford.
Roberts, J. D. and Caserio, M. C. (1965). "Basic Principles of Organic Chemistry". Benjamin, Philadelphia.
Sarkanen, K. V. and Ludwig, C. H. (Eds) (1971). "Lignins". Wiley-Interscience, New York.
Timasheff, S. N. and Fasman, G. D. (Eds) (1969). "Structure and Stability of Biological Macromolecules". Marcel Dekker, New York.
Vernon, L. P. and Seeley, G. R. (Eds) (1966). "The Chlorophylls". Academic Press, New York.
West, E. S., Todd, W. R., Mason, H. S. and Van Bruggen, J. T. (1966). "Textbook of Biochemistry". MacMillan, London.

6

Biochemical Systems in Living Organisms

INTRODUCTION

An actively living cell may be formally considered to consist of a number of closely linked chemical reactions (or in the dormant spore having a potential operating capacity) operating within a close specifically structured skeletal system which may of course play a major part in the chemistry. The chemical processes are concerned essentially, at least in the simple cell, with the elaboration of new cell material and the production of energy. In the net process both functions derive from simple exogenous chemicals such as water, ammonia, amino acids, monosaccharides, etc.

There appear to be a great many chemical reactions operating even in the simple cell, but in spite of the complexity of these reactions they are dominated largely by the production and functioning of the well-defined classes of chemical substances outlined in Chapter 5, namely the polysaccharides, the proteins and the nucleic acids and their precursor molecules, monosaccharides, α-amino acids and nucleotides. These three groups of molecular types are common to all life forms and it is difficult to imagine any living system which does not contain at least two of the three types, especially the nucleic acids and the proteins, although since nucleic acids are polymers of specific monosaccharide derivatives, a source of the latter type of material is always essential. Among the many other molecules concerned with life processes the rather vaguely defined group of lipids is undoubtedly the most important. Theories which describe living systems without nucleic acids have been outlined in an elegant manner and will be discussed later, but generally the nucleic acids would appear to be essential as central controlling agents and proteins as catalysts (enzymes) for most of the reactions. Polysaccharides are common as skeletal units in both prokaryotic and eukaryotic cells, but alternative structures derived from proteins or nucleic acids quite possibly could and frequently do take their place.

It would appear reasonable that evolution of living systems proceeded hand in hand with an evolution of the associated biochemical systems (and there is in fact some evidence for this) and extrapolation to zero time might produce a simple chemical origin for the ordered living system. Certainly concepts of this type have proved appealing to many workers and have led to theories of life origins which will be discussed later.

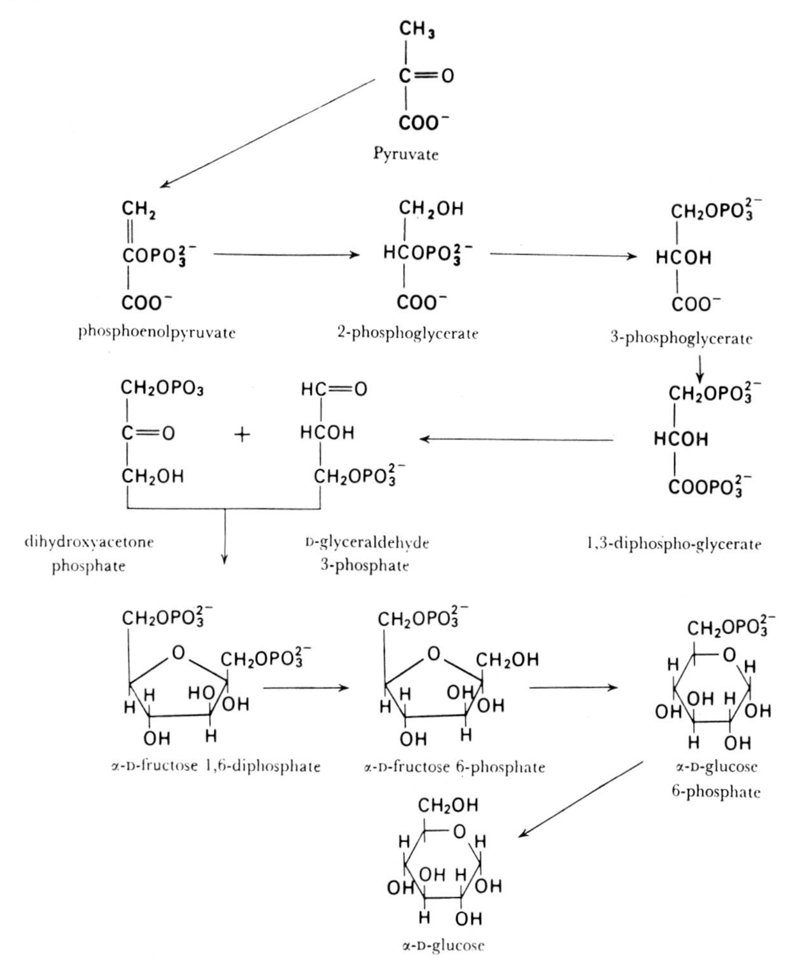

Fig. 6.01. Formation of glucose from pyruvate

BIOSYNTHESIS OF MONOSACCHARIDES

In different living systems pathways of monosaccharide biosynthesis differ appreciably although the basic underlying chemistry has generally much in common. D-Glucose is a dominant central intermediate in monosaccharide biosynthesis although the capacity of living systems to form free D-glucose is relatively limited. All cells, however, appear to be capable of forming glucose-6-phosphate by the central pathway of

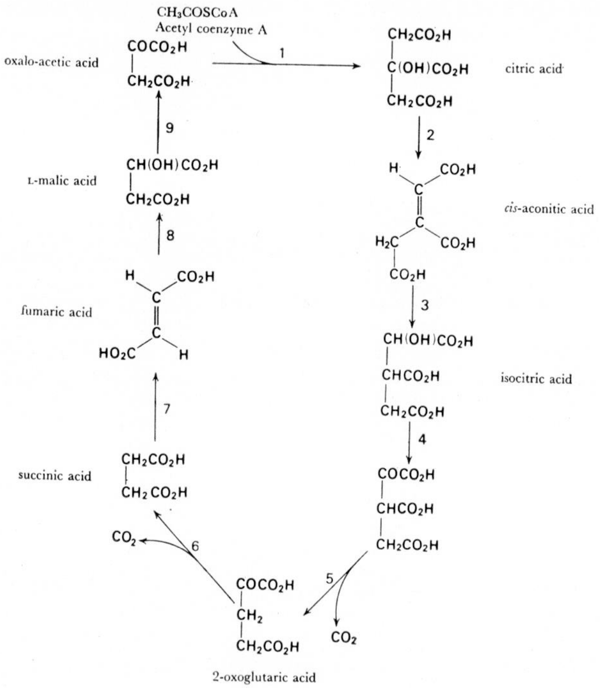

Fig. 6.02. The tricarboxylic acid cycle

gluconeogenesis, namely pyruvate → glucose 6-phosphate (Fig. 6.01). Nevertheless only photosynthetic and chemosynthetic autotrophs are able to form net amounts of glucose-6-phosphate *de novo* by reduction of carbonic acid. In animal and other non-photosynthetic organisms a cyclical pathway (the tricarboxylic acid cycle) (Fig. 6.02) operates; in this pathway ingested glucose or some glycogenic α-amino acids are converted into pathway intermediates which may ultimately be reconverted into glucose via pyruvate. That amino acids can in many instances form a source of monosaccharides is of particular interest and could possibly limit the necessity for exogenous monosaccharides in a primeval system. Pathways leading to starch and cellulose, however, appear to be universal although they differ considerably in degree. On the other hand, formation of many extracellular polymers including some wall materials of bacteria and fungi are highly specific processes.

PHOTOSYNTHESIS

This is a process occurring in plants and photosynthesizing organisms whereby the energy of the sun is used to transform carbon dioxide and

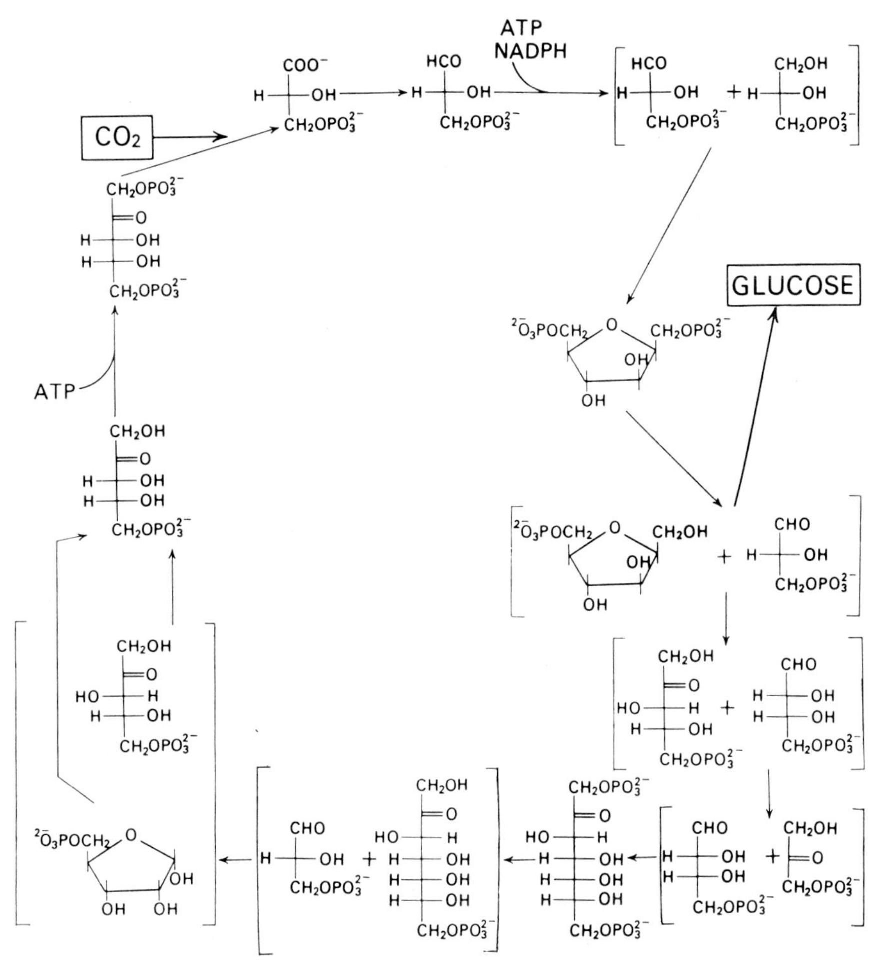

Fig. 6.03. The conversion of CO_2 to glucose during photosynthesis

water reductively into monosaccharides. There are various mechanisms whereby this reaction may occur (see Chapter 7), but a major cycle involving several interlocked reactions of numerous simple intermediates is the Calvin pathway (Fig. 6.03); other pathways undoubtedly exist. Thus in photosynthetic bacteria, glucose is probably formed by a reversal of the normal monosaccharide degradative pathway which leads to net formation of tricarboxylic acid cycle intermediates.

Nevertheless, in spite of the apparent diversity of biochemical reactions leading to monosaccharides, whether from pyruvate or other tricarboxylic acid cycle intermediates, or by some reductive photosynthetic pathway, the broad chemistry involved is remarkably similar, having intermediates in all cases consisting of simple hydroxy-aldehydes or -ketones or acids, with carbon chains not exceeding seven nor less than three. Indeed the basic reactions common to all the pathways (cf. Figs 6.01, 6.02 and 6.03) may be simply summarized as follows:

$$C_3 + C_3 \rightarrow C_6$$
$$C_4 + C_3 \rightarrow C_7$$
$$C_6 + C_3 \rightarrow C_4 + C_5$$
$$C_7 + C_3 \rightarrow C_5 + C_5$$

Fig. 6.04

This of course is simply another way of indicating that the enzymes involved in these particular reactions form common groups or families within species of living systems.

Ribose-5′-phosphate which is required for nucleotide synthesis is produced from glucose-6-phosphate by an overall oxidation scheme which involves the co-enzyme NADP, and may be briefly written:

$$\text{Glucose-6-phosphate} + 2\text{NADP} \rightarrow \text{D-ribose-5}'\text{-phosphate} + CO_2 + 2\text{NADPH}$$

Fig. 6.05

It is especially interesting to note that ribulose-1,5-diphosphate related to ribose phosphate is regenerated at the end of each Calvin cycle (Fig. 6.03) and must be regarded as a necessary component of the cycle which in fact is generally written as follows:

$$\text{6-ribulose-1,5-diphosphate} + 6\ CO_2 + 18\ \text{ATP} + 12\text{NADPH}$$

Fig. 6.06

The net result of course would exclude the ribulose derivative.

BIOSYNTHESIS OF DI- AND POLYSACCHARIDES

Starches and cellulose

Post-war research work, especially by Leloir and his co-workers, has shown that the active glycosyl group which is involved in most glycosyl biosynthetic reactions including syntheses of di- and polysaccharides involves a nucleoside 5′-diphosphate sugar derivative (NP_2-sugar) which is formed from a nucleoside 5′-triphosphate (NP_3) and the particular sugar in the presence of a pyrophosphorylase enzyme:

$$\text{N-P}_3 + \text{sugar-1-P} \longrightarrow \text{NP}_2\text{-sugar} + \text{PP}_i$$

e.g.

$$\text{Uridine-P}_3 + \alpha\text{-D-glucose-1-P} \longrightarrow \underset{\text{(UDPG)}}{\text{Uridine-P}_2\text{-glucose}} + \text{PP}_i$$

Fig. 6.07

UDPG is the active glucose donor in mammals. In plants the analogous ADPG is the main glucose donor formed by the similar reaction:

$$\text{Adenine-P}_3 + \alpha\text{-D-glucose-1-P} \longrightarrow \underset{\text{(ADPG)}}{\text{Adenine-P}_2\text{-glucose}} + \text{PP}_i$$

Fig. 6.08

During the biosynthesis of starch (glycogen) the glucose group of UDPG (ADPG) is transferred under enzyme control to the terminal glucose unit of an amylose chain as follows:

$$(\text{glucose})_n + \text{(A)UDPG} \longrightarrow (\text{glucose})_{n+1} + \text{(A)UDP}$$

Fig. 6.09

The specific enzyme (glycogen synthetase) which controls this reaction has a requirement for a small amount at least of a preformed polymer (primer) of an α-$(1 \rightarrow 4)$ polyglucose containing at least four glucose residues but the efficiency increases markedly with increase in primer chain length. In a similar manner branching enzymes give rise to the $1 \rightarrow 6$ linkage found in both glycogens and starches (amylopectin moiety).

Cellulose is formed by a similar sequence of reactions involving either GDP-, CDP- or ADP- glucose and chitin formation involves the analogous UDP-N-acetylglucosamine.

The various naturally occurring disaccharides are similarly formed by reaction of a nucleoside diphosphate sugar with either a second sugar or a phosphorylated sugar, e.g.:

UDP-G + Fructose 6-P ⟶ UDP + Sucrose 6′-P

or

UDP-G + Fructose ⟶ UDP + sucrose

+ UDP-galactose $\xrightarrow{\text{D-glucose}}$ UDP + lactose (Gal-Glu)

Fig. 6.10

Nitrogenous di- and Polysaccharides

The general biosynthetic routes to di- and polysaccharides outlined under (a) apply equally to the biosynthesis of nitrogenous analogues derived from nitrogenous monosaccharides such as glucosamine and its derivatives. Thus hyaluronic acid is produced by an alternating reaction of UDP-glucuronic acid with UDP-N-acetylglucosamine. The sialic (or N-acylneuraminic) acids are derived from glucosamine by the following sequence of reactions:

N-acetylglucosamine ⟶ N-acetylmannosamine $\xrightarrow{\text{ATP}}$ N-acetylmannosamine 6-P

$\xrightarrow{\text{phosphoenolpyruyate}}$ N-acetylneuraminic acid 9-P

N-acetylneuraminic acid (NANA) $\xrightarrow{\text{CTP}}$ CMP-NANA

Fig. 6.11

The CMP-NANA now donates NANA to the terminal sugar unit of the polysaccharide chain in gangliosides and glycoproteins with elimination of CMP:

.... sugar + CMP-NANA ⟶ .... sugar-NANA + CMP

The nitrogenous saccharides are as we have seen previously (Chapter 5) important constituents of bacterial cell walls. Early studies by Park showed that uridine nucleotides accumulated in cultures of bacteria which had been inhibited by sub-liminal amounts of penicillin using techniques analogous to those used with sulphanilamide drugs; this work led to an understanding of purine nucleotide biosynthesis. Later work by Strominger and co-workers were able to relate these observations to an understanding of the chemical structures of the walls and their biosynthesis. The peptidoglycan (Murein) of Gram-positive bacterial cell walls is produced in four major steps:

1. UDP-N-acetylglucosamine + phosphoenolpyruvate

   ↓

   UDP-N-acetylmuramic acid

   ↓ L-Ala, D-Glu, L-Lys and D-Ala, ATP

   (UDP-N-acetylmuramyl pentapeptide)

2. UDP-N-acetylmuramyl pentapeptide

   (a) Undecaprenyl phosphate
   $Me_2C{:}CHCH_2\ (CH_2CMe{:}CHCH_2)_9CH_2CMe{:}CHCH_2O\text{-}P$
   (b) UDP-N-acetylglucosamine
   ↓ (c) ammonia and ATP

   $(\text{Murein})_{n+1}$

3. $(\text{Murein})_{n+1}$ + $\xrightarrow{\text{Gly t-RNA}}$ $(\text{Murein})_{n+1}$-pentapeptide

4. $(\text{Murein})_{n+1}$pentapeptide ⟶ cross linked $(\text{Murein})_{n+1}$
   ↘ D-Ala

The last step is inhibited by penicillin.

Fig. 6.12

BIOSYNTHESIS OF α-AMINO ACIDS AND PROTEINS

Amino acids

There are twenty α-amino acids which are required for protein synthesis and their biosynthesis is clearly of considerable importance. Whereas higher animals can only synthesize about half their amino-acid requirements (the remaining so-called essential amino acids must be obtained from an external source), higher plants and many prokaryotic organisms (e.g. *E. coli*) can synthesize all the amino acids using various nitrogen sources including ammonia, nitrate, nitrite, or, in the case of nitrogen-fixing legumes, nitrogen, although a requirement for a reduced form of nitrogen is most usual. On the other hand some micro-organisms, e.g. *Leuconostoc mesenteroides*, require sixteen of the amino acids before they can grow and like similar organisms will normally only grow on the amino-acid rich deposits found in decaying biological material.

Although the biosynthesis of each amino acid involves an individual reaction sequence and specific enzymes there are nevertheless many points of similarity. Thus many of the intermediate reaction products (pyruvate and other α-keto-acids) are derived from the carbohydrate cycles and of course most amino-acids consequently provide by a reversal of these processes valuable sources of carbohydrate cycle intermediates and hence ultimately of hexoses. For many of the amino acids (especially the so-called essential amino acids) the pathways of biosynthesis may vary substantially within species, and especially between prokaryotic and eukaryotic cells.

Aspartic Acid and Glutamic Acid

These are the most ubiquitous of the amino acids which occur in proteins.

Glutamic acid is produced from ammonia and ketoglutaric acid:

$$NH_3 + HO_2C.CO.CH_2CH_2CO_2H + NADPH + H^+ \rightleftharpoons HO_2C.CH(NH_2)CH_2CH_2CO_2H + NADP^+$$

Fig. 6.13

and aspartic acid by a similar reaction of oxaloacetic acid:

$$NH_3 + HO_2CCOCH_2CO_2H + NADPH + H^+ \rightarrow HO_2CCH(NH_2)CH_2CO_2H$$

or alternatively by a transamination reaction involving glutamic acid:

$$HO_2CCH(NH_2)CH_2CH_2CO_2H + HO_2CCOCH_2CO_2H \rightarrow HO_2CCOCH_2CH_2CO_2H + HO_2CCH(NH_2)CH_2CO_2H$$

Fig. 6.14

Asparagine and Glutamine

The amides are produced from the acids by formation of an acyl phosphate and amination of this

$$HO_2CCH(NH_2)CH_2CH_2CO_2H + ATP \longrightarrow$$

$$HO_2CCH(NH_2)CH_2CH_2CO_2PO_3^- \xrightarrow{NH_3}$$

$$HO_2CCH(NH_2)CH_2CH_2CONH_2$$

An alternative pathway includes a transamination step:

$$HO_2CCH(NH_2)CH_2CO_2H + HO_2CCH(NH_2)CH_2CH_2CONH_2$$

$$\downarrow$$

$$HO_2CCH(NH_2)CH_2CONH_2 + HO_2CCH(NH_2)CH_2CH_2CO_2H$$

Fig. 6.15

Proline

Proline is produced from glutamic acid by a reaction sequence involving formation of glutamic semialdehyde:

$$\text{Glutamic acid} \xrightarrow{\text{NADH}} OHC.CH_2CH_2CH(NH_2)CO_2H \longrightarrow$$

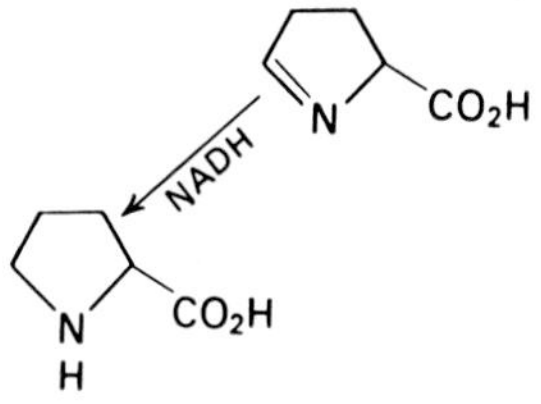

CO₂H

NADH

CO₂H

N
H
(Pro.)

Fig. 6.16

Alanine, Isoleucine, Valine and Leucine

These are derived from pyruvate. Alanine is normally produced by reductive amination of pyruvate or by a transamination reaction:

$$CH_3COCO_2H + NH_3 \xrightarrow{2H} CH_3CH(NH_2)CO_2H$$

or

$$CH_3COCO_2H + HO_2CCH(NH_2)CH_2CH_2CO_2H \longrightarrow CH_3CH(NH_2)CO_2H + HO_2CCOCH_2CH_2CO_2H$$

Fig. 6.17

Valine similarly results from α-ketoisovaleric derived from pyruvate and acetaldehyde:

$$(CH_3)_2CHCOCO_2H \xrightarrow{NH_3} (CH_3)_2CHCH(NH_2)CO_2H$$

Fig. 6.18

Isoleucine by a variation from keto-β-methylvaleric acid:

$$CH_3CH_2CH(CH_3)COCO_2H \xrightarrow{NH_3} CH_3CH_2CH(CH_3)CH(NH_2)CO_2H$$

Fig. 6.19

and *Leucine* is similarly formed from α-ketoisocaproic acid derived in turn by a sequence of reactions from α-ketoisovaleric acid:

$$(CH_3)_2CHCH_2COCO_2H \longrightarrow (CH_3)_2CHCH_2CH(NH_2)CO_2H$$

Fig. 6.20

Glycine and Serine

Serine results from 3-phosphoglyceric acid by oxidation and transamination:

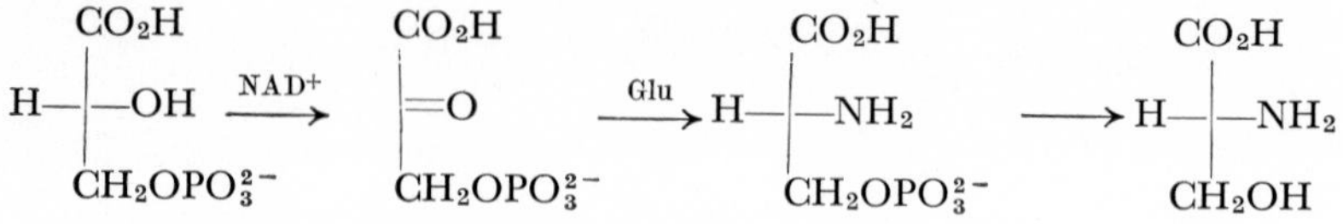

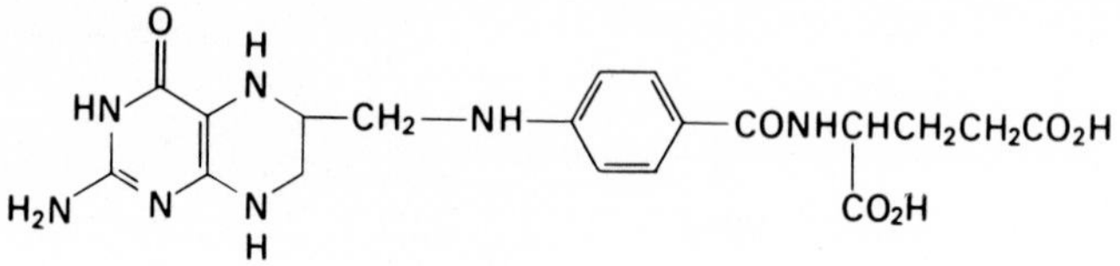

$$\begin{array}{c} CO_2H \\ H-\!\!|\!\!-OH \\ CH_2OPO_3^{2-} \end{array} \xrightarrow{NAD^+} \begin{array}{c} CO_2H \\ |\!\!=O \\ CH_2OPO_3^{2-} \end{array} \xrightarrow{Glu} \begin{array}{c} CO_2H \\ H-\!\!|\!\!-NH_2 \\ CH_2OPO_3^{2-} \end{array} \longrightarrow \begin{array}{c} CO_2H \\ H-\!\!|\!\!-NH_2 \\ CH_2OH \end{array}$$

O, HN, H, N, H₂N, N, N, H
$-CH_2-NH-C_6H_4-CONHCH(CO_2H)CH_2CH_2CO_2H$

[Tetrahydrofolic acid (THF)]

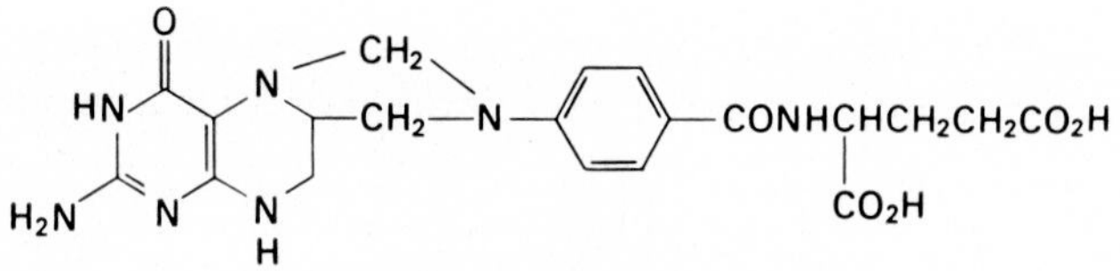

O, HN, N, CH_2, H₂N, N, N, H
$-CH_2-N-C_6H_4-CONHCH(CO_2H)CH_2CH_2CO_2H$

[N^5,N^{10}-methylene (THF)]

Fig. 6.21

and glycine by hydrogenolysis of serine with tetrahydrofolate (FH_4):

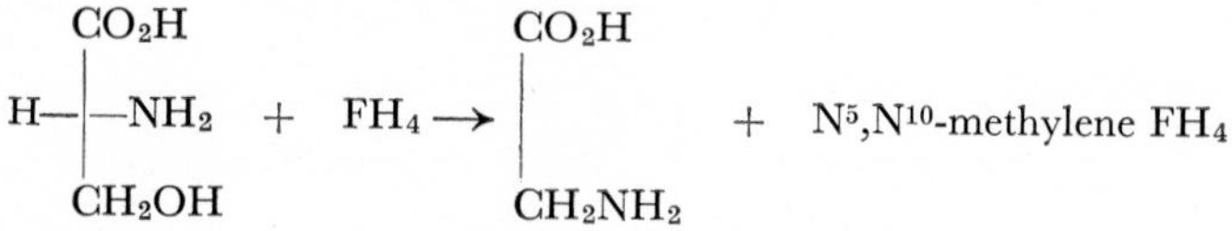

$$\begin{array}{c} CO_2H \\ H-\!\!|\!\!-NH_2 \\ CH_2OH \end{array} + FH_4 \longrightarrow \begin{array}{c} CO_2H \\ | \\ CH_2NH_2 \end{array} + N^5,N^{10}\text{-methylene } FH_4$$

Fig. 6.22

THF (Fig. 6.22) and related compounds are intimately concerned with numerous 1-carbon unit transfers in many biochemical sequences including purine nucleotide biosynthesis.

Methionine and threonine

These amino acids are produced from homoserine which in turn results from aspartic acid (see Fig. 6.23).

$$HO_2CCH_2CH(NH_2)CO_2H \rightarrow OHCCH_2CH(NH_2)CO_2H \rightarrow$$

(Aspartic semialdehyde)

$$HOCH_2CH_2CH(NH_2)CO_2H$$

(homoserine)

$$^{2-}O_3POCH_2CH_2CH(NH_2)CO_2H \rightarrow CH_3CH(OH)CH(NH_2)CO_2H$$

and

$$HOCH_2CH_2CH(NH_2)CO_2H \xrightarrow[\text{CoA}]{\text{Succinyl}}$$

$$HO_2CCH_2CH_2CO{-}OCH_2CH_2CH(NH_2)CO_2H \xrightarrow{\text{Cys}}$$

$$HO_2CCH(NH_2)CH_2{-}S{-}CH_2CH_2CH(NH_2)CO_2H$$

$$\longrightarrow HSCH_2CH_2CH(NH_2)CO_2H \xrightarrow{N^5{-}\text{MeTHF}} CH_3SCH_2CH_2CH(NH_2)CO_2H$$

(homocysteine) (Met)

Fig. 6.23

Cysteine

This amino acid results from a somewhat devious replacement of the hydroxyl of serine by an SH group the sulphur atom coming from methionine:

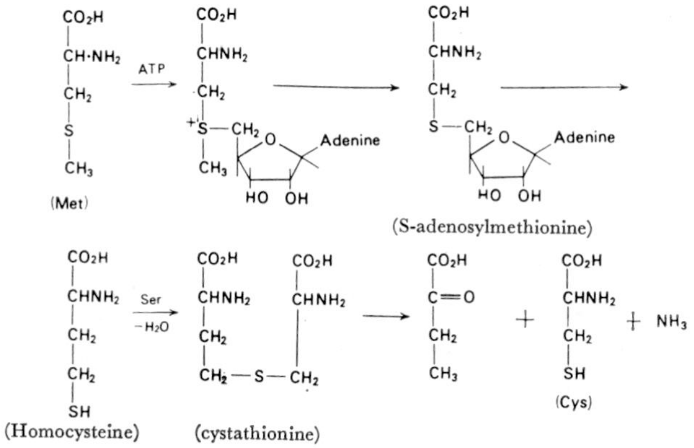

$CO_2H{-}CH{\cdot}NH_2{-}CH_2{-}S{-}CH_3$ (Met) $\xrightarrow{\text{ATP}}$ $CO_2H{-}CHNH_2{-}CH_2{-}\overset{+}{S}(CH_3){-}CH_2$–ribose(HO, OH)–Adenine $\longrightarrow$ $CO_2H{-}CHNH_2{-}CH_2{-}S{-}CH_2$–ribose(HO, OH)–Adenine $\longrightarrow$

(S-adenosylmethionine)

$CO_2H{-}CHNH_2{-}CH_2{-}CH_2{-}SH$ (Homocysteine) $\xrightarrow[-H_2O]{\text{Ser}}$ $CO_2H{-}CHNH_2{-}CH_2{-}CH_2{-}S{-}CH_2{-}CHNH_2{-}CO_2H$ (cystathionine) $\longrightarrow$ $CO_2H{-}C{=}O{-}CH_2{-}CH_3$ + $CO_2H{-}CHNH_2{-}CH_2{-}SH$ (Cys) + NH_3

Fig. 6.24

Lysine

The amino acid is produced via diaminopimelic acid in bacteria and higher plants and from α-aminoadipic acid in many fungi. In each case carbohydrate cycle intermediates are involved:

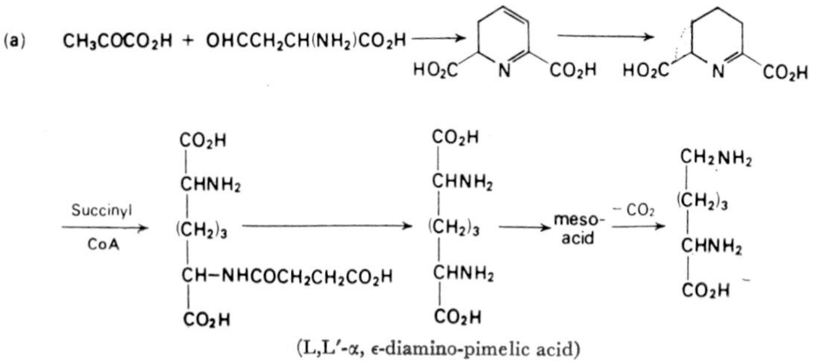

(a) $CH_3COCO_2H + OHCCH_2CH(NH_2)CO_2H \longrightarrow$ [dihydropyridine ring with HO_2C and CO_2H] $\longrightarrow$ [tetrahydropyridine ring with HO_2C and CO_2H]

$\xrightarrow[\text{CoA}]{\text{Succinyl}}$ $CO_2H{-}CHNH_2{-}(CH_2)_3{-}CH{-}NHCOCH_2CH_2CO_2H{-}CO_2H$ $\longrightarrow$ $CO_2H{-}CHNH_2{-}(CH_2)_3{-}CHNH_2{-}CO_2H$ $\longrightarrow$ meso-acid $\xrightarrow{-CO_2}$ $CH_2NH_2{-}(CH_2)_3{-}CHNH_2{-}CO_2H$

(L,L′-α, ε-diamino-pimelic acid)

Fig. 6.25(a)

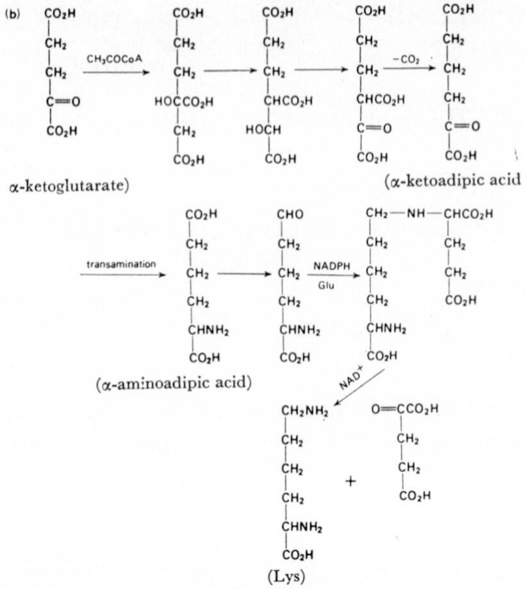

(b) α-ketoglutarate) $\xrightarrow{CH_3COCoA}$ ... $\xrightarrow{-CO_2}$ (α-ketoadipic acid

$\xrightarrow{\text{transamination}}$ (α-aminoadipic acid) $\rightarrow$ $\xrightarrow[Glu]{NADPH}$ $\xrightarrow{NAD^+}$ (Lys) + $O{=}CCO_2H(CH_2)_2CO_2H$

Fig. 6.25(b)

Arginine

This amino acid results in net amount by a sequence of reactions which involve the amino acid ornithine:

$$\text{Glu} \rightarrow \text{N-Acetyl Glu} \rightarrow \text{N-Acetyl Glu-}\gamma\text{-phosphate}$$

$$\xrightarrow{NADPH} OHC.CH_2CH_2CH(NHCOCH_3)CO_2H$$

$$\longrightarrow H_2N(CH_2)_3CH(NHCOCH_3)CO_2H$$

$$\longrightarrow H_2N(CH_2)_3CH(NH_2)CO_2H$$

(ornithine)

$$\xrightarrow{NH_2COOPO_3^{2-}} H_2NCONH(CH_2)_3CH(NH_2)CO_2H$$

(citrulline)

$$\xrightarrow[ATP]{Asp} HO_2CCH_2CH(CO_2H)NH.C(:NH)NH(CH_2)_3CH(NH_2)CO_2H$$

(arginosuccinic acid)

$$\downarrow$$

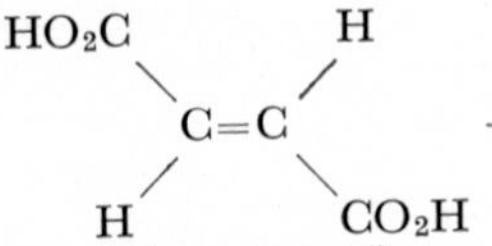

$$HO_2C(H)C{=}C(H)CO_2H + H_2N.C(:NH)NH(CH_2)_3CH(NH_2)CO_2H$$

(Fumaric Acid) (Arg)

The use of Asp as a source of amidine nitrogen in this reaction is reminiscent of reactions in the pathway of biosynthesis *de novo* of purine nucleotides (see later).

Fig. 6.26

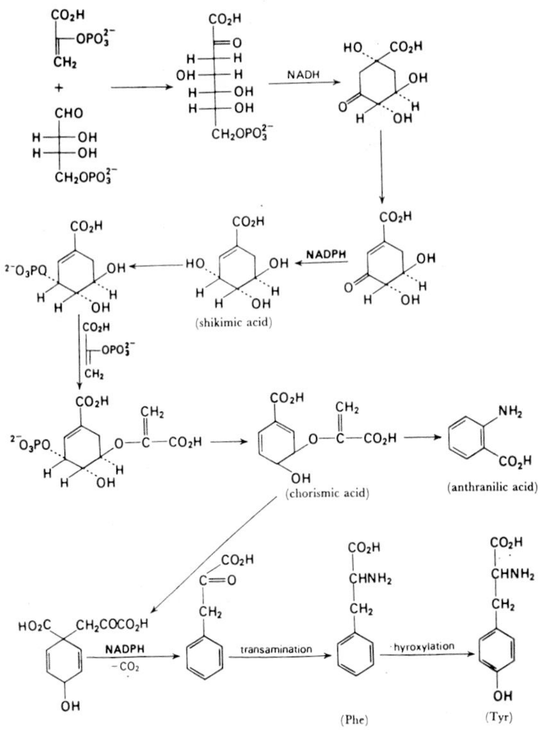
NADH
NADPH
(shikimic acid)
(chorismic acid)
(anthranilic acid)
NADPH
$-CO_2$
transamination
hyroxylation
(Phe)
(Tyr)

Fig. 6.27

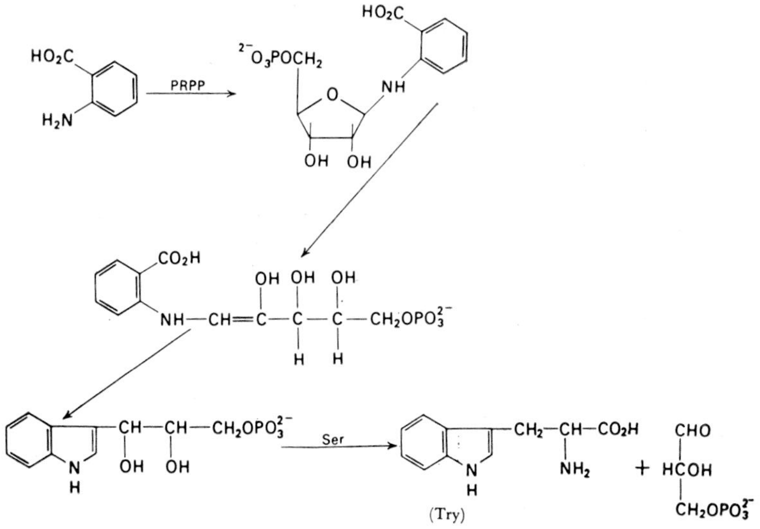
PRPP
Ser
(Try)

Fig. 6.28

Phenylalanine, Tyrosine and Tryptophane

These are the aromatic (benzenoid) amino acids which are produced from carbohydrate cycle precursors proceeding via cyclohexane derivatives which include a central intermediate shikimic acid (Figs. 6.27, 6.28).

Histidine

The biosynthesis of this amino acid involves a reaction sequence somewhat similar to that leading to tryptophane and involves nucleotide intermediates:

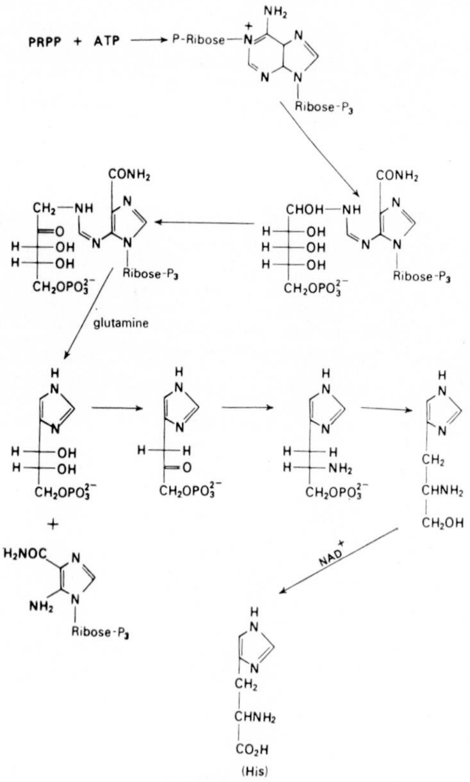

Fig. 6.29

Proteins

The synthesis of proteins in nature is intimately bound up with nucleic acids. The processes by which a particular protein is produced from a DNA (gene) of part of a chromosome in a cell is described in Chapter 7.

THE GENETIC CODE

The mRNA in the ribosomes functions in conjunction with appropriate amino acid tRNA's and enzymes, to produce proteins with predetermined amino acid sequences (see Chapter 7). We have clearly implied a relationship between nucleic acid base sequence stretching back to DNA of the chromosomal material and amino-acid sequence in the derived polypeptide. This relationship, the genetic code, is one of the more important discoveries which have resulted from the original Watson-Crick double helical hydrogen bonded structure for DNA.

In 1961 Nirenberg and Matthaei prepared a sample of poly-U (polyuridine 5′-phosphate with 3′-5′ internucleotidic linkages) by polymerization of UDP in the presence of the enzyme polynucleotide phosphorylase. The poly-U was incubated with a suspension of messenger depleted *E. coli* ribosomes containing also all the α-amino acids which in separate tubes were individually labelled (^{14}C). After a suitable incubation period the resultant polypeptides were precipitated with acid and analysed. It was found that only one radioactive amino acid, namely phenylalanine, had been incorporated into the polypeptide and moreover that the so-formed polypeptide was poly-phenylalanine. In other words, a sequence of U in the nucleic acid results in the formation of a sequence of Phe in the polypeptide. Similar work with a large variety of biosynthetic nucleic acids soon established the existence of a code (the genetic code) which related a triplet base sequence in a nucleic acid with a particular α-amino acid in the derived polypeptide (Table 6.01).

Further work, especially by Khorana and his co-workers, resulted in the chemical synthesis of all the possible di- and trinucleotides from the four bases' nucleotides and these were enzymically polymerized to messenger nucleic acids of known (triplet) sequence. From these materials, by experiments of the type described above, the base sequences of the triplets were determined for all the amino acids.

There are various features about the code which bear special comment:

(1) The code is degenerate (an unfortunate term). There is more than 1 triplet sequence of bases coding for one amino acid (see Table 6.01). All the amino acids with the exception of Met and Try have more than one code. On the other hand, a given triplet never codes for more than one amino acid.

(2) The triplet code must be correctly set at the beginning, i.e. there

are no commas or other means of distinguishing sequences other than starting at one end and working forwards.

(3) Three of the 64 (4^3) possible triplets do not code for any known amino acid. They have (again unfortunately) been termed nonsense triplets and apparently function as chain terminating units.

Table 6.01. The Genetic Code

2nd → / ↓1st	U	C	A	G	↓ 3rd
U	PHE	SER	TYR	CYS	U
	PHE	SER	TYR	CYS	C
	LEU	SER	STOP	STOP	A
	LEU	SER	STOP	STOP	G
C	LEU	PRO	HIS	ARG	U
	LEU	PRO	HIS	ARG	C
	LEU	PRO	GLU	ARG	A
	LEU	PRO	GLU	ARG	G
A	ILEU	THR	ASPN	SER	U
	ILEU	THR	ASPN	SER	C
	ILEU	THR	LYS	ARG	A
	MET	THR	LYS	ARG	G
G	VAL	ALA	ASP	GLY	U
	VAL	ALA	ASP	GLY	C
	VAL	ALA	GLU	GLY	A
	VAL	ALA	GLU	GLY	G

There have been many experiments which confirm the genetic code both *in vitro* and *in vivo*. Thus mutation of the tobacco mosaic virus by exposure to nitrous acid results in conversion of the primary NH_2 group of both purines (A or G) or pyrimidines (C-) to the hydroxyl (OH) group. In particular A → hypoxanthine and C → U. The resulting changes produced in the amino acid sequences may be observed in isolated polypeptides and are as predicted by the code. Similarly the universality of the code has been confirmed by many experiments using mammalian tRNAs and bacterial ribosomes (see chapter 7).

BIOSYNTHESIS OF NUCLEOTIDES AND NUCLEIC ACIDS

The nucleic acids are derived by polymerization of their monomeric mononucleotide precursors, generally in the 5′- di- or tri-phosphate form. The mechanism of formation of the simple nucleotides in the living cell is of special interest since these, or very similar sequences of chemical reactions, must have existed in the first primeval cell. In a sense therefore these particular metabolic paths afford a unique and specific link with the past. At the same time the formation of the specific types of organic intermediaries in the pathway would be minimal requirements of any abiogenic synthetic pathway.

Mononucleotides may be biochemically formed in two ways: (a) by exchange and interchange of the ribosyl unit with preformed bases or other nucleotides.

$$(1)\ P{-}S\overset{\beta}{-}base_1 + base_2 \rightleftharpoons P{-}S\overset{\beta}{-}base_2 + base_1$$

$$(2)\ P{-}S\overset{\beta}{-}base_1 + P{-}S\overset{\beta}{-}base_2 \rightleftharpoons P{-}S\overset{\beta}{-}base_2 + P{-}S\overset{\beta}{-}base_1$$

$$(3)\ P{-}S\overset{\alpha}{-}PP + base \rightarrow P{-}S\overset{\beta}{-}base + PP_i$$

Fig. 6.30

(b) by the *de novo* route from simple chemical substances.

The last route is perhaps of greater interest in the context of this thesis and the particular sequences of biochemical reactions leading to the pyrimidine and the purine nucleotides are outlined in Figs 6.31 and 6.32 respectively. There are a number of points about these pathways which require special comment:

(i) They are apparently universal since examination of the pathways in plant, animal, bacterial and fungal sources indicates that the same types of reactions are occurring.

(ii) The *de novo* syntheses proceed with D-ribose only. It may be noted that deoxyribonucleotides are produced only by a reduction of the corresponding ribonucleotides by a mechanism which we need not go into here and do not form directly from simple chemical precursors. The ribonucleotides in this sense therefore appear to be of more fundamental and primitive origin. One might reasonably conclude that ribonucleotides function in nucleic acid, and protein biosynthesis may have preceded their subsequent evolutionary reduction to deoxyderivatives.

(iii) In the *de novo* biosynthesis of the purine nucleotides, with the exception of the D-ribose moiety, the only chiral centre introduced is derived from L-aspartic acid, at two positions in the pathway (Fig. 6.32) where the same enzyme is in fact involved. It is interesting to note that this particular use of the aspartic acid nitrogen, as a source of a carboxamide group (in AICAR), could well be imagined to be replaced in a more primitive model by ammonia. Ammonia can certainly take the place of L-glutamine in the conversion of PRPP→PRA.

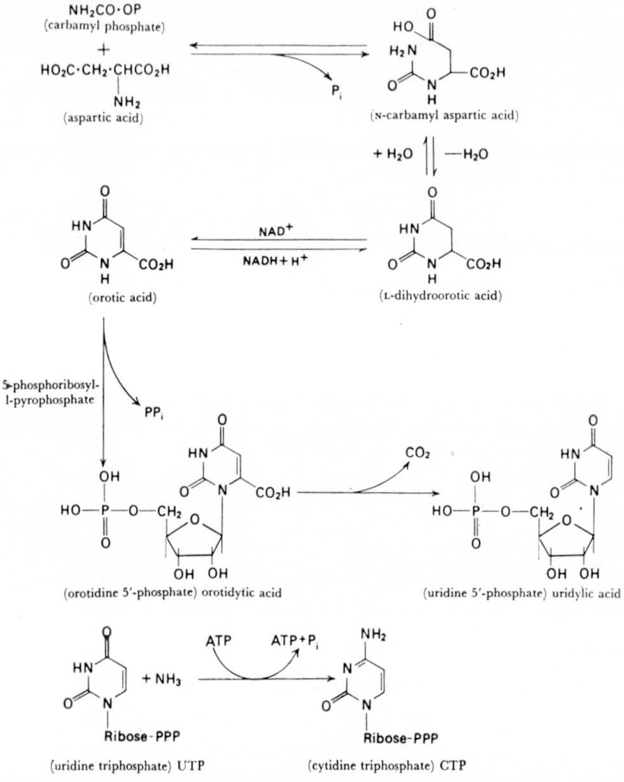

$NH_2CO{\cdot}OP$ (carbamyl phosphate) + $HO_2C{\cdot}CH_2{\cdot}CH(NH_2)CO_2H$ (aspartic acid) ⇌ (N-carbamyl aspartic acid) + P_i

(N-carbamyl aspartic acid) ⇌ (L-dihydroorotic acid) ($+H_2O$ / $-H_2O$)

(orotic acid) ⇌ (L-dihydroorotic acid) (NAD^+ / $NADH + H^+$)

(orotic acid) + 5-phosphoribosyl-1-pyrophosphate → (orotidine 5′-phosphate) orotidytic acid + PP_i

(orotidine 5′-phosphate) orotidytic acid → (uridine 5′-phosphate) uridylic acid + CO_2

(uridine triphosphate) UTP + NH_3 → (cytidine triphosphate) CTP (ATP → ATP+P_i)

Fig. 6.31. Biosynthesis of pyrimidine nucleotides

(iv) In the *de novo* biosynthesis of purine nucleotides all the bonds formed, with one exception, are between carbon and nitrogen. This is particularly interesting since such reactions are certainly more readily achieved in non-living systems than is the formation of carbon to carbon bonds. It is also of interest that the one specific carbon–carbon linkage formed in this sequence of reactions has in fact been readily duplicated in an *in vitro* experiment in the absence of enzymes, namely by direct carboxylation of aminoimidazole ribotide (AIR) with bicarbonate under quite mild (optimum temperature 60°) conditions (Shaw and co-workers). The same reaction proceeds slowly even at room temperature.

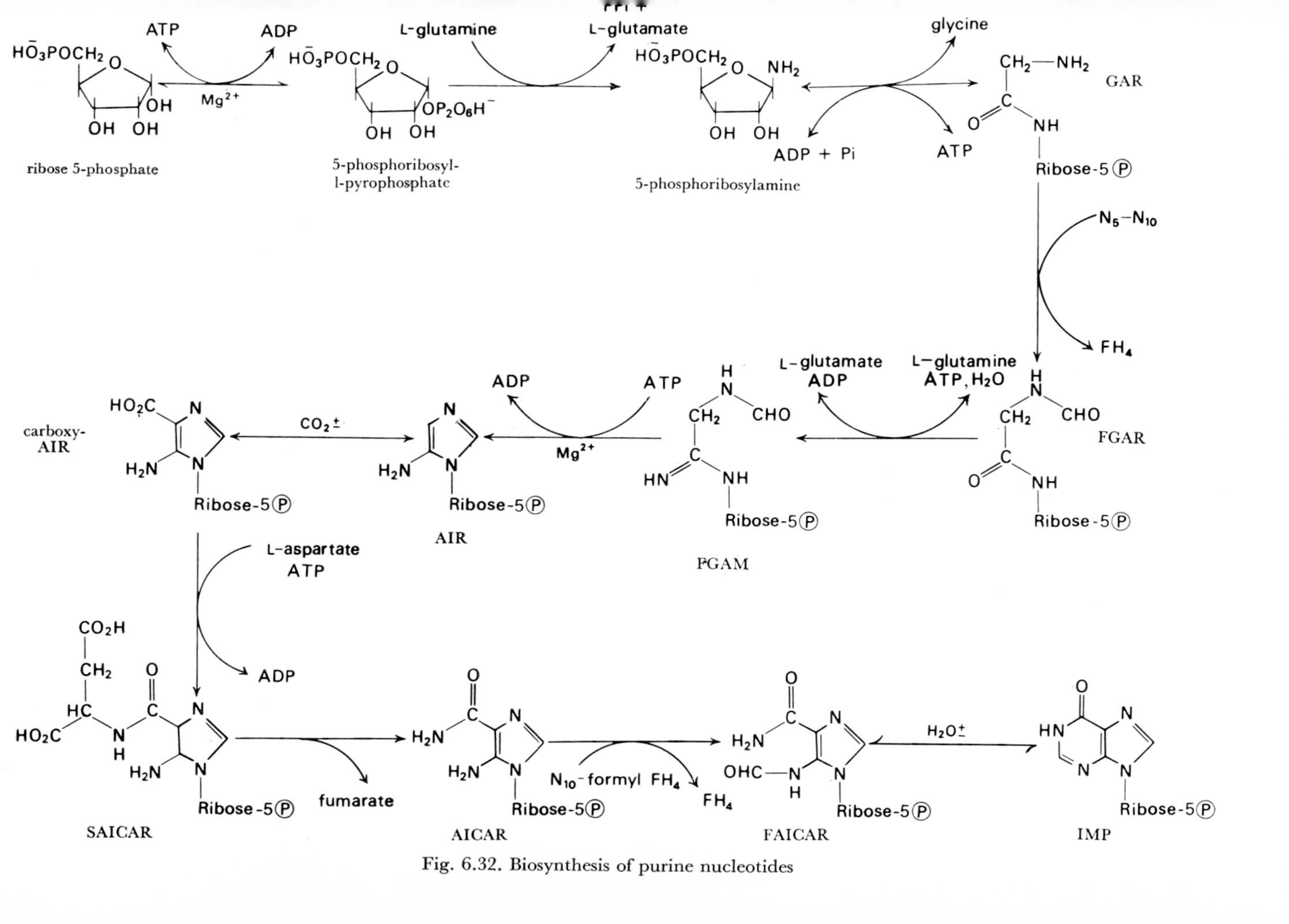

Fig. 6.32. Biosynthesis of purine nucleotides

Fig. 6.32—*continued*

(v) The biosynthesis of pyrimidine nucleotides is in some ways quite different from that of the purine derivatives. Thus the ribosyl unit is introduced at a much later stage when the pyrimidine ring has been formed. There are more chiral centres involved, although these are derived also from L-aspartic acid but they are less readily dispensed with and require more stereo-specific control. The whole process seems less primitive and it is tempting to assume that the original primeval cell contained only purines and no pyrimidines. If this were so, present knowledge of the genetic code would restrict the number of amino acids to four—glycine (GGG, GGA), lysine (AAA, AAG), glutamic acid (GAA, GAG) and arginine (AGA, AGG).

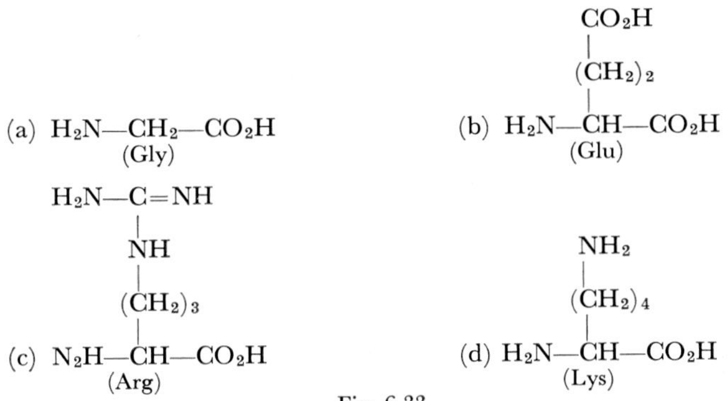

(a) $H_2N—CH_2—CO_2H$ (Gly)

(b) $H_2N—CH(CH_2CH_2CO_2H)—CO_2H$ (Glu)

(c) $N_2H—CH—CO_2H$ with side chain $(CH_2)_3—NH—C(=NH)—NH_2$ (Arg)

(d) $H_2N—CH—CO_2H$ with side chain $(CH_2)_4—NH_2$ (Lys)

Fig. 6.33

This is a particularly interesting sequence of amino acids containing as it does one neutral (Gly), one acidic (Glu) and two basic amino acids with a gradual increase in the number of methylene groups in the side chain from O(Gly) to 4(Lys), and in addition all are aliphatic compounds. Proteins derived from such amino acids would clearly have a substantial armoury of functional groups to operate a wide range of enzyme activities which are well known to involve many hydrogen transfer reactions between acidic and basic groups, i.e.:

$$—CO_2^- + HN^+— \rightleftharpoons —CO_2H + N—$$

or more precisely

$$CO_2^- + S + HN^+ \rightleftharpoons CO_2...\overset{H}{\underset{+}{S}}...N \overset{H_2O}{\rightleftharpoons} CO_2H + N + S \text{ modified}$$

In contrast, nucleic acids derived from pyrimidines alone would contain perhaps rather less interesting amino acids from a functional point of view, namely-Ser, Phe, Leu and Pro.

Enzymic Polymerization of Mononucleotides

Several enzymes are now known which are able to polymerize various types of nucleoside ribo- and deoxyribo-phosphates to their respective polymers and it would appear certain that many more such enzymes will be found with a variety of specialized functions in this area.

Polynucleotide Phosphorylase

This enzyme, the first useful polymerizing enzyme, was isolated by Ochoa and Grunberg-Manago in 1955 and is apparently only found in bacteria. The enzyme catalyses the polymerization of nucleoside diphosphates and requires Mg^{++} and a primer of RNA to which new units may be added. The reaction is reversible and the major function of the enzyme may be to break down mRNA in the cell rather than to act as a synthesizing enzyme. Nevertheless the enzyme has proved invaluable in providing a route to biosynthetic RNAs and so enabling the first principles of the genetic code to be established.

DNA-Polymerase

Discovered by Kornberg and co-workers in 1956, this enzyme catalyses the synthesis of DNA from 2-deoxynucleoside 5′-triphosphates and requires Mg^{++} and some preformed DNA; it only works when all 4 nucleoside triphosphates are present. The reaction is also reversible and *in vivo* the enzyme may function to hydrolyse DNA molecules. It adds about 1000 nucleotide residues per min per mol of enzyme and can produce DNA with molecular weights of several millions. The enzyme has been found in animal, plant and bacterial cells. In eukaryotic cells it is largely present in the nucleus. The preformed DNA is required both as a primer and as a template and the bigger the primer material the better the enzyme works and the greater the amount of DNA formed. On the template DNA the enzyme builds a complementary strand of DNA. This enzyme will also incorporate into the DNA some unnatural purines or pyrimidine nucleoside triphosphates, but only when there is no interference with the hydrogen bonding process between the bases in the template strands. Thus dITP can replace dGTP and Br-dTP and Br-dCTP may replace dTTP and dCTP respectively.

DNA-Ligase

This particular enzyme is able to join two DNA molecules together by the 5′- and 3′- terminal ends of two segments of DNA both of which are in double helical association. Figure 6.34 shows the two steps.

The enzyme may use this ability to produce a cyclic DNA, similar to those which occur in bacterial chromosomal material, from an acyclic molecule, and it may also join breaks in damaged DNA, or recombine genes during crossing over in eukaryotic cells.

A combination of DNA polymerase and DNA-ligase was used in 1968 by Kornberg and Sinsheimer to produce a synthesis of the bio-

(1) $NAD^+ + E \rightleftharpoons E\text{—}AMP + NMN + (PP_i)$
(or ATP) (nicotinamide mono nucleotide)

(2) $E\text{—}AMP + 5'\text{—}P \text{ end} + 3'\text{—}OH \text{ end} \rightleftharpoons 5'\text{—}3' \text{ phosphodiester} + E + AMP$

Fig. 6.34

logically active form of a circular double stranded virus DNA derived from the ΦX174 bacteriophage.

RNA-Polymerase

Discovered in 1959 by Weiss, Hurwitz and Stevens. This enzyme produces RNA from nucleoside 5′-triphosphates. It requires Mg^{++} and a double stranded DNA-template primer. The four nucleoside 5′-triphosphates are all required and the RNA produced is complementary to the DNA primer material. The enzyme will also incorporate certain unnatural bases. Thus 5-fluoro-UTP and 5-Br-UTP may replace TTP and ITP can replace GTP.

RNA-Replicase

This enzyme was discovered by Spiegelman in 1961. It can use a viral RNA as a template for formation of new RNA and is induced in the infected host cell by the invading virus.

BIOSYNTHESIS OF LIPIDS

Introduction

Although the majority of naturally occurring lipids are essentially polymers (perhaps better described as *n*-mers) of acetic acid and generally of low molecular weight there are two distinct routes to their formation. The first involves the more or less direct conversion of acetate (or the derived and presumably more active malonate) to the *n*-mer material and the second involves the initial formation of a branched chain C_6 unit (mevalonate) from acetate and the further elaboration of this as active C_5 units into various lipids whose structures generally reflect the branched structure. The first group of compounds include normal straight chain compounds such as hydrocarbons, fatty acids and fats, long-chain alcohols, waxes and numerous derivatives. The second group includes rubbers, many terpenoid compounds, carotenoids and related steroids which arise by cyclization of acyclic intermediates. There are of course many hybrid types which arise from a combination of the two main routes.

n-Merization of Acetic Acid and Derivatives

The polymethylenes $CH_3(CH_2)_nCH_3$ and various related lipids including fatty acids, long-chain alcohols, waxes and other esters with and without additional functional groups including double and triple bonds, hydroxyl, thiol, epoxide groups, etc., are produced biosynthetically essentially by an *n*-merization of acetic acid and concomitant or subsequent reduction, dehydration, hydrogenation, oxidation or some similar chemical modification of the initial products.

In principle, for example:

$$3\ CH_3COOH \longrightarrow CH_3CO\ CH_2CO\ CH_2CO_2H \xrightarrow{8H} CH_3\ (CH_2)_4CO_2H$$

Fig. 6.35

The active form, which operates to produce these results, consists of a thiol ester formed from acetic acid and a small molecular weight (*ca* 10,000) protein named the acyl carrier protein. A serine residue at position 36 in this protein is attached in a phosphodiester link to a pantothenic acid side chain which in turn becomes acylated on its terminal SH group by either an acetyl or a malonyl group. The molecular unit involved in this particular acylation is either acetyl Co-A or the derived malonyl Co-A produced by reaction of the acetyl derivative with carbon dioxide in the presence of ATP:

$$CH_3CO\text{—}S\text{—}CoA + CO_2 \xrightarrow[Mn^{++}]{ATP} HO_2CCH_2CO\text{—}SCoA + ADP + P_i$$

$$\begin{array}{l} \text{Protein} \quad\quad O \\ \quad \vdots \quad\quad\quad\quad \| \\ \text{—}CH_2\text{—}O\text{—}P\text{—}OCH_2C(CH_3)_2CHOH\text{—}CONH(CH_2)_2CONH(CH_2)_2\text{—}SH \\ \quad\quad\quad\quad\quad | \\ \quad\quad\quad\quad OH \quad\quad\quad\quad\quad\quad\quad\quad\quad\quad\quad\quad\quad\quad \equiv ACP\text{—}SH \end{array}$$

Fig. 6.36. Acyl carrier protein (ACP) prosthetic group

The first stages of chain formation involve formation of acetoacetyl-S-ACP:

$$\begin{array}{l} CH_3CO\text{—}S\text{—}ACP + HO_2C\text{—}CH_2\text{—}CO\text{—}S\text{—}ACP \longrightarrow \\ \quad\quad CH_3COCH_2CO\text{—}S\text{—}ACP + CO_2 + ACP\text{—}SH \end{array}$$

Fig. 6.37

so that the evolved carbon dioxide used to convert acetate to malonate does not appear in the final product.

The acetoacetyl derivative by reduction (involving NADPH), dehydration of the resulting alcohol and further reduction of the crotonyl derivative yields *n*-butyryl-S-ACP:

$$\begin{array}{l} CH_3COCH_2CO\text{—}S\text{—}ACP \longrightarrow CH_3CHOHCH_2CO\text{—}S\text{—}ACP \longrightarrow \\ \quad\quad CH_3CH{:}CHCO\text{—}S\text{—}ACP \longrightarrow CH_3(CH_2)_2CO\text{—}S\text{—}ACP \end{array}$$

Fig. 6.38

The chain may now be extended in a similar fashion ultimately to form longer chain fatty acids, although most living systems have a propensity to complete the chain process at either palmitic ($C_{15}H_{31}CO_2H$) or stearic ($C_{17}H_{35}CO_2H$) acids, which suggests that the enzymes concerned can only handle a limited chain length extension process. The overall reaction equation leading to palmitic acid may be written:

$$8CH_3CO{-}CoA + 14NADPH + 14H^+ + 7ATP \longrightarrow C_{15}H_{31}CO_2H + 8CoA + 14NADP^+ + 7ADP + 7P_i + 6H_2O$$

Fig. 6.39

The fats (acylated glycerols) are produced from glycerol phosphate in two steps involving initial formation of a phosphatidic acid by acylation of glycerol-1-phosphate with two fatty acyl CoA molecules

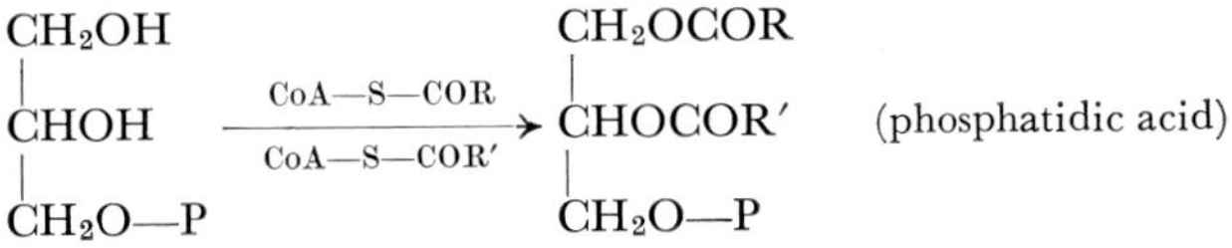

$$\begin{array}{l} CH_2OH \\ | \\ CHOH \\ | \\ CH_2O{-}P \end{array} \xrightarrow[CoA{-}S{-}COR']{CoA{-}S{-}COR} \begin{array}{l} CH_2OCOR \\ | \\ CHOCOR' \\ | \\ CH_2O{-}P \end{array} \quad \text{(phosphatidic acid)}$$

Fig. 6.40

followed by hydrolysis to the diacylglycerol and subsequent acylation of this with a further acyl CoA molecule:

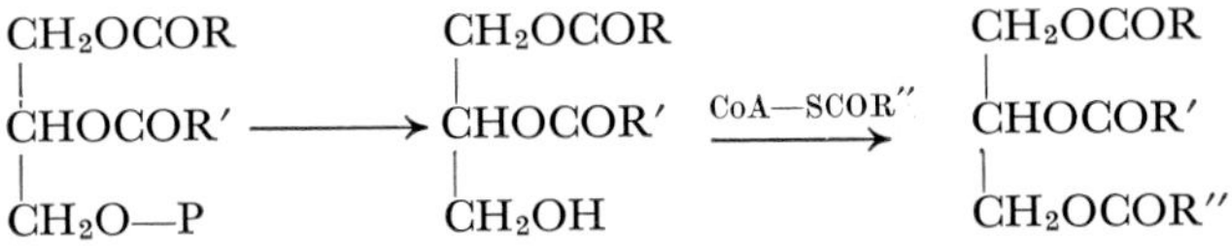

$$\begin{array}{l} CH_2OCOR \\ | \\ CHOCOR' \\ | \\ CH_2O{-}P \end{array} \longrightarrow \begin{array}{l} CH_2OCOR \\ | \\ CHOCOR' \\ | \\ CH_2OH \end{array} \xrightarrow{CoA{-}SCOR''} \begin{array}{l} CH_2OCOR \\ | \\ CHOCOR' \\ | \\ CH_2OCOR'' \end{array}$$

Fig. 6.41

Numerous phosphoglycerides derived from phosphatidic acid exist. They are mostly produced by initial formation of a cytidine diphosphate diacyl glycerol by reaction of CTP with the phosphatidic acid derivative:

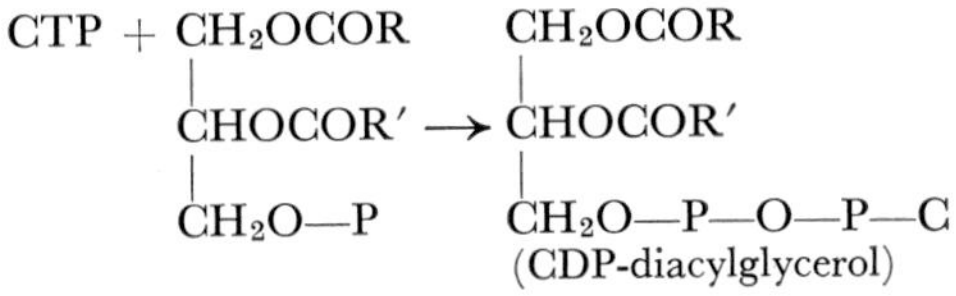

$$CTP + \begin{array}{l} CH_2OCOR \\ | \\ CHOCOR' \\ | \\ CH_2O{-}P \end{array} \longrightarrow \begin{array}{l} CH_2OCOR \\ | \\ CHOCOR' \\ | \\ CH_2O{-}P{-}O{-}P{-}C \end{array}$$

(CDP-diacylglycerol)

Fig. 6.42

The CDP-diacyglycerol can then react with numerous substances to produce phosphoglycerides by displacement of a CMP unit.

e.g.

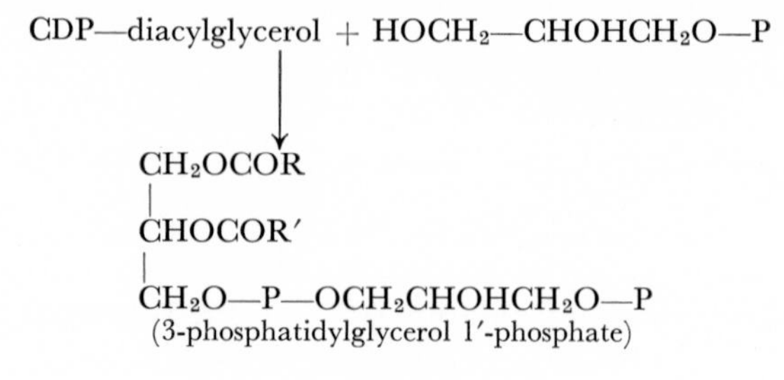

Fig. 6.43

Pathways Involving Mevalonate

Mevalonic acid is produced from acetate by a simple series of reactions: which differ slightly from those involved in fatty acid biosynthesis.

$$CH_3CO—S—CoA + CH_3COCH_2CO—SCoA$$
$$\rightarrow HO_2CCH_2C.CH_3(OH).CH_2CO_2H$$
$$\rightarrow HO_2CCH_2C.CH_3(OH)CH_2CH_2OH \rightarrow CO.CH_2.C.CH_3(OH)CH_2.CH_2$$

(mevalonic acid) (mevalonolactone; ring closed through —O—)

Fig. 6.44

The mevalonate unit is then converted into active C_5-pyrophosphate units which undergo *n*-merization processes with production of numerous types of lipid molecules (Fig. 6.45).

Extension of processes of this type leads to natural rubbers

$$(—CH_2.CMe{:}CHCH_2—)_n,$$

carotenoids such as β-carotene (Fig. 6.46(a)) and by cyclization reactions to triterpenes and steroids (Fig. 6.46(b)).

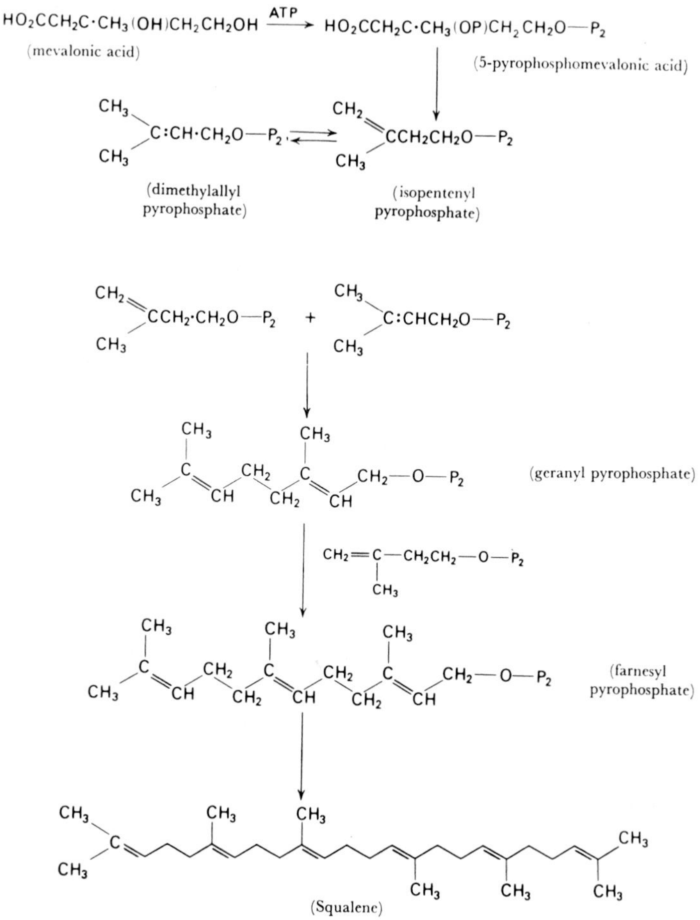

$HO_2CCH_2C{\cdot}CH_3(OH)CH_2CH_2OH \xrightarrow{ATP} HO_2CCH_2C{\cdot}CH_3(OP)CH_2CH_2O{-}P_2$

(mevalonic acid) (5-pyrophosphomevalonic acid)

$(CH_3)_2C{:}CH{\cdot}CH_2O{-}P_2 \rightleftharpoons CH_2{=}C(CH_3)CH_2CH_2O{-}P_2$

(dimethylallyl pyrophosphate) (isopentenyl pyrophosphate)

$CH_2{=}C(CH_3)CH_2{\cdot}CH_2O{-}P_2 + (CH_3)_2C{:}CHCH_2O{-}P_2$

$\downarrow$

$(CH_3)_2C{=}CH{-}CH_2{-}CH_2{-}C(CH_3){=}CH{-}CH_2{-}O{-}P_2$ (geranyl pyrophosphate)

$\downarrow$ $CH_2{=}C(CH_3){-}CH_2CH_2{-}O{-}P_2$

$(CH_3)_2C{=}CH{-}CH_2{-}CH_2{-}C(CH_3){=}CH{-}CH_2{-}CH_2{-}C(CH_3){=}CH{-}CH_2{-}O{-}P_2$ (farnesyl pyrophosphate)

$\downarrow$

(Squalene)

Fig. 6.45. Biosynthesis of squalene

(a)

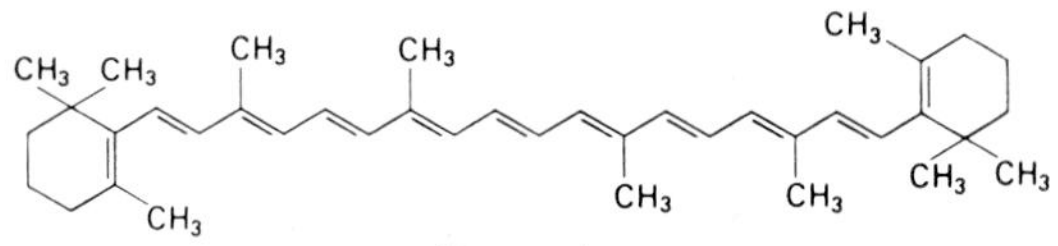

(β-carotene)

(b)

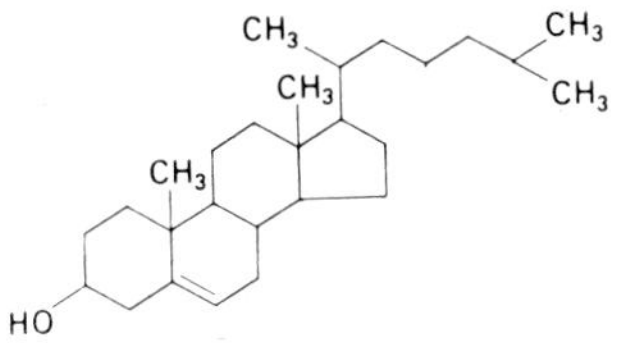

(cholesterol)

Fig. 6.46

SUGGESTED FURTHER READING Chapter 6

Books

Bernhard, S. (1968). "The Structure and Function of Enzymes". Benjamin, Philadelphia.

Calvin, M. and Bassham, J. A. (1962). "The Photosynthesis of Carbon Compounds". Benjamin, Philadelphia.

Calvin, M. and Jorgenson, M. (Eds) (1968). "Bio-organic Chemistry". Freeman, San Francisco.

Chargaff, E. and Davidson, J. N. (Eds) (1955). "The Nucleic Acids", Volumes I–III. Academic Press, New York.

Davidson, J. N. (Ed.) (1972). "Biochemistry of Nucleic Acids". Wiley.

Florkin, M. and Mason, H. S. (Ed.) (1962). "Comparative Biochemistry", Volumes I–VII. Academic Press, New York.

Goodwin, T. W. (Ed.) (1965). "Chemistry and Biochemistry of Plant Pigments". Academic Press, London.

Goodwin, T. W. (Ed.) (1968). "The Metabolic Roles of Citrate". Academic Press, London.

Greenberg, D. M. (Ed.) (1967). "Metabolic Pathways". Academic Press, New York.

Hutchinson, D. W. (1964). "Nucleotides and Coenzymes". Wiley, New York.

Lehninger, A. L. (1970). "Biochemistry". Worth, New York.

Meister, A. (1965). "Biochemistry of the Amino Acids". Academic Press, New York.

West, E. S., Todd, W. R., Mason, H. S. and Van Bruggen, J. T. (1966). "Textbook of Biochemistry". MacMillan, London.

Woese, C. R. (1967). "The Genetic Code". Harper and Row, New York.

Articles

Akhtar, M., and Wilton, D. C. (1970). Enzyme Mechanisms. *Ann. Rep. chem. Soc.* 557–574.

Crick, F. H. C. (1968). The Origin of the Genetic Code. *J. mol. Biol.* **38,** 367–379.

Green, D. E. and Allman, D. W. (1968). Biosynthesis of Fatty Acids. *In* "Metabolic Pathways", Vol. 2. (Ed. D. M. Greenberg). Academic Press, New York.

Hartman, S. C. and Buchanan, J. M. (1959). Purine Nucleotide Biosynthesis. *Ann. Rev. Biochem.* **28,** 365.

Holley, R. W. *et al.* (1965). Structure of a Ribonucleic Acid. *Science, N.Y.* **147,** 1462–1465.

Kornberg, A. (1948). Nucleotide Pyrophosphate and Triphosphate Nucleotide Structure. *J. biol. Chem.* **174,** 1051–1052.

Kornberg, A. (1967). Biosynthesis of DNA. Regul. Nucleic Acid Protein Biosyn., Proc. Int. Symp., Luteren, The Netherlands.

Kornberg, A., Sinsheimer, R. L. and Goulian, M. A. (1968). Enzymatic Synthesis of DNA. 24 Syntheses of Infectious Phage XΦ 174 DNA. *Proc. natn. Acad. Sci.* **58,** 2321–2328.

Kornberg, A. (1969). Active Centre of DNA Polymerase. *Science, N.Y.* **163,** 1410–1418.

Nirenberg, M. W., and Matthaei, J. H. (1961). The Dependence of Cell-free Protein Synthesis in *E. coli* upon Naturally Occurring or Synthetic Polyribonucleotides. *Proc. natn. Acad. Sci.* **47,** 1588–1602.

Nirenberg, M. W. and Leder, P. (1964). RNA Codewords and Protein Synthesis. *Science, N.Y.* **145,** 195.

Ochoa, S. and Grunberg-Manago, M. (1955). Enzymic Synthesis and Breakdown of Polynucleotides-Polynucleotide Phosphorylase. *J. Am. chem. Soc.* **77,** 3165–3166.

Park, J. T. (1952). Isolation and Structure of the Uridine 5′-pyrophosphate Derivatives which Accumulate in *Staphyloccus aureus* When Grown in Presence of Penicillin. *2nd Congr. Intern. Biochem., Chim Biol. VII, Symposium Mode d'Action des Antiboitiques* **31–39.**

Sanger, F. (1959). The Chemistry of Insulin. *Ann. Rep. chem. Soc.,* **45,** 283–292.

Spiegelman, S. *et al.* (1968). The Mechanism of RNA Replication. *Quant. Biol.* **34,** 101–124.

Staunton, J. (1970). Biosynthesis. *Ann. Rep. chem. Soc.* 535–556.

Vandecasteele, J. P. (1971). Bacterial Metabolism and its Regulation. *Rev. Inst. Francais, Petrole* **26,** 221–261.

Weiss, S. B. and Nakamoto, T. (1961). Net Synthesis of Ribonueleic Acid With a Microbial Enzyme Requiring Deoxyribonucleic Acid and Four Ribonucleoside Triphosphates. *J. biol. Chem.* **236,** 18–20.

7

Living Systems

INTRODUCTION: THE DEFINITION OF LIFE

Sooner or later we must ultimately explain or define in reasonably precise terms what we mean by the terms "life" or "living system", to provide at least a frame of reference against which we may set our various physical, chemical, biochemical, biological and geochemical data. Aristotle believed that nowhere on a line drawn from the smallest atom to the most complex of living things was it possible to say where non-life ended and life began. Many scientists of today would agree with this viewpoint and suggest that matter simply differs according to its degree of organization.

Recent and startling advances in molecular biology and biochemistry have led to quite rigid definitions of life couched in purely physical or chemical terms. For instance—"Life is the structure replication of enzymes ensured by exactly reproduced nucleic acid molecules." The rigidity of such definitions, however, is their weakness since they fail to describe the system and only describe one or more of its functions.

From simple examples of little organized matter such as a hydrogen molecule one can progress in a seemingly orderly and logical fashion through more complex and more ordered molecules as in crystalline lattices through even more complex molecules such as proteins and nucleic acids to nucleoprotein adducts which exist in nature (viruses) to more complex adducts (viruses with membranes but see later) and more complex simple bacteria (rickettsia, psittacosis) and finally to the incredibly complex differentiated and integrated cells and tissues of plants and animals.

Such concepts, of course, are in some ways strongly reinforced by the undoubted truths of Darwinian type biological evolution in which more complex living systems are seen to develop from apparently simpler living systems. At the same time it has been comparatively easy conceptually to extrapolate evolutionary sequences of these types backwards to include the chemicals from which they are derived, to produce a simple general theory not only of the nature of life but of its origins.

At the other end of the spectrum of ideas we have different views about the nature of living systems, in which a dominant central (ego-centric?) role is played by the human species. These are, of course, the various supernatural concepts, which represent in a sense the intuition of many of the most advanced life forms on this planet and as such cannot be excluded. They attempt to combine life, the universe and the nature of things and in this last they have perhaps to some extent an edge over the more rigid narrower concepts outlined earlier.

We define the living system in a much simpler manner, namely as a machine. This can hardly claim to be novel, but we believe that the machine concept should be accepted in a much more literal sense than has generally been the case. The living system machine is a chemical one. This particular concept is not always stressed. In the sense that the living system possesses extraordinary properties such as the ability (in mammals at least) to think, reason and control environment, there is no conceptual difficulty in providing alternative systems capable of such properties.

Thus it requires no great imagination to elaborate some system of sub-atomic particles or of fields, so ordered that they have abilities to control, direct and manipulate their environment and, like advanced forms of chemical living systems, to think and to reason logically. One might define such a form of life as physical or cosmological. There is no evidence for the existence of any such organized form of matter but in some ways it is almost to be expected.

In addition, of course, there is no particular reason why ours should be the only form of chemical life. The sole ability of carbon to form such a wide and diverse group of compounds among all the elements virtually ensures that any reasonably complex chemical living system would have to be carbon-based. But this said, there are innumerable possible macromolecular variations completely unrelated to the nucleic acids and/or proteins which formally at least could lead to viable systems.

However, on this planet at least only chemical living systems are characterized by the universality of the various chemical systems from which they are constituted. Thus the nature and functions of the major macromolecules are universal, as are the biosynthetic pathways from which they and their monomeric precursors are derived. In addition the macromolecules have combined to produce the biological "monomer" unit of life, namely the simple cell, which is characterized by its relatively small size and by the way it has been elaborated. There are two quite distinct types of cell, the prokaryotic and the eukaryotic cells, which make up the known living systems. In addition there are nucleoprotein complexes—the viruses—which appear to occupy a position in the hinterland between the living and the non-living systems.

PROKARYOTIC AND EUKARYOTIC CELLS

Introduction

Recent advances in molecular biology, and especially the use of electron microscopy and histochemical staining techniques, have clearly shown that there exist two fundamentally different types of simple

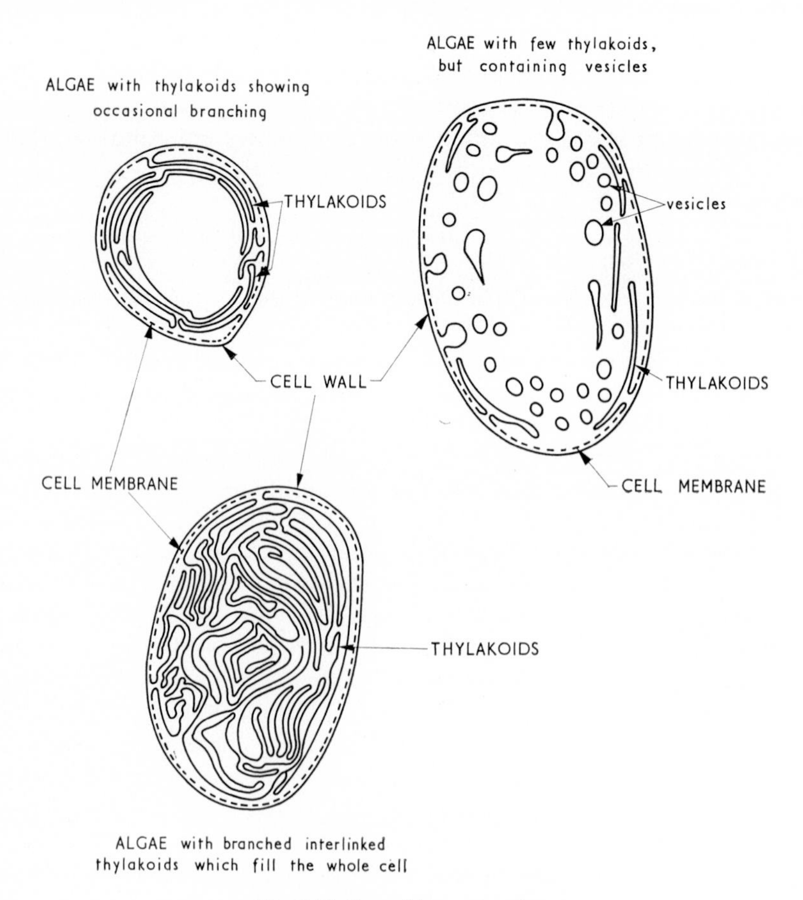

Fig. 7.01. Some blue-green algae

cells into which all living systems may be classified. These were first termed Prokaryotic and Eukaryotic cells by E. C. Dougherty in 1957. The names are derived from a description of the nuclear material in each type of cell—prokaryon for the moneran nucleus (from *pro-* = before and *karyon* = kernel), and eukaryon for the nucleus of higher organisms (from *eu* = well and *karyon*). The prokaryotes are restricted to the blue-green algae (Fig. 7.01), the bacteria (Fig. 7.02) and the actinomycetes and the eukaryotes include all other cells. Between the two

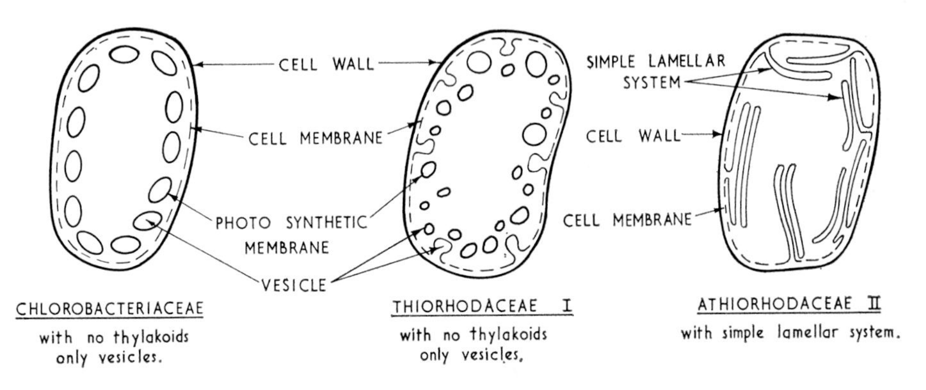

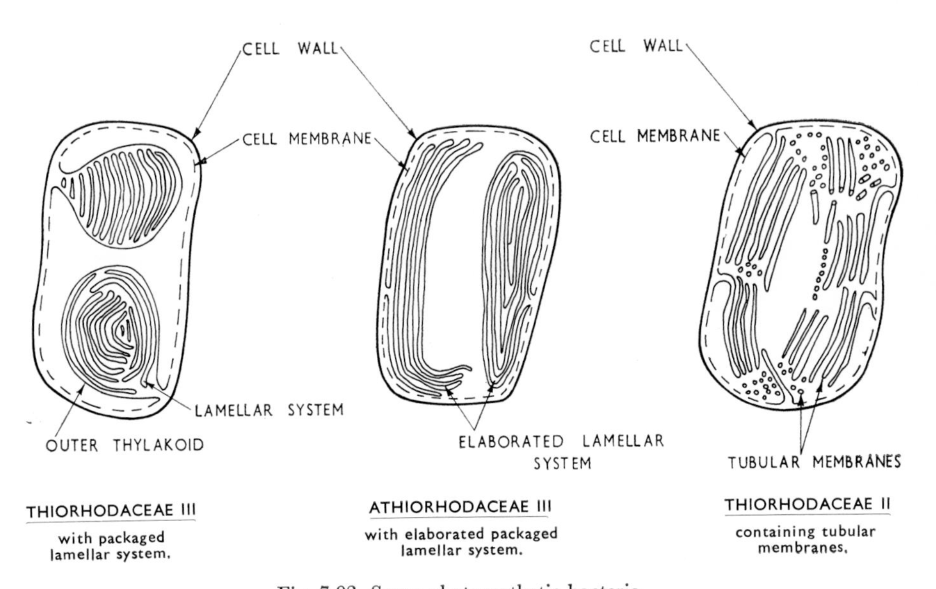

Fig. 7.02. Some photosynthetic bacteria

there is accepted to exist a major morphological discontinuity—perhaps the deepest division in the whole of biological science. There are several clearly defined differences between the two cell types.

The eukaryotic cell (Figs 7.03, 7.04 and 7.05) is distinguished by the way its contents are neatly packaged in discrete functioning units,

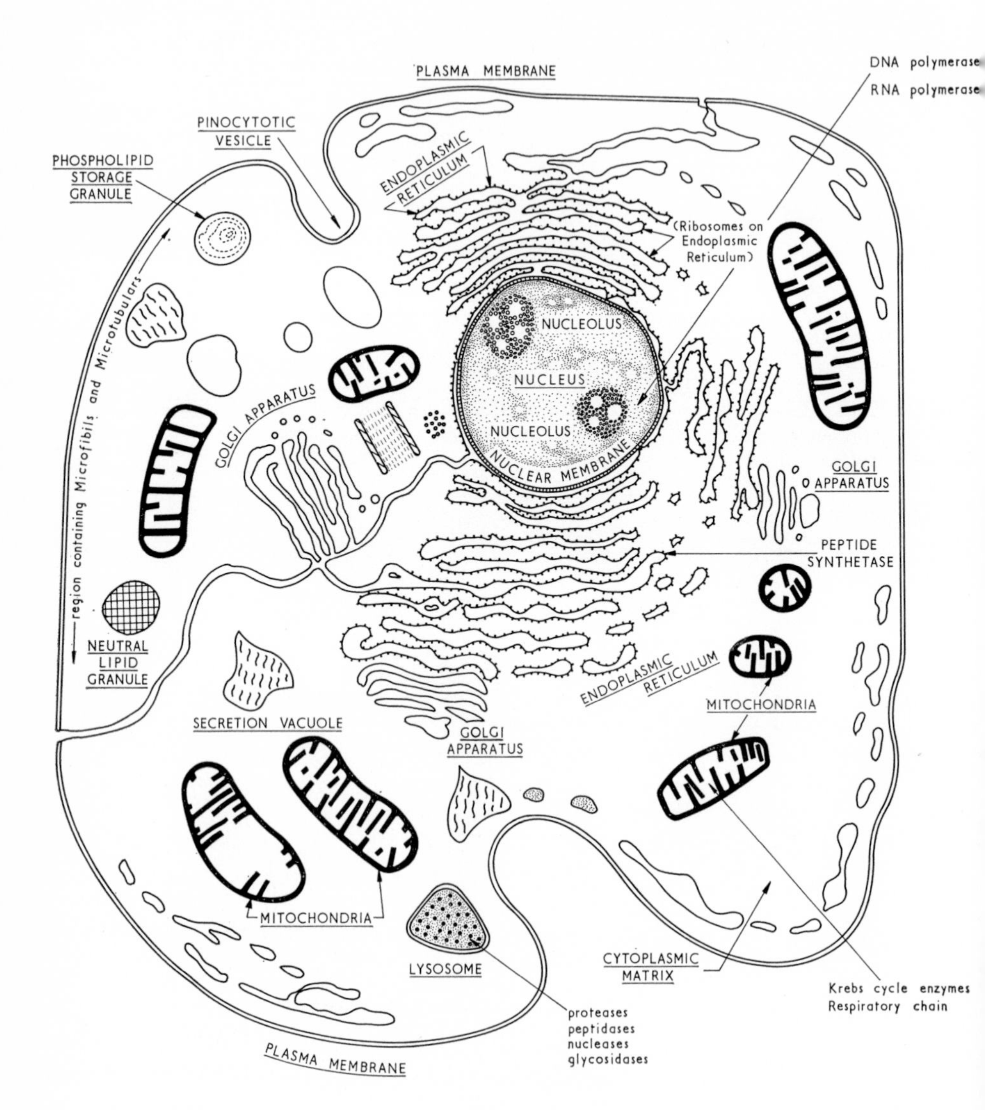

Fig. 7.03. Schematic diagram of an animal cell showing various cell components and functions. Modified from various authors including J. Brachet (ed.). "The Living Cell". *Scientific American*, 1961.

which appear to float in a sea of streaming (at times) cytoplasm. The cell contains a complex nucleus structure (4–6 μm diameter) which is bounded by a membrane, which in turn possesses large porous openings. DNA in the nucleus is associated with basic proteins (histones) and other materials, and forms the multi complex of chromosomes. In

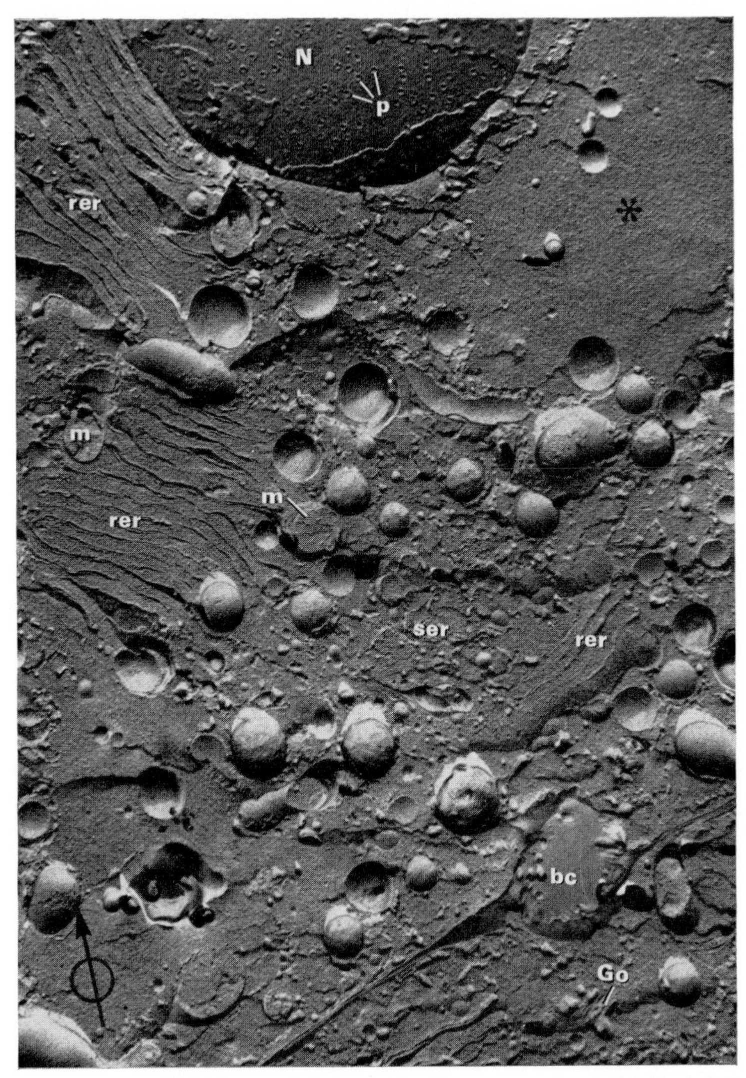

Fig. 7.04. Eukaryotic cell (rat liver hepatocyte) (a) Freeze etched material. Encircled arrow indicates direction of shadow cast (m = mitochondrion; ser = smooth endoplasmic reticulum; rer = rough endoplasmic reticulum; Go = Golgi complex; N = nucleus; p = nuclear pores; bc = bile canaliculus).

addition, however, the nucleus also contains RNA, especially in the nucleolus region. There are present several hundred (up to 1000 in animal cells) organelles called mitochondria, which occupy perhaps one-fifth of the cell volume. They are roughly spherical bodies about 1 μm in diameter and contain an extended membrane, a matrix and various granules. They are concerned with oxidation of carbohydrates

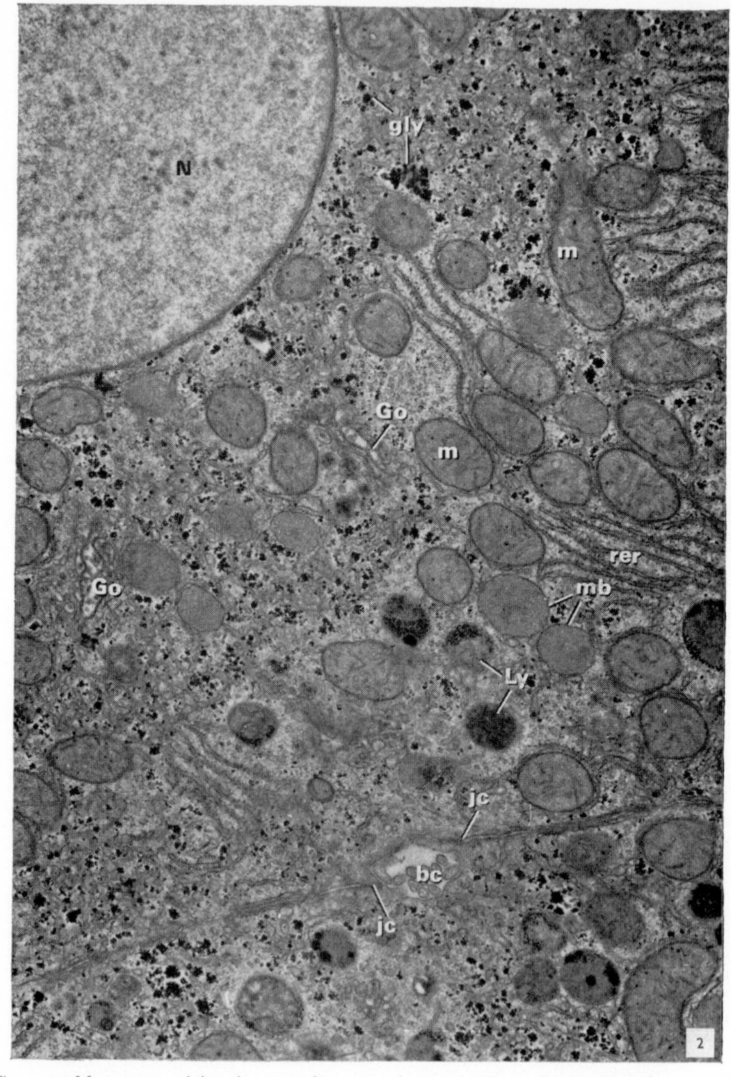

(b) Comparable area to (a) using an electron microscope (ly = lysosome; gly = glycogen; jc = junctional complex; mb = microbody).
Photographs by kind permission of L. Orci (1971). *J. Ultrastruct. Res.* **35,** 1.

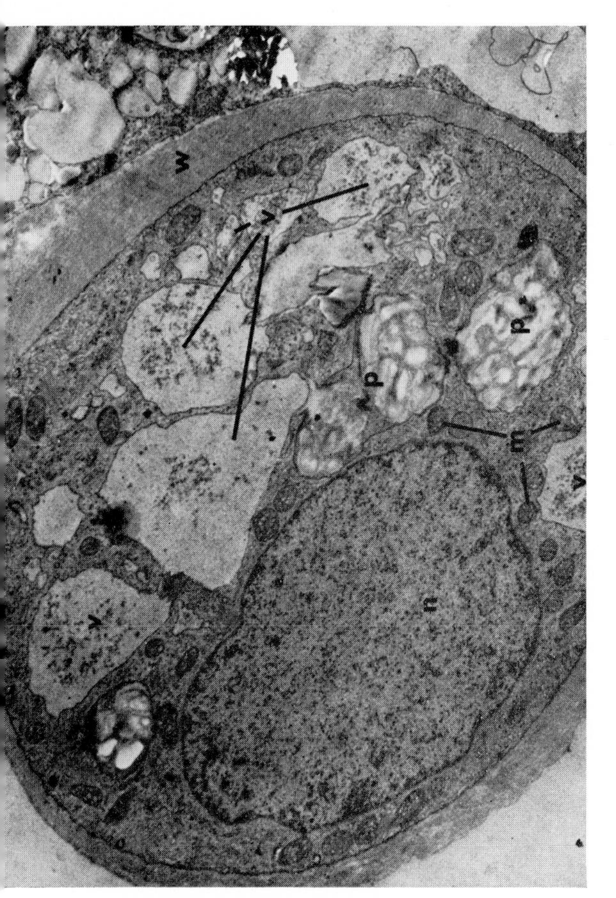

Fig. 7.05 (right). Eukaryotic cell. Guard cell of *Opuntia ficus-indica* var. *elongata*. (n = nucleus; m = mitochondria; v = vacuoles; p = plastids; w = wall). Photograph by kind permission of W. W. Thomson (1970). *Am. J. Bot.* **57,** 309.

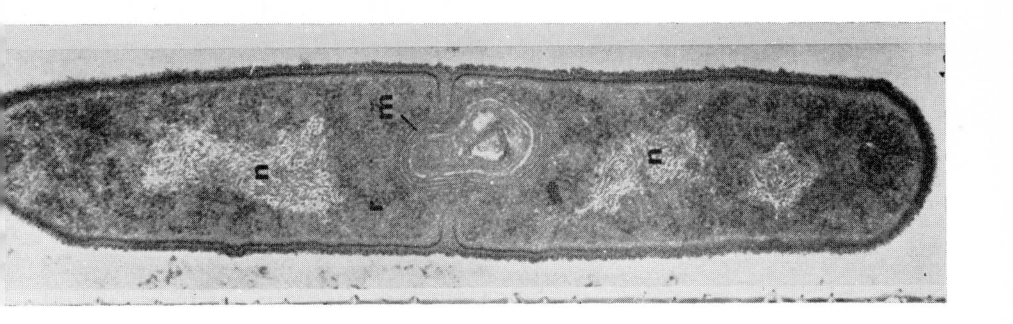

Fig. 7.06 (left). Prokaryotic cell (*Bacillus licheniformis*), (n = the DNA region; r = ribosomes; m = mesosomal membranes). Photograph by kind permission of P. J. Highton (1970). *J. Ultrastruct. Res.* **31,** 247.

and other materials and generally may be regarded as the power house of the cell. They also contain DNA and are capable of replication. Higher plant cells also contain chloroplasts, which are somewhat analogous to mitochondria, and indeed when they are present the number of mitochondria is reduced in bulk correspondingly. The chloroplasts are also packaged in a membrane, contain enzymes for various (especially oxidative, including photosynthetic) functions, contain their own distinctive DNA and are also capable of replication. Ribosomes occur as discrete bodies along a rough endoplasmic reticulum that is an extension of the cell membrane into the cytoplasm. It is here that major protein synthesis occurs. The Golgi apparatus are vesicles which are surrounded by membranes and generally appear in a flattened form. They are concerned with cell membrane formation and function. The lysosomes are small (0·2–5 μm diameter) bodies neatly packaged by membranes and contain hydrolases which are concerned with digestion of invasive matter. Other primitive organelles also occur, including peroxisomes (0·5 μm diameter) concerned with oxidative reactions and again surrounded by a membrane. The whole eukaryotic cell is bounded by a membrane the nature of which is reasonably well understood in many fungi and in higher plants, but little is known about its chemistry in animal cells although it apparently contains protein and lipids.

The prokaryotic cell (Fig. 7.06), however, contains no nucleus, no mitochondria, no chloroplasts or any such organelle. The genophore consists of a single (usually circular) strand of DNA and the ribosomes are suspended in the rigid gel-like (non-streaming) cytoplasm associated with the other membrane which may form invaginations into the central cell body.

Mitochondria

Occurrence and Structure

These are discrete membrane-bound yet flexible bodies which occur in eukaryotic cells but not in prokaryotic cells. They were first discovered by Kolliker in 1857 as the so-called "sarcosomes" of muscle, but it is only over the past few decades that they have been recognized as the power house of the cell where simple molecules, especially carbohydrates and fats, are combusted with production of energy. They consist of small, generally globular bodies (Fig. 7.07) about the size (1 μm diameter) of an average bacterium, distributed evenly throughout the cytoplasm. In animal cells (e.g. rat hepatocytes) there are about 800 mitochondria occupying about one-fifth of the cytoplasmic volume, and the numbers appear constant and characteristic for a specific cell but may vary at different stages of development. By contrast in plant cells there are usually smaller numbers (100–200), occupying a corresponding smaller part of the cytoplasm and their

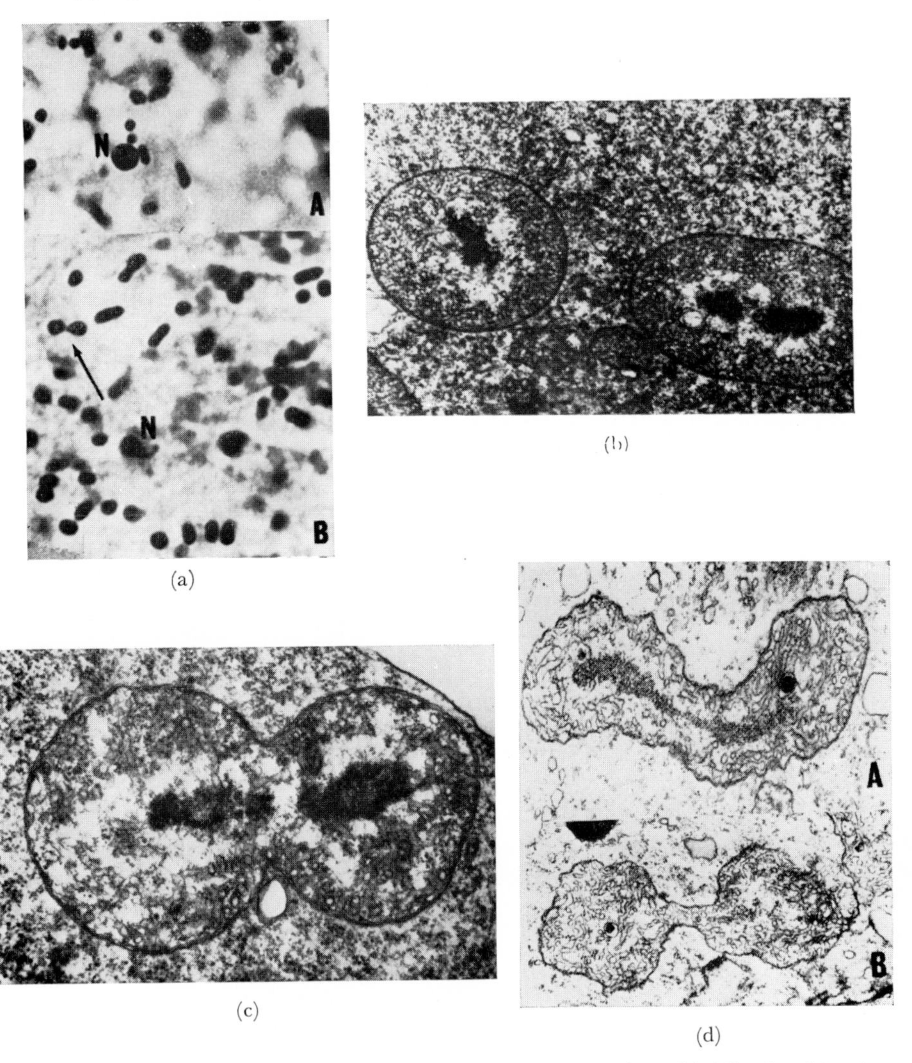

Fig. 7.07. Mitochondria from plasmodia of *Physarum polycephalum*. (a) Mitochondria, *A* untreated, *B* after actinomycin treatment. (b) Two mitochondria from untreated plasmodium. (c) Dividing mitochondria from untreated plasmodium. (d) Mitochondria from plasmodium treated with actinomycin: *A* Elongated mitochondria showing delayed division of nucleoid. *B* Mitochondria with narrow constriction. Photographs by kind permission of E. Guttes (1969). *Experientia* **25,** 66.

place is invariably taken in large measure by the somewhat analogous chloroplasts; in the light chloroplasts offer an alternative energy-producing catalytic system via photosynthesis, the mitochondria taking over during the dark. Nevertheless the general structure and function of mitochondria in all eukaryotic cells whether of plant or animal source, in spite of some variation in shape, are remarkably similar.

The mitochondrion has two main membranes. The outer one is a smooth bounding membrane similar in all eukaryotic cells, whereas the inner membrane is featured by the complex invaginations (cristae) which function to increase the surface area substantially (Fig. 7.08).

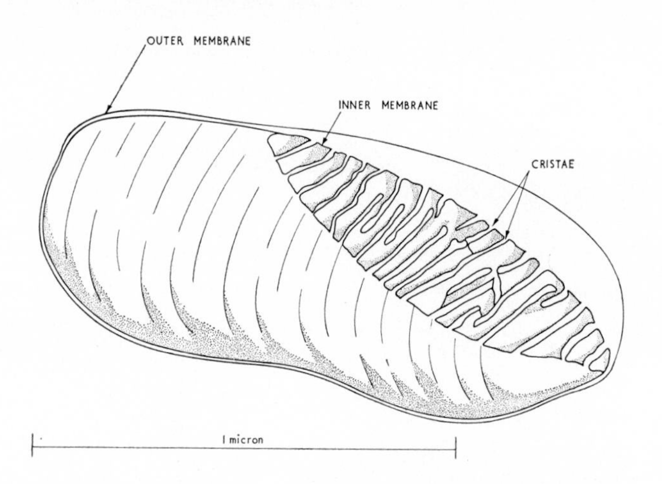

Fig. 7.08. A mitochondrion showing arrangement of membranes

and the detailed structure of these vary greatly in different cells although the net effect is similar. This inner membrane may be compared with the bacterial cytoplasmic membrane and there are many similarities, although there are differences also. In general, cells which are associated with high work load and hence require large amounts of energy (e.g. heart cells, 200–250 m^2/g, muscle, 400 m^2/g) generally possess many tightly packed cristae (200–250 m^2/g), whereas in other cells (liver cells) they may be more sparse and irregular (40 m^2/g). Indeed the cristae may additionally undergo conformational changes rapidly according to the amount of work they are required to perform.

The outer membrane is about 70 Å thick and may be separated from the inner membrane by treatment with detergent. It is then obtained as a transparent envelope, which contains largely protein but is especially rich in lipids, including cholesterol and phosphatidyl inositol. The inner membrane when spread out is about 50–55 Å thick and its surface is covered with small knoblike particles attached by tiny stalks to the membrane, although the protrusions only appear after phosphotungsate fixing for electron micrograph preparations and this may partly exaggerate the structure.

Occupying the interstices between the internal cristae is the matrix, a structurally rigid gel whose amount increases inversely with the amount of inner membrane material. The matrix contains a satellite DNA which differs in structure from that of the nuclear DNA, and amounts to 0·1–0·2% of the total cellular DNA, mitochondrial ribosomes on which protein synthesis can occur, and other inclusions of little known function.

The outer membrane readily allows passage of most small molecules, but the inner membrane is highly selective and only permits passage of water and some small neutral molecules and fatty acids. However, it is possible for certain substances to cross the membrane by an exchange diffusion process which involves some carrier material (or permease). An important example involves the passage of ADP, which can only enter the inner compartments of the mitochondrion if a molecule of ATP comes out. The carrier involved appears to be an enzyme-like protein. Other carriers are also involved in the transport of substances such as glutamate, phosphate and some dicarboxylic acids. It is interesting to note that most of the co-enzymes required by the mitochondrion, including NAD, NADP and NADH, CoA and the nucleoside mono-, di- and triphosphates, are not transportable across the membranes (thus NADH is not oxidized in the presence of mitochondria although NAD^+-requiring malate is, by intromitochondrial oxidation), hence it is clear that the mitochondrion has a distinct and separate source of these types of compounds from the general cytoplasmic pool, and the comparison with the whole prokaryotic cell is again obvious.

Function of Mitochondria

Much of the protein of the mitochondrion (25–30%) is concerned with enzymic reactions, whereas the remainder in association with lipids and no doubt other chemical materials apparently plays a largely structural role. The ability to separate the mitochondrial constituents by detergent or salt treatment and differential density gradient centrifugation has enabled assessments of the nature and site of enzymic activities to be carried out.

Outer membrane

Although this membrane with its high lipid content plays a largely structural role, nevertheless several enzyme systems are associated with it. These include NADH–cytochrome *c* oxidoreductase (rotenone insensitive), cytochrome b_5, acyl-CoA synthase and monoamine oxidase.

The Inner Membrane

The inner parts of the mitochondrion are undoubtedly the major source of the numerous biochemical reactions. There is present in the matrix both DNA (generally cyclic) and ribosomes and this enables mitochondria to undergo cell division and protein synthesis. This in fact occurs and mitochondria generally undergo division after cell division has ceased. In addition they occur packaged in sperm cells, spores, seeds, etc., since the cells do not have the genetic material to reproduce them. They can also produce their own proteins. The mechanisms of both types of reactions appear to be virtually identical with concepts of replication and protein synthesis in prokaryotic cells and will be dealt with later.

The most important functions of the mitochondria other than production of the materials required for self assembly (nucleotide and DNA biosynthesis) and enzyme synthesis are perhaps those concerned with energy production and control. The more important reactions involved here include the conversion of ADP to ATP and the secretion of the latter into the cytoplasm, the combustion of carbohydrates by way of the Krebs tricarboxylic acid cycle, and fatty acid oxidation. The enzymes (flavoproteins, cytochromes) concerned with oxidative reactions are located in the plane of the inner membranes, whereas the knob-like sub-units contain the so-called F_1 coupling factor for oxidative phosphorylation which is concerned with the conversion of ADP to ATP. When this protein (mol wt 280,000) is removed from mitochondria, (a) they can no longer carry out oxidative phosphorylation and (b) the spheres have disappeared. Addition of purified F_1 factor to the depleted mitochondria results in (a) return of ability to convert ADP to ATP and (b) reappearance of the globular sub-units attached to the membranes.

The rate at which oxidative reactions proceed is governed essentially by the ratios of ADP to ATP and P_i in the cytoplasm.

$$P_i + ADP \rightarrow ATP + H_2O$$

Thus the rate of oxygen consumption is at a maximum when the ADP and P_i concentrations are high in the medium and that of ATP is low. When the concentration of ATP is high by comparison, the respiratory rate may be little more than 5% of the maximum. Maximum rates may occur with ADP concentrations as low as 0·02 mM. This results in a respiratory control mechanism. Thus when a muscle works hard its cytoplasmic ATP breaks down to ADP. This is rapidly adsorbed via the permease carrier into the mitochondria with a subsequent large increase in oxygen consumption. So long as the work continues so the cycle repeats. At the cessation of work, ATP again accumulates in the extramitochondrial environment and activity diminishes.

The mitochondrial matrix also contains several enzymes, including those required for enzyme and nucleic acid synthesis, tricarboxylic acid cycle enzymes and appropriate pools of the various co-enzymes and fellow travellers including NAD and derivatives, mono- and other nucleotides, etc., which cannot gain access from the extra-mitochondrial environment.

Although there is a barrier between cytoplasmic NADH and intramitochondrial NADH, nevertheless it is possible by a so-called shuttle reaction for the mitochondria to effect indirectly oxidation of cytoplasmic NADH. This is achieved by the intermediacy of glycerol phosphate as follows:

Extramitochondrial

Dihydroxyacetone phosphate + NADH→H + L-glycerol-3-phosphate + NAD

The glycerol phosphate enters the mitochondrion:

Intramitochondrial

Glycerol-3-phosphate + Flavoprotein→dihydroxyacetone phosphate + Flavoprotein H_2

The dihydroxyacetone phosphate leaves the mitochondrion and the reduced flavoprotein passes its reducing properties ultimately to oxygen via a sequence of transfers involving various coenzymes.

No mitochondria are visible in yeast cells when grown on a glucose medium under aerobic conditions, although the oxidative enzymes are clearly operating. During aerobic growth, on the other hand, mitochondria soon appear, so apparently the units for their assembly are readily available. Detailed information about the morphogenesis of these mitochondria is not fully available, i.e. whether from a simpler uncoiled or extended complex or by *de novo* assembly from macromolecules.

The ability of yeast to dispense with mitochondria has made it a special object for study in this field. In addition treatment of yeast with acridine dyes results in inhibition of mitochondrial development and formation of a mutant yeast known as *petite* (*colonie*) mutants. These contain few, if any, mitochondria, and although respiratory deficient they are able to oxidize glucose especially by the Embden-Meyerhof pathway.

Summary

The mitochondria may in many ways be considered to be separate partly independent (to an extent not known) intracellular structures, each possessing its own machinery for replication, transcription and translation, some of which elements are specific to the genetic system of the eukaryotic cell. In this sense and in others they have properties very similar to those of the typical prokaryotic cell.

Chloroplasts

Photosynthesis may be regarded as involving two reaction steps. Firstly the use of light energy both to reduce $NADP^+$ and to phosphorylate ADP to ATP. Secondly a dark reaction in which the derived NADPH and ATP are used to reduce carbon dioxide to various carbohydrates and related compounds.

In eukaryotic cells the photosynthetic apparatus is sited in discrete membrane-bound flexible bodies termed the chloroplasts, whereas in prokaryotic cells the apparatus is sited in the cell membrane which may invaginate to form specialized sub-units. In this sense therefore there is a clear and distinct parallel between the chloroplasts and the mitochondria.

Chloroplasts are generally larger than mitochondria with diameters about 8 μm and consequently there are smaller numbers (*ca* 40) per cell. The chloroplasts are one of a group of membrane-bound organelles called the plastids and are distinguished by containing chlorophyll. Other plastids may contain pigments (chromoplasts) or be unpigmented (leucoplasts). Some of the chromoplasts store carotenoid pigments and the leucoplasts include the amyloplasts which store starch grains. The plastids are similar and, like the mitochondria, contain DNA which is different from the nuclear DNA, and are generally capable of (nuclear partially or complete independent) replication.

The chloroplasts are usually ellipsoids or flattened discs (Figs 7.09 and 7.10), but other more involved ribbon-like (as in *Spirogyra*), stellate (in *Zygnema*) or irregular forms (*Oedogonium*) also occur. The numbers present vary from one per cell in eukaryotic algae such as *Chlorella* to more than 100 in *Euglena* and about 40 in higher plant cells.

They generally occur in those parts of the plant body which are exposed to light. Their structure is similar to that of the mitochondria. They possess an outer somewhat fragile membrane and an inner folded membrane which encloses a rigid gel matrix called the stroma traversed by a series of membranous sub-units. These take the form of flattened vesicles called thylakoid discs which are stacked in groups called grana (Fig. 7.11).

The number of thylakoid discs and derived grana varies substantially according to the plant series. In general the more evolved forms have increasingly complex grana conformations. The red algae characteristically contain single thylakoid discs which run the entire length of the chloroplast. In the Cryptophyta the thylakoids occur in pairs closely compressed together and also run the whole length of the plastid. The most common arrangement in the eukaryotic algae, however, are the triple thylakoids. In other algae multiple thylakoids predominate (Chlorophyta) and there may be as many as 100 triple thylakoid bands in *Nitella*. The thylakoids may coalesce but generally stack like piles of coins (Fig. 7.12). In the higher plants the internal membranes of the chloroplasts are very complex but are characterized by the stacks of grana lamellae.

A major function of chloroplasts involves the light-induced phosphorylation of ADP to ATP in complete analogy with the oxidative phosphorylation of ADP in the mitochondrion. In plant photosynthesis

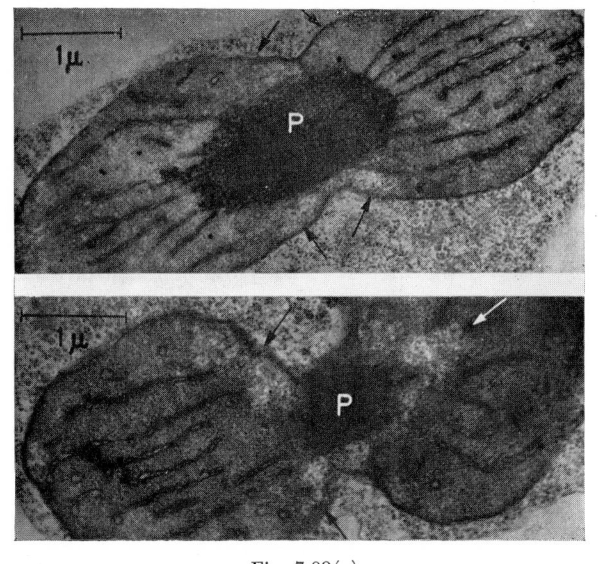

Fig. 7.09(a)

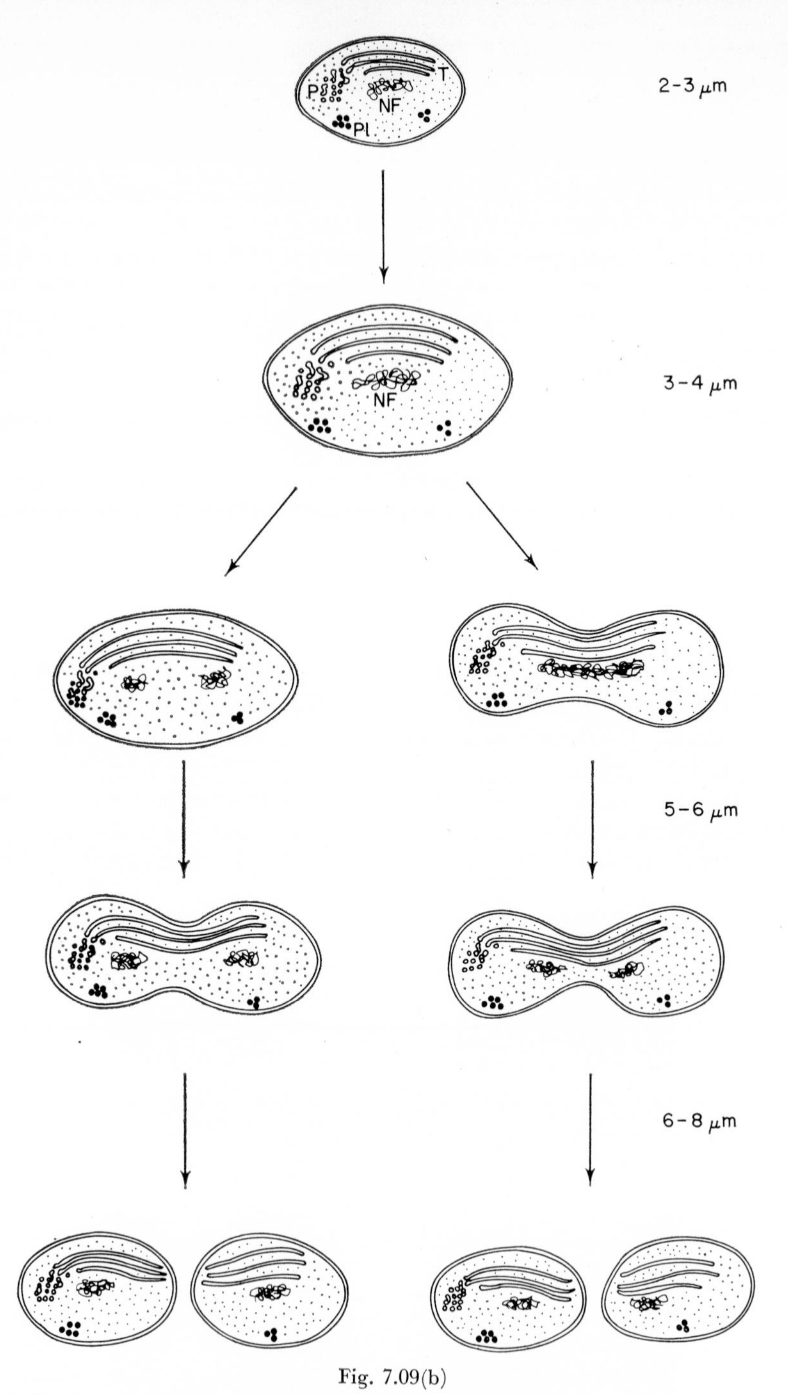

Fig. 7.09(b)

Fig. 7.09. The etioplast (a) A constriction zone of the etioplast, showing two poorly contrasted DNA-bearing zones (arrowed). (b) Diagrammatic representation of the division of the DNA-bearing plastid zone (NF) in the etioplast of *Hordeum vulgare*, before and during the division of the plastid. (P = prolamellar body; T = thylakoid; Pl = plastoglobule.) Photograph and diagram by kind permission of B. Sprey (1968). *Planta* **78,** 115.

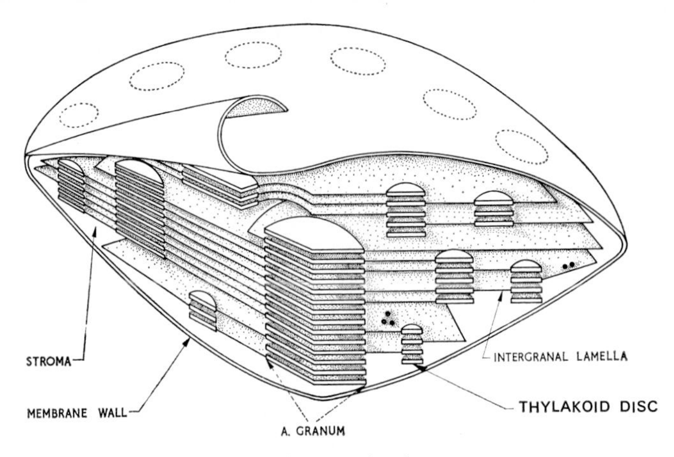

Fig. 7.10. Chloroplast showing inner structure

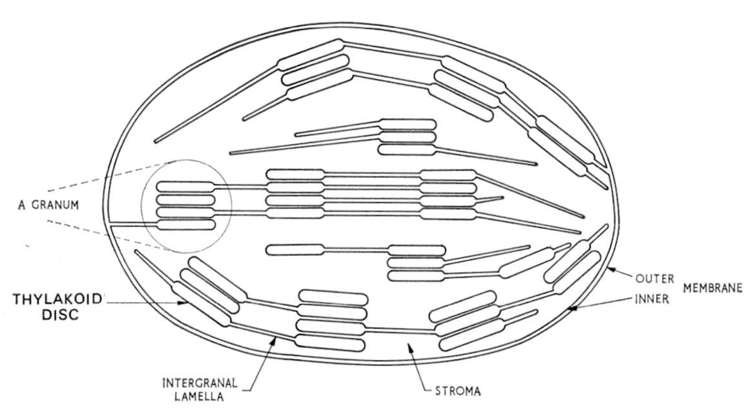

Fig. 7.11. Diagrammatic representation of developing chloroplasts showing arrangement of thylakoid discs and membranes

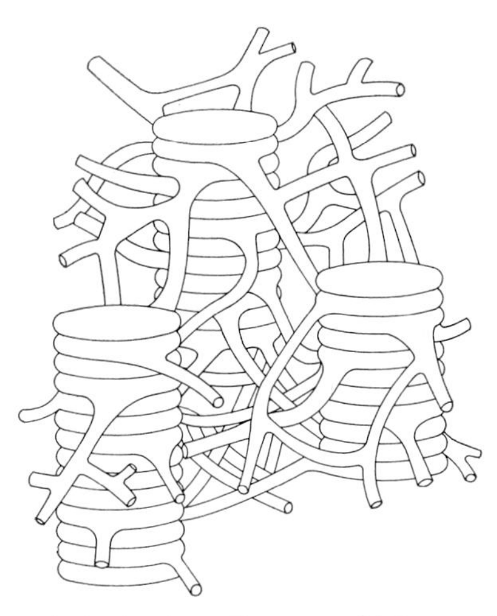

Fig. 7.12. Chloroplast: structure of grana and intergranal lamella tubes

there appear to be two types of light-stimulated reactions. One of these is activated by light of long wavelength (680 nm) and one by light of shorter wavelength (<600 nm). The first of these has been termed photosystem I. It is associated with chlorophyll *a* and does not result in oxygen evolution. This system is present in all photosynthesizing cells including the photosynthetic prokaryotic bacteria. The second photosystem II requires a second type of chlorophyll (*b*, *c* or *d*) and additional phycobilin pigments and is accompanied by oxygen evolution. This system only occurs in the oxygen-evolving eukaryotic cells and is not present in photosynthetic bacteria. Photosystem I contains about 200 chlorophyll molecules and 50 carotenoid molecules; pigment of photosystem II contains 200 chlorophyll molecules, 200 chlorophyll *b*, *c* or *d* and there may be other pigments including phycobilins and xanthophylls (oxidized carotenoids) present. The two systems may be separated from one another by differential density gradient centrifugation of detergent lysed chloroplasts. It is believed that the complex of the two systems occurs in the thylakoid discs.

The photo-reduction of $NADP^+$ to NADH is not fully understood but involves proteins including ferredoxin, an iron and sulphur-containing electron acceptor protein (molecular weight 11,600—spinach ferredoxin) which like bacterial ferredoxins does not contain methionine, histidine or tryptophane. Ferredoxin alone in the reduced state cannot reduce $NADP^+$, but an enzyme–ferredoxin–NADP–oxido-reductase (a flavo-protein) has been isolated in crystalline form from spinach chloroplasts and will catalyse the reduction in the dark. Other electron carriers which occur in chloroplasts include various new cytochromes, a blue copper-containing protein (plastocyanin), various quinones including vitamin K (Fig. 7.13) and plastoquinone (Fig. 7.14), related to CoQ of mitochondria (Fig. 7.15).

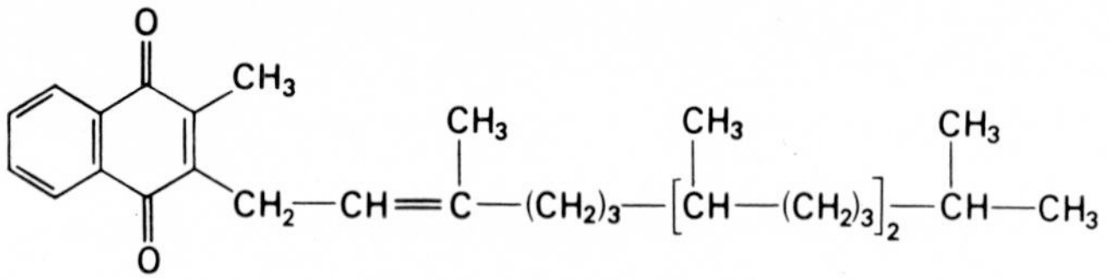

Fig. 7.13. Vitamin K_1

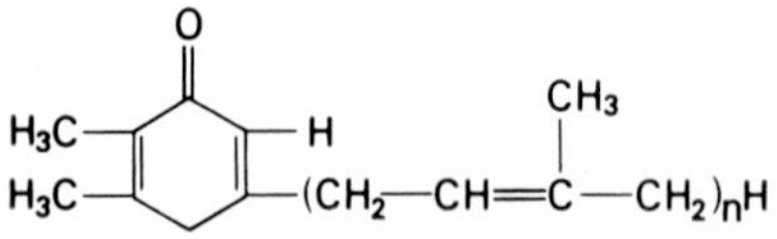

Fig. 7.14. Plastoquinone

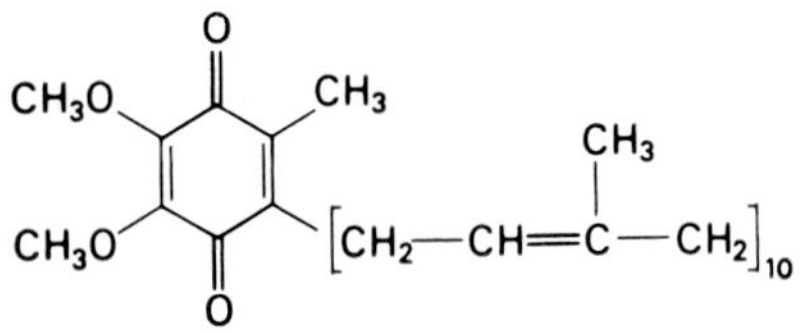

$$\mathrm{CH_3O},\ \mathrm{CH_3O},\ \mathrm{CH_3}\text{-benzoquinone}-[\mathrm{CH_2}-\mathrm{CH}=\overset{\mathrm{CH_3}}{\overset{|}{\mathrm{C}}}-\mathrm{CH_2}]_{10}$$

Fig. 7.15. Coenzyme Q_{10} (Co Q_{10})

In photosynthetic prokaryotic cells there are no chloroplasts and the enzyme and coenzyme apparatus for the reactions are bound to the infolded outer membrane as chromatophores. No oxygen is evolved from these systems and only photosystem I is therefore present.

Ribosomes—and the Endoplasmic Reticulum (Figs 7.03 and 7.04)

Ribosomes are macromolecular ribonucleoprotein complexes which occur as free floating small particles in prokaryotic cells, but in eukaryotic cells they are larger and are generally bound to the endoplasmic reticulum. They also occur in the mitochondria and chloroplasts of eukaryotic cells in a manner reminiscent of that in prokaryotic cells and in the cell nucleus. During the early 1950s they were shown to be the site of protein biosynthesis, by Zamecnik and co-workers.

The cytoplasm of eukaryotic cells in the intra-organelle spaces is traversed by a complex series of membranous structures known as the endoplasmic reticulum. Much of this forms flattened vesicles with an inner grana or cisternae; these interconnect to form channels which traverse the cytoplasm and through which molecules may pass to the cytoplasmic membrane. In the rough-surfaced endoplasmic reticulum the outer surface is lined with adherent ribosomes. The ribosomes are frequently connected by an RNA (messenger RNA) molecule into organized arrays of up to 100 or more called polyribosomes which in chick embryos have been shown to form ordered crystal-like shapes. In cells such as reticulocytes where the protein synthesized is of a limited nature (globins) and largely for internal use, the ribosomes occur free as polysomes in the cytoplasm, but when, as is more usual, protein is required to be transported to the cytoplasmic membrane and hence excreted, attachment to the endoplasmic reticulum is usual. In other parts of the eukaryotic cell similar smooth surfaced vesicles occur which do not possess adherent ribosomes. Both the smooth and rough endoplasmic reticulum contain glycoproteins and appear to have several as yet not fully understood functions, some of which include protein, lipid and cell membrane synthesis.

The ribosomes consist of RNA (prokaryotic cells 60–65%, eukaryotic cells 50%) and protein (prokaryotic cells 35–40%, eukaryotic cells 50%). In all organisms studied they fall into two distinct groups. Ribosomes derived from either prokaryotic cells or from mitochondria chloroplasts and other chromo- and leucoplasts of eukaryotic cells have sedimentation values of about 70s corresponding to a molecular weight of about $2{\cdot}8 \times 10^6$. The ribosomes of eukaryotic cells other than those present in the specific organelles have a sedimentation value of 80s (mol. wt $4{\cdot}0 \times 10^6$). In addition the ribosomes are all found to consist of two unequal size sub-units. The 70s ribosomes have sub-units with sedimentation values 50s (mol. wt $1{\cdot}8 \times 10^6$) and 30s (mol. wt $1{\cdot}0 \times 10^6$). In *E. coli* the 50s unit contains RNA molecules (23s and 5s in size) and about 30 polypeptide chains. The 30s unit contains an RNA molecule (1×10^6) and about 20 polypeptide chains.

The 80s ribosomes have sub-units of about 60s and 40s, but some variation occurs among animals, plants and fungi. Animal ribosomes contain sub-units 28s (variable mol. wt from about $1{\cdot}4$–$1{\cdot}75 \times 10^6$ sea urchin to man according to evolution of species) and 18s RNA, whereas plants contain 25s and 16s or 18s RNA. It is of special interest that apparently both eukaryotic plants and prokaryotic bacteria have conserved their rRNA size, whereas in animals the larger of the RNA units has increased with increasing evolution. The reason for this is not known but may be connected with increased demands for integration and differentiation of tissues.

The prokaryotic cell contains about 10^4 ribosomes and in *E. coli* this accounts for about 25% of the cell. In eukaryotic organisms there are about 10^6–10^7 ribosomes per cell. The prokaryotic cell may synthesize about 5 ribosomes per sec and the rapidly growing eukaryotic cell dividing once in 24 h requires 10–100 ribosomes per sec. The ribosomal RNA (rRNA) which tends to be substantially *o*-methylated is produced by the nucleolus in eukaryotic cells and there is a well-known relationship between the size and density of the nucleolus and the number of ribosomes per cell. Anucleolate embryos of *Xenopus laevis* were unable to produce ribosomes and the anucleolate mutant was found to lack DNA complementary to the rRNA. In addition, further evidence for the implication of the nucleolus in rRNA production comes from the identification of precursors of rRNA in isolated nucleoli and from electron micrographs of nucleoli from *Triturus viridescens* (newt) in which growing chains of the ribosomal precursor nucleoprotein are seen on a deoxynucleoprotein chain. In both prokaryotic and eukaryotic cells the rRNA is synthesized as molecules which are larger than the final product (about 40% larger in mammals), the excess material being discarded.

Ribosomes and Protein Synthesis

In the early 1950s Zamecnik and co-workers using ^{14}C-labelled amino acids showed that cell-free homogenates of rat liver, when supplemented with ATP and other chemicals, were able to synthesize proteins. The radioactive material was found after differential centrifugation to reside in the ribosomes attached to the endoplasmic reticulum from which they could be freed by treatment with substances such as deoxycholate. The isolated ribosomes, when incubated with

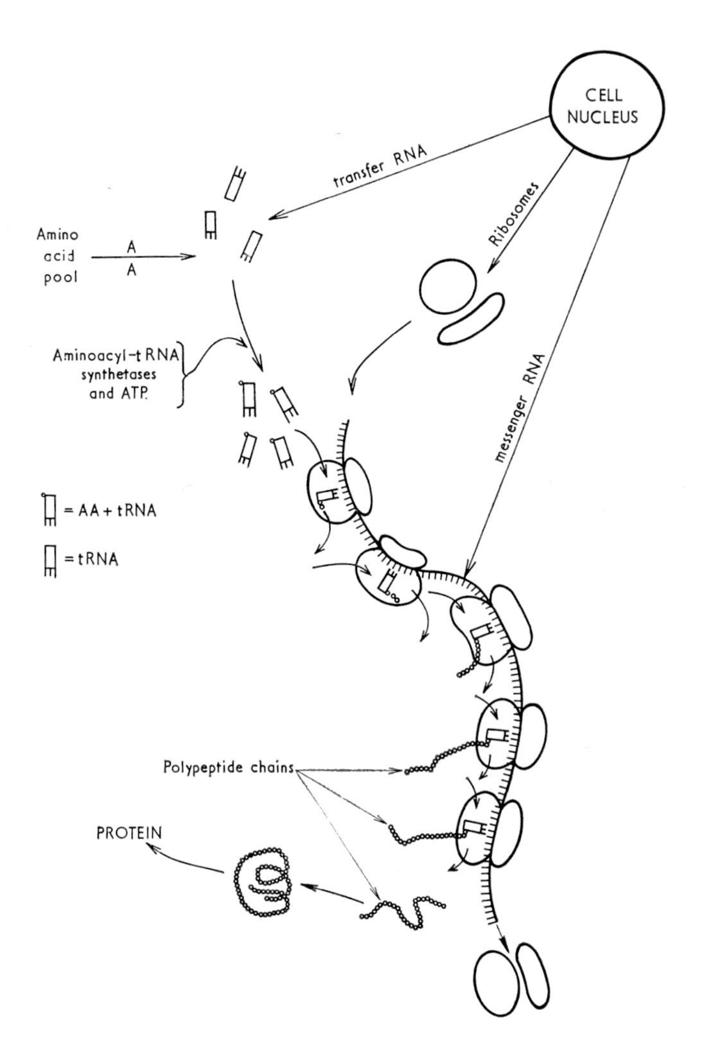

Fig. 7.16. Protein synthesis

Mg^{2+}, ATP and a soluble fraction of the rat liver were able to incorporate amino acids into polypeptides. These general results were soon found to be generally applicable to a wide variety of other cells both prokaryotic and eukaryotic. The required soluble fraction was found to contain two main essential factors, one of which is the tRNA characteristic of the specific amino acid and the other an enzyme which will produce an ester of the amino acid with the terminal adenosine moiety (at position 3′) of its tRNA. The resultant aminoacyl tRNA is attached to the ribosome and may partake in peptide bond formation. The subsequent condensation polymerization which leads to the formation of a polypeptide may be divided into three phases: initiation of the chain, propagation and termination.

Chain Initiation (Fig. 7.16)

The ribosome first must dissociate into its two main sub-units (in the case of prokaryotic cells the 50s and 30s units) by an interaction with three protein initiation factors F_1, F_2 and F_3 which appear attached to the 30s unit. This complex then becomes attached to the mRNA and to an initiation tRNA which in *E. coli* is normally the protected N-formylmethionine derivative (fMet-tRNA). Finally this as yet inactive initiation complex associates with the larger (50s) sub-unit with liberation of the initiation factors (which may be used again) and formation of the fully functional ribosome. The specific order of attachment apparently ensures that the first amino acid (fMet) is attached at the correct (terminal) position.

Chain Propagation

Each ribosome can make a whole polypeptide chain. It does so effectively by moving along the mRNA molecule and reading the message which has of course been encoded from the DNA of the genophore. Each mRNA molecule may contain at any one time many ribosomes forming a polysomal unit in prokaryotic cells, especially in which synchronous protein synthesis can occur at several centres. In addition several different proteins may be coded by a single mRNA chain. The steps involved in the propagation are shown in Fig. 7.17.

The second amino acid (of the final polypeptide) binds to the appropriate codon (base triplet in the mRNA) as its tRNA derivative. This step requires both GTP and a protein T factor and the stereo-chemistry must be such that the aminoacyl groups are close in space. The next phase involves interaction of the primary amino group of the unprotected amino ester with the first protected formyl derivative, with the aid of an enzyme (peptidyl transferase) present in the 50s unit. The

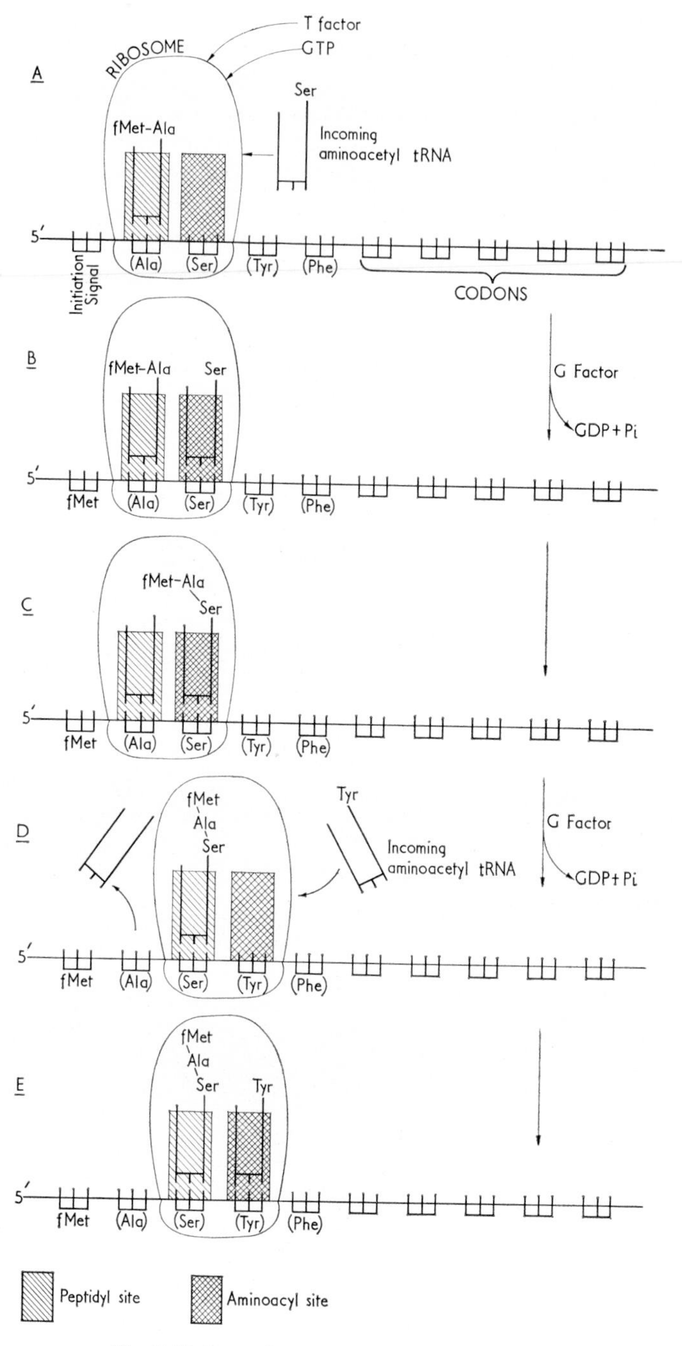

Fig. 7.17. Step-wise elongation of peptide chain

M

first (now free tRNA, termed $tRNA_p$) remains bound to the mRNA. Next follows a reaction in which the peptidyl-tRNA produced above displaces the tRNAp from its site accompanied therefore by a simultaneous movement of mRNA which moves along by one codon. This movement apparently takes place through a sort of enclosed groove in the ribosomal unit since a large part of the mRNA is not attacked by ribonuclease under these conditions; it is presumably screened from such attack. The process described is then repeated many times.

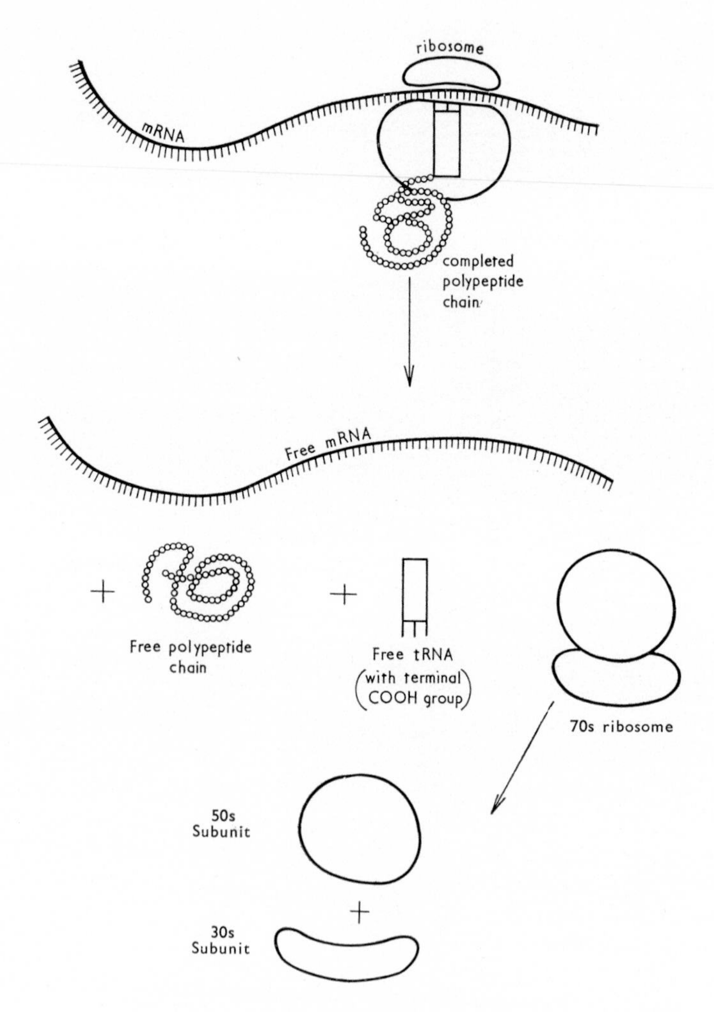

Fig. 7.18. Termination step of peptide synthesis

Chain Termination (Fig. 7.18)

There are three codons in the mRNA which do not code for any amino acids and apparently function as chain-terminating units by preventing attachment of any tRNA at that point. At this stage the polypeptide formed is attached both via the carboxy terminal amino-acid to its tRNA and by hydrogen bonds (codon-anticodon) to the mRNA. A special protein release factor serves at this stage to catalyse hydrolysis of the terminal ester link and the polypeptide separates as does the ribosome from the mRNA.

Nucleus

The nucleus is characteristic of eukaryotic cells. It consists of a flexible material bounded by a double membrane. The nuclear diameter is 4–6 μm and the membranes are each about 40–60 Å thick. In addition there is a clear space between the membranes of varying size but roughly 200 Å across. The whole surface of the membrane is covered with large holes or pores (diameter 500 Å) formed by fusion of the inner and outer membranes and occupying about 10% of the nuclear surface area, in mammalian cells at least. These may at times be closed by a plug of some unknown material. In addition there appears to be a direct, albeit transitory, connection of the nucleus via the internal cannular spaces of the endoplasmic reticulum with the external cell membrane, especially with rapidly dividing cells. Thus certain acridine dyes are capable of staining (by intercalation) the nuclear material without affecting other intracellular or intraorganellar materials. During cell division the nuclear membranes disintegrate into small pieces which may form vesicles in the cytoplasm. The nuclear body as described above is hence usually termed the "interphase nucleus". These may subsequently be used again after cell division in the formation of new nuclear membrane. Within the nucleus there also resides one or more nucleoli, irregular shaped bodies which contain RNA and protein, as much as 80% of the dry weight consisting of protein. There appears to be no membrane around the nucleolus, but it has a fine structure consisting of fibres and particles against a background protein network. The main constituents of the nucleus are the chromosomes which consist of DNA associated with RNA and with various proteins, especially the basic histones and acidic proteins. There is no nucleus in prokaryotic cells and the genophore consists of a single molecule of (usually circular) DNA which is suspended in the gel-cytoplasm.

Function of the Nucleus and Nucleolus

The nucleus and its contained DNA is very much concerned with cell replication. The chromosomes have not been too well resolved in electron micrographs, but it would appear that they may be best regarded as linear structures composed of a single enormously long molecule of DNA (albeit folded and coiled) associated along its whole length with basic and acidic proteins. It is generally thought that the histones act as derepressors of replication, and related acidic proteins containing large amounts of aspartic and glutamic acid operate to remove the basic protein coat and free the DNA. They may thus be considered to be derepressors. The chromosomes also contain RNA which appears to be associated with the DNA at certain specific points only and may also have repressor activity.

The evidence for the linear structure of the genophore in eukaryotic cells is not complete but comes from, on the one hand, genetic analysis of *Drosophila* (fruit fly), mouse, maize and *Neurospora* and other organisms, indicating a linear arrangement of genes along each chromosome. Other evidence comes from electron micrographs of chromosomes which have been spread out by treatment with detergents and by removal of proteins with proteinases. They may also be extended manually up to 10 times their length by micromanipulation. When extended in such manners the chromosomes are revealed as extended linear structures which contain more dense regions at regular intervals. These dense regions are known as the *chromomeres*. Experiments with *Chironomus* larvae suggest that pieces of DNA isolated therefrom must have extended across several chromomere regions, indicating that the chromomeres represent a more dense folding or coiling of the DNA at particular regions, and not joints or new materials. There is certainly no evidence at present for any joining material throughout the chromosome strand.

The nucleolus is associated with a specific site on the chromosome strand known as the *nucleolar organizer*.

The chromomeres have long been associated with the concept of a unit of genetic activity (gene) and this concept seems to be receiving increasing support. The chromomeres, however, appear to be too large to operate as a single gene. This anomaly has been overcome by a suggestion by Callan and Lloyd from studies of the lampbrush chromosome derived from developing amphibian eggs (especially the newt *Triturus*). These form long strands (200 μm long) with symmetrical lateral loops (Fig. 7.19) which arise from the dense regions on the main strand. Staining indicates presence of RNA and in addition the fluffy appearance of the loops disappears after treatment with RNAase, although the loop remains intact. The chromosomes are apparently continuous and linear and readily extensible. It is suggested that each loop consists of a number of linearly arranged copies of a single gene. At each chromomere there exists a "master" copy and information from this is transferred to each of the "slave" copies.

It is also suggested that the slave copies alone (not the master copy)

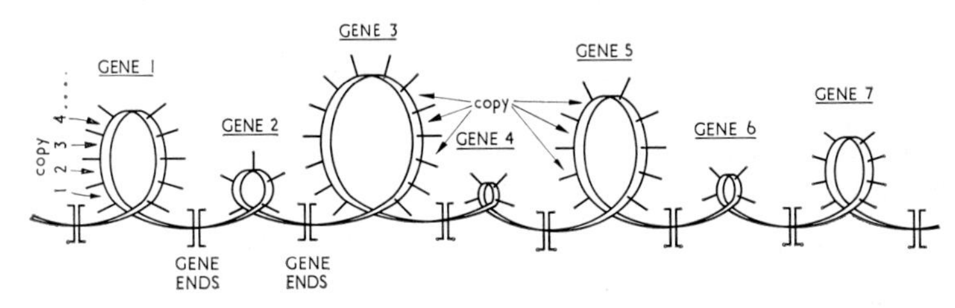

Fig. 7.19. Chromomere C_1 C_2 C_3C_4 C_5 C_6 C_7

take part in RNA production and indeed the RNA from different "puffs" (enlarged sections) of polytene chromosomes (multistranded) have each a characteristic base composition. The duplicate slave copies presumably would help increase the net amount of RNA synthesis. Addition of histone causes inhibition of RNA synthesis and retraction of the loops into the main body of the chromosome. Folding, coiling and other similar architectural and configurational changes in chromosomes are (not unexpectedly) produced by pH changes and by exposure

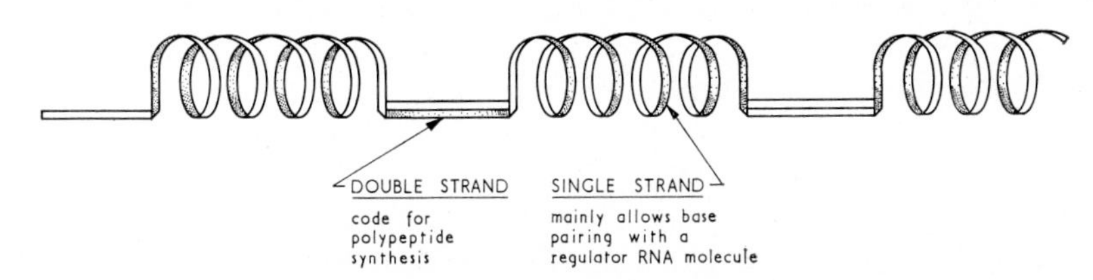

Fig. 7.20. Globular coiled chromomere sites in eukaryotic DNA

to specific metal ions, especially Ca^{++}. The nucleolar organizer appears to be similar to the other chromomeres and to be specifically concerned with rRNA synthesis—perhaps of complete ribosomes, small numbers of which have occasionally been noted in the nucleus.

It has been suggested (Crick, 1971) that the globular coiled chromomere sites in eukaryotic DNA which are recognized by regulator molecules are mainly single stranded DNA (Fig. 7.20). This would allow regulators (RNA?) to bind in a complementary fashion. The interbands of fibrous DNA code for polypeptide synthesis and will be double stranded.

DNA Content of Eukaryotic Cell Nuclei

One might suppose that the more highly evolved the species the more DNA would be required to accomplish the enhanced work load required. Boivin and Vendrelys first showed that mammalian cells from different tissues contain the same amount of DNA and that sperm cells contain half this amount. This type of observation strongly suggested the genetic function of DNA. However, when the DNA contents of nuclei of cells of different species are compared it becomes clear that the DNA content cannot be related completely to the position of the organism on the evolutionary scale (Table 7.01). In the case of prokaryotic cells the simple circular DNA genophore is required to produce perhaps 1000–2000 enzymes, and various other repressors, and structural proteins required for membranes, etc., mounting in all to, say, 6000 genes. The total DNA requirement allowing 600 base pairs per gene would be (600×6000) $3{\cdot}6 \times 10^6$ base pairs (cf. Table 7.02), corresponding to a length of about 1·2 mm. The length of the *E. coli* genophore is in fact about 1·4 mm and the DNA content is consequently approximately as expected. The DNA content of the more complex fungi *Neurospora* and *Aspergillus* and the yeasts is about 10 times greater, which roughly fits the sort of enhanced requirement in such organisms, produced by sexual reproduction. Similarly the fruit fly, with twice the amount of DNA as a fungus, again emphasizes the greater complexity. However, whereas mammals have about 40 times the DNA of the fruit fly, certain amphibia, including the primitive lung fish *Protopterus*, have 30 times more DNA than mammals! Similarly in higher plants there are even greater amounts of DNA per cell and these vary enormously even within the same species of plant and even the same genus show major variations. Thus *Vicia faba* has 7 times the DNA of *V. sativa* in spite of having the same chromosome number. The ratio of DNA in 12 measured *Vicia* showed that they fell into three groups in a rough ratio of 1 : 2 : 4. Similar ratios in several Ranunculaceae examined were 1 : 8 : 12 : 16 : 20 : 24 : 40. These variations have been extended to include *Mesostomum* (flatworm) species with the same chromosome number but an 11-fold difference in DNA content per nucleus.

If one were to accept DNA content as a measure of evolutionary progress man would fall somewhere between a shark and a mouse, or alternatively about the same as a common lupin. It would at first sight appear that many living systems have far more DNA (especially when

Table 7.01. Estimated DNA content of some eukaryotic cells

Species	Amount of DNA per cell ($\times 10^{-10}$ g)
HIGHER PLANTS	
Lilium longiflorum	22·0
Tradescantia ohioensis	12·0
Vicia faba	0·6
Vicia pannonica	0·21
Vicia grandiflora	0·08
Anemone tetrasepala	0·55
Anemone blanda	0·21
Anemone virginiana	0·12
Anemone parviflora	0·07
Zea mays	0·08
Lupinus albus	0·005
Aquilegia (hybrid)	0·001
VERTEBRATES	
"Congo eel" (*Amphiuma*)	9·0
Lungfish (*Protopterus*)	1·0
Newt (*Triturus*)	0·6
"Mud puppy" (*Necturus*)	0·49
Frog (*Rana*)	0·16
Toad (*Bufo*)	0·075
Shark (*Carcharias*)	0·065
Man (*Homo*)	0·06
Dog (*Canis*)	0·06
Horse (*Equus*)	0·06
Mouse (*Mus*)	0·055
Alligator	0·05
Carp (*Cyprinus*)	0·035
Fowl (*Gallus*)	0·025
Shad (*Eucinosomus*)	0·009
INVERTEBRATES	
Crab (*Plagusia*)	0·015
Sea Urchin (*Echinometra*)	0·009
Snail (*Tectarius*)	0·007
Jelly fish (*Cassiopeia*)	0·0035
Fruit fly (*Drosophila*)	0·0018
FUNGI	
Ustilago maydis	0·0006
Neurospora crassa	0·0005
Aspergillus nidulans	0·0005
Saccharomyces cerevisiae	0·00025
BACTERIA	
Escherichia coli	0·00006

Table 7.02. Estimated DNA content of some cells and viruses

Species	Number of nucleotide base pairs (× 1 million)	DNA per cell (× 10^{-12} g)
Bacteriophage 2	0·07	0·00008 per virion[1]
Bacteriophage T4	0·22	0·00024 per virion[1]
Bacteria	2	0·002–0·06
Fungi	20	0·02–0·17
Sponges	100	0·1
Molluscs	1100	1·2
Birds	2000	2·0
Fishes	2000	2·0
Higher plants	2300	2·5
Reptiles	4500	5·0
Mammals	5500	6·0
Amphibia	6500	7·0

[1] Virus.

one accepts the linear chromosome) than they can use for genetic purposes. These apparent anomalies may be explained in several ways. Thus the gene duplication (master slave concept) may be greater in some organisms than in others, leading to large amounts of DNA without any increase in complexity. This could occur in organisms requiring a more rapid synthesis of RNA. Also it seems certain that only a proportion (perhaps a quite small amount—10%) of DNA has a genetic function and the rest plays other roles many of which are as yet unknown but will be concerned with control.

Replication of DNA

The double helical structure of DNA (Chapter 5) provides a physicochemical basis for a template function of the genophore which includes replication of DNA and transcription into complementary RNA. At the same time reversal of transcription from RNA to DNA is possible in tumour viruses. The so-called central dogma of molecular biology would now be simply stated that no information transfer occurs from protein to nucleic acid but only the reverse, i.e.

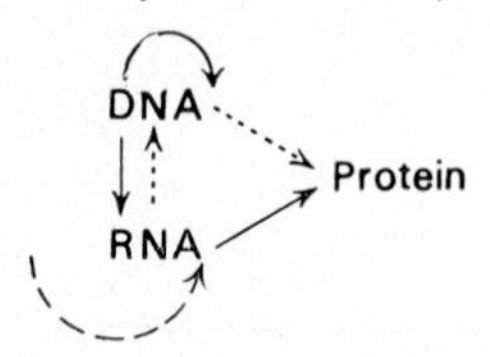

In their original note on the double helical structure for DNA Watson and Crick noted: "It has not escaped our notice that the double helical structure suggests the manner in which DNA is replicated."

There are in fact different possible ways in which the replication may proceed:

(1) Conservative replication—without strand separation.
(2) Semi-conservative replication—with strand separation.
(3) Dispersive replication, in which parent chains break at intervals.

Experiments using ^{15}N labelling of DNA (via NH_4Cl), originally by Meselson and Stahl in 1957–1958, allowed separation of "heavy" and "light" DNA and showed that in *E. coli* cells, DNA replicated in the

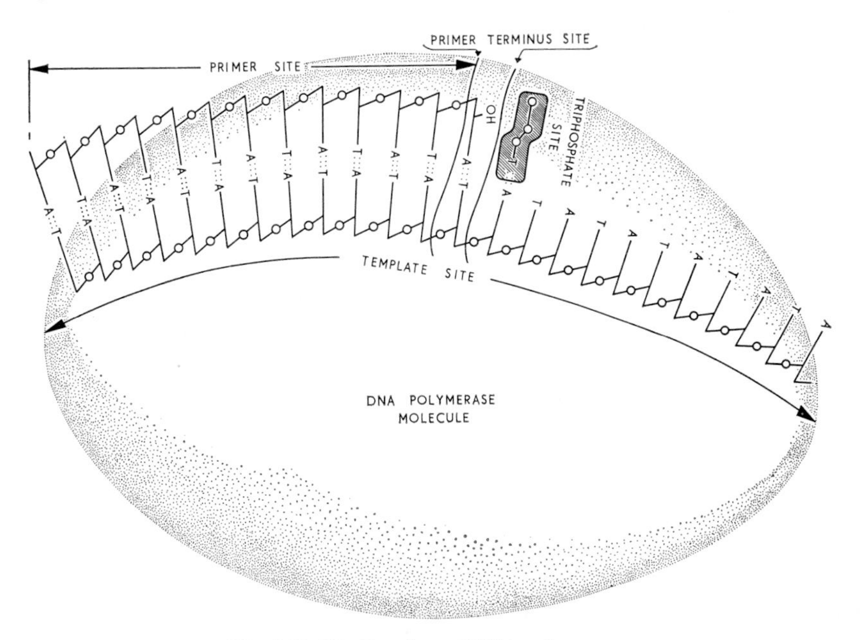

Fig. 7.21. Binding sites of DNA polymerase

semi-conservative manner. Similar replication also occurs in eukaryotic cells.

The actual replication (Fig. 7.21) requiring a DNA polymerase, a small amount of pre-existing DNA and an appropriate pool of mono-nucleotide precursors, may be envisaged to occur in the manner illustrated in Fig. 7.22. The pre-formed DNA is probably required as a template on which the synthesis of a complementary chain can begin. Also it has been shown that the newly synthesized DNA chain runs in the opposite direction to the template DNA chain. Also the polymerase makes internucleotide linkages in the 5′–3′ direction.

The replication of a simple circular viral (ΦX174-bacteriophage) DNA was finally achieved in the test tube by Goulian, Kornberg and Sinsheimer in 1968 by the action of a DNA polymerase and DNA ligase. The circular duplex DNA is typical of the prokaryotic cell and

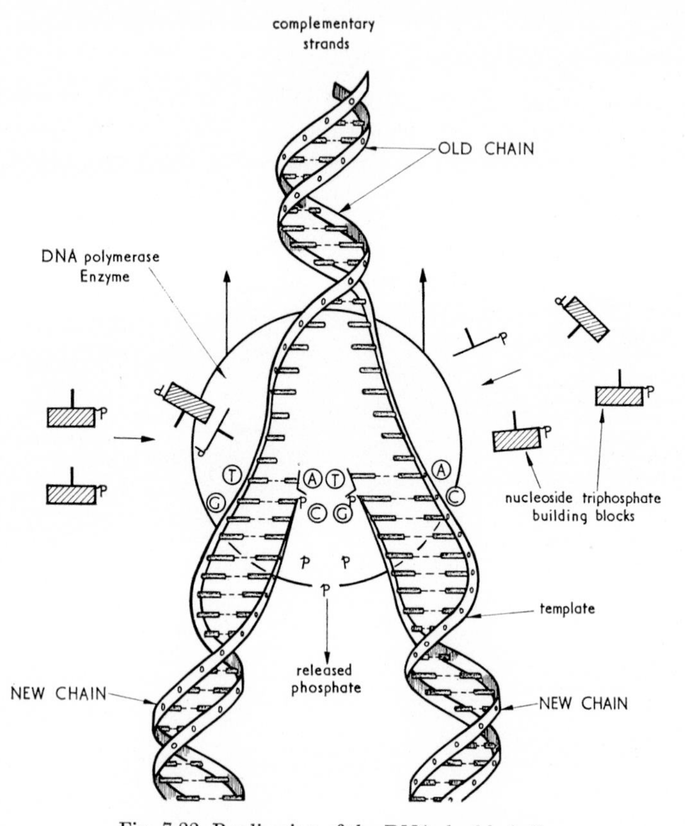

Fig. 7.22. Replication of the DNA double helix

its replication begins normally at one point to produce two circular duplex chromosomes or 4 strands in total. The test tube experiments began with a circular single-stranded duplex ^{3}H labelled ΦX174 DNA from which was produced (by the host *E. coli*) a complementary and attached "heavy" duplex strand made "heavy" by using 5-bromodeoxyuridine, which readily takes the place of thymidine in the polymerization process. The double duplex was partially hydrolysed with DNAase to give mixtures of various closed and open chain duplex forms of the light (^3H) original material and "heavy" (Br-dU) prepared material. The purified (by density centrifugation) "heavy" material was used as a template with pure DNA polymerase and normal mononucleotide triphosphates to give an acyclic "normal" complementary duplex strand which was finally cyclized with the

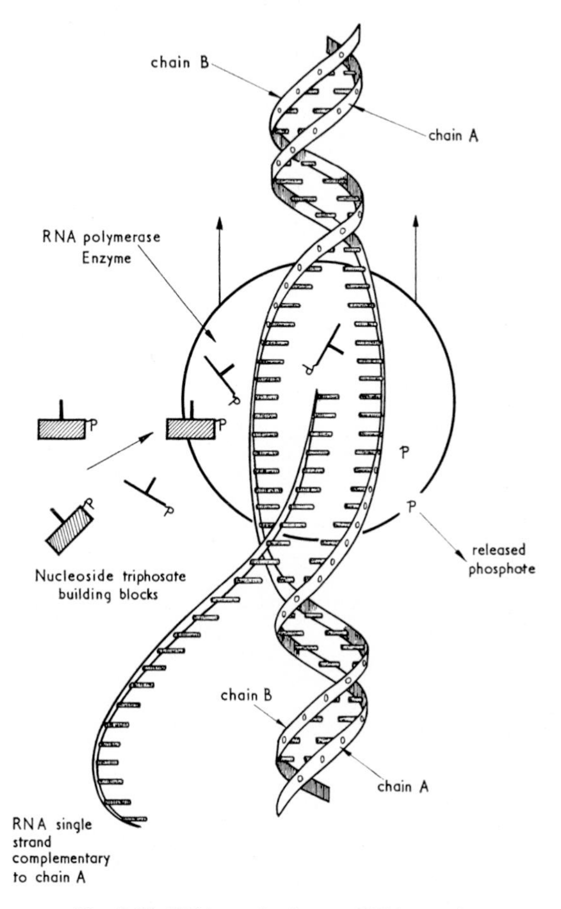

Fig. 7.23. RNA synthesis on a DNA template

DNA ligase to produce material identical to the natural DNA. The virus incidentally in some unknown manner only "selects" one of the duplex strands for its final assembly.

Replication is not confined to DNA. Thus mRNA is transcribed in the nucleoli from DNA, but in addition RNA may replicate RNA in certain cells infected by RNA-containing viruses. In these systems apparently DNA is not involved and only intact viral RNA (and not host cell RNA) is acceptable as primer. It seems likely that more examples of RNA replication will arise in due course, perhaps in eukaryotic cells. The general replication mechanism via RNA is apparently similar to that involving DNA (Fig. 7.23).

Lysosomes

Lysosomes (Figs 7.03 and 7.24) are small single membrane-bound vesicles (0·25–0·5 μm diameter) which were first detected in animals

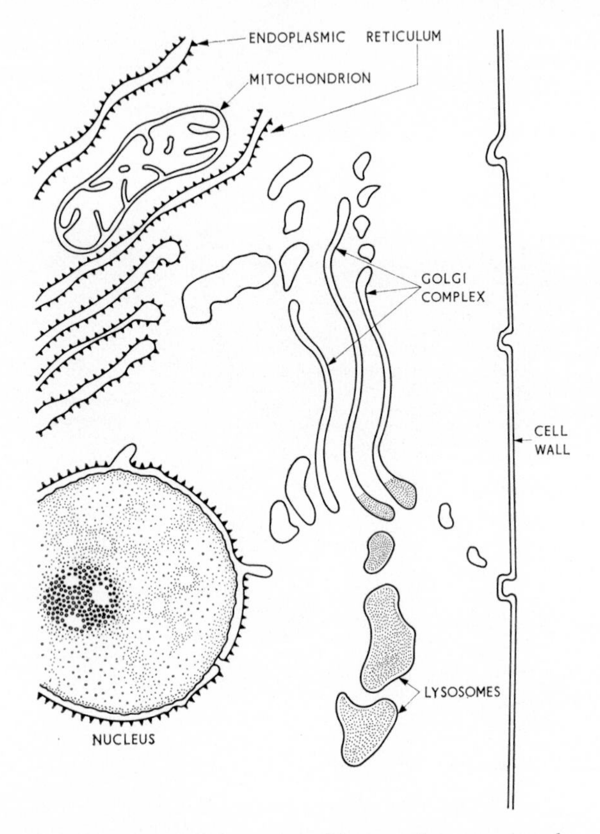

Fig. 7.24. Possible formation of lysosomal storage vacuole

by de Duve and co-workers in 1949 and later identified by electron microscopy in rat liver cells by Novikoff in 1955. They appear to exist in all animal cells and probably in plant cells also but do not occur in prokaryotic cells. They contain several (at least twelve) enzymes which are capable of controlling the hydrolysis of all the naturally occurring polymers such as proteins, polysaccharides and nucleic acids. The hydrolases must clearly be isolated from the general cytoplasm, otherwise they would presumably cause irreversible damage. In this sense they have been called "suicide bags". They occur in large quantity in those cells which have specialized digestive functions such as the protozoa which injest food material by a process known as pinocytosis to form an enclosed phagosome. This unit fuses with a lysosome particle to produce a digestive vacuole in which the large polymeric materials are broken down into their smaller monomer units. These in turn may be excreted, leaving behind a vacuole which may contain undigested material and presumably enzyme residues. The vacuole may be ejected from the cell in a manner opposite to that which occurs during pinocytosis and reminiscent of the way in which a virus leaves the cell: a sort of exocytosis.

In addition to their general scavenging functions lysosomes are also implicated in autodigestion of part of the cell contents during starvation and in addition have specialized functions including regression of the tail of the tadpole during the later stages of metamorphosis. At this time the tail cells contain very large quantities of lysosomes.

Golgi Complex

The Golgi apparatus (first described by Golgi in 1898) is present only in the eukaryotic cell. There may be 1–20 Golgi bodies in a cell (Fig. 7.03). In animal cells they usually consist of stacks of flattened vesicles of varying shapes (Fig. 7.25) and are generally considered to be

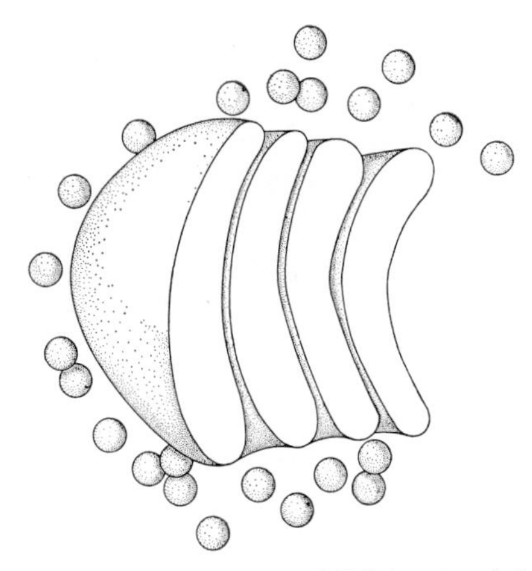

Fig. 7.25. Diagrammatic representation of Golgi region (after W. A. Jensen)

derived from the smooth endoplasmic reticulum. In plant cells the analogous structures are termed dictyosomes which, however, have a more random distribution than the Golgi bodies which normally clump together.

The functions of the Golgi apparatus, indeed of most of the endoplasmic reticulum, is only partly understood. The Golgi bodies are, however, concerned with the secretion, transport, and possibly, in some circumstances, the formation of cell wall materials and related compounds which in many cases may be transported to the extracellular position by an exocytosis (or reversed pinocytosis). Thus the polysaccharide scales of the alga *Chrysochromulina* (containing ribose and galactose) have been shown by Manton (1966) to be formed *de novo* in the Golgi apparatus and subsequently transported to the outer wall. Similar wall formation has been observed in the alga *Pleurochrysis* by Brown (1969). In animal cells, on the other hand, frequently the materials transported are apparently proteins which it is generally assumed are produced *de novo* on the more or less adjacent ribosome-covered rough endoplasmic reticulum. Such compounds include collagen secretion within Golgi vesicles and their subsequent discharge by exocytosis to the extracellular spaces. Zymogen granules and hormones such as norepinephrine and melanin type pigments are also secreted into the Golgi bodies and subsequently discharged. There is also evidence of a more general function in eukaryotic cells, namely the production of lysosomes.

Peroxisomes (Glyoxysomes)

Peroxisomes are small (0·5 μm diameter) bodies bound by a single membrane which occur in eukaryotic cells from a wide variety of sources. In rat liver they form about 2·5% of the total protein (6·5 mg/g wet liver) and in number they are about one-quarter the number of mitochondria. Their functions are not fully understood but they appear to be concerned with three major processes at least: (1) they contain a catalase which is of special importance for the reduction of hydrogen peroxide to water; (2) they are concerned with oxidative metabolism but apparently have no mechanism for ATP synthesis by oxidative phosphorylation. They may take part in the oxidation of NADH; (3) they are concerned with oxidation of α-amino acids to α-keto acids and in this sense are perhaps involved in glyconeogenesis. It has been shown that the entire glyoxylate cycle is specifically located in the peroxisomes of germinating seedlings where there is conversion of fat into carbohydrates. In general they give the appearance of having a primitive oxidative role in cell metabolism and it has been suggested that they might represent an early type of oxidative metabolism which has arisen from an ancestral anaerobic environment. In this sense they could be regarded as an evolutionary antecedent of the mitochondrial tricarboxylic acid cycle whose functions have become less important as the more efficient systems take over. Such concepts are, however, fraught with danger and tend to be derived from fixed concepts about the evolution of cells. Thus it could equally be argued that the peroxisomes have arisen as specialized (and hence more advanced evolutionary) structures to take care of certain specialized oxidative reactions in the eukaryotic cell, which for some reason are not readily handled by the alternative respiratory enzyme chains of, for example, the mitochondria.

Miscellaneous Inclusions

The term vacuole is a somewhat generalized description for a variety of ill-defined cytoplasmic inclusions which are commonly present in plant cells where they may be very large, especially in older cells, causing the remaining cell contents to spread out in a flattened array around the cell perimeter. The vacuoles contain aqueous solutions of various types of salts, organic acids, proteins and pigments, many of which may eventually crystallize out. They are presumably partly concerned with segregation of waste products, especially where extracellular secretion may be difficult.

In addition to the lysosomes and peroxisomes there are other so-called micro bodies which occur in eukaryotic cells. These include *lomasomes* which are analogous to prokaryotic mesosomes and have been detected in the region between the cell membrane and the hyphal wall. They may be involved in fungal wall synthesis. *Woronin bodies* have also only been seen in fungal hyphae and may be concerned with blocking certain pores. Some of these particular inclusions are not dissimilar to the orbicules present in the specialized tapetal cells of anthers of higher plants and concerned with sporopollenin (wall) formation.

The above-mentioned prokaryotic mesosomes are invaginations of the prokaryotic membrane which may form ovoid vesicle-containing convoluted membranes. On the other hand, the invaginated membrane may fill the whole cell in a series of stacked and coiled membranes as in *Azotobacter* (Gelman *et al.*, 1967). They appear to function as mitochondrial equivalents, in DNA replication, and in wall enzyme formation, although the difficulty of isolating structures such as these has hindered their study.

VIRUSES (Figs 7.26 and 7.27)

One of the more difficult aspects of the virus concept is how to define it. Fraenkel-Conrat states that viruses are defined as follows: "Particles

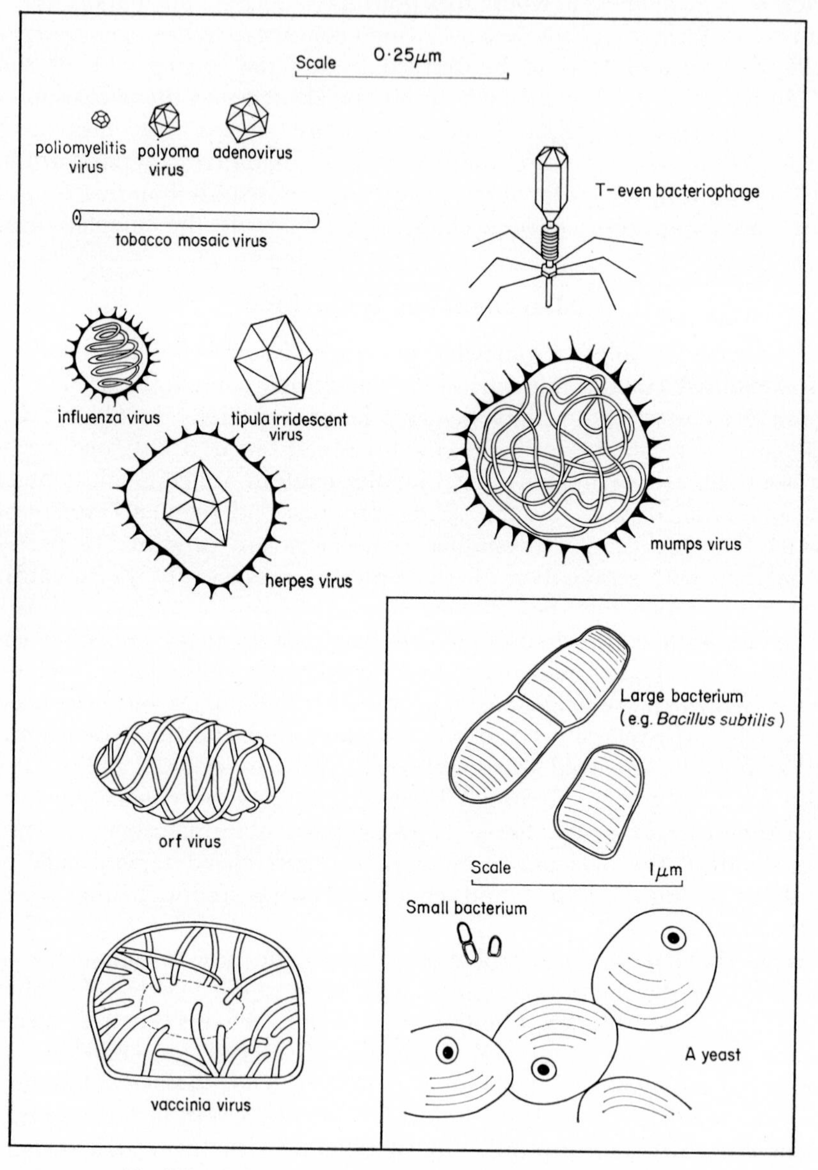

Fig. 7.26. Relative size and shapes of viruses and simple cells

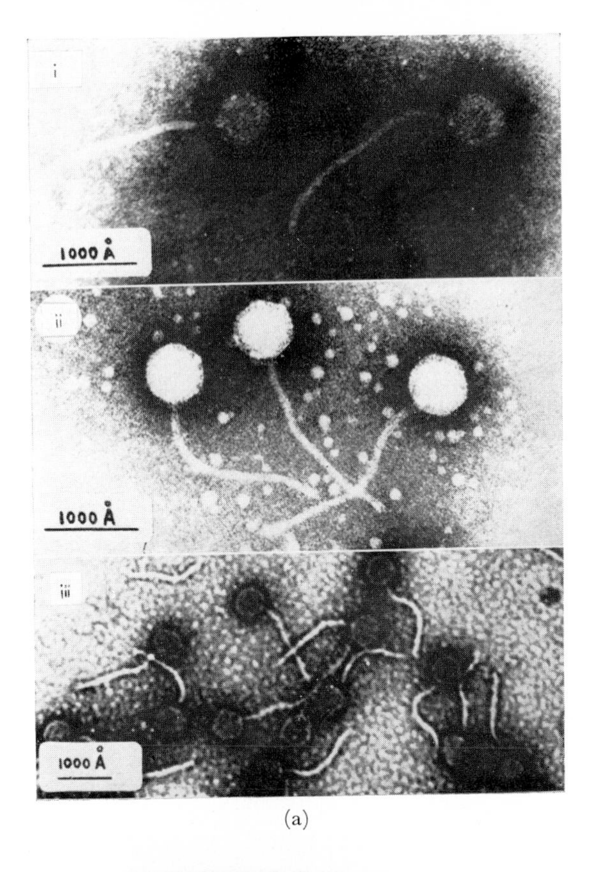

(a)

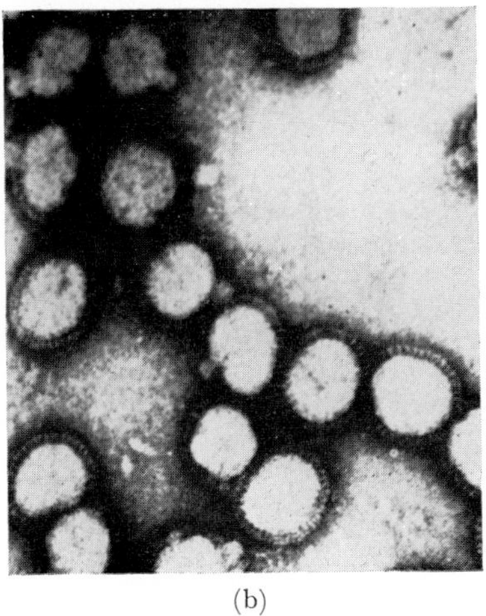

(b)

Fig. 7.27. Electron micrographs of some viruses (a) N-phages: (i) Phage Nl; (ii) Phage N6; (iii) Phage N5. Each phage was stained with 2% phosphotungstic acid. (b) Influenza virions. Stained with sodium phosphotungstate and showing the characteristic layer of surface spikes. Photographs (a) kindly supplied by N. Davidson 1970). *Virology* **40,** 102. Photograph (b) kindly supplied by R. W. Compans (1970). *Virology* **42,** 880.

made up of one or several molecules of DNA or RNA, are usually but not necessarily covered by protein, which are able to transmit their nucleic acid from one host cell to another and to use the host's enzyme apparatus to achieve their intracellular replication by superimposing their information on that of the host cell; or occasionally to integrate their genome (genophore) in reversible manner in that of the host and thereby to become cryptic or to transform the character of the host cell." The virus unit is generally called the virion and the total protein coat the capsid which may be composed of a number of sub-units known as capsomeres (Fig. 7.28).

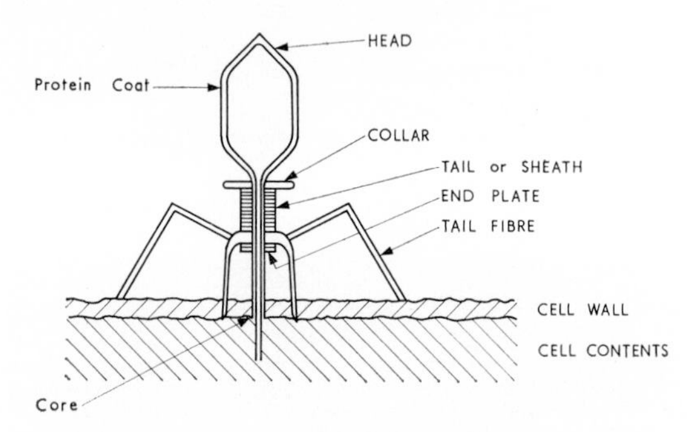

Fig. 7.28. Structure of a virus (bacteriophage T4)

In some ways this may not be an entirely or completely satisfactory definition since many viruses have a close association with chemical substances other than nucleic acids and proteins and this is excluded from the definition (however see later). Some of the problems associated with virus classification and definition appear to have arisen in direct relationship to their original discovery, that is as infective agents which passed through filters of mesh size sufficiently fine to retain known bacteria. The earliest observations in this area appear to be those of Iwanowsky (1892), who noted that the agent causing mosaic disease in tobacco plants passed the finest filters. The concept of a new infective agent was developed further by Beijerinck (1898), but it was not until Stanley (1935) showed that the agent in mosaic disease could be obtained in an infective para-crystalline form, that the identity of the virus (at least this one) as a chemical complex as distinct from a living system became fully established. Nevertheless in spite of these distinctions, in relatively recent times many organisms originally considered to be viruses have been re-classified as simple very small prokaryotic cells. In spite of the confusion surrounding definition and classification to which we are neither competent nor willing to add, there are various statements one can make about the virus concept which seem to be generally acceptable, although in the certain knowledge that only a minute fraction of the likely virus population of this planet has been observed, one hesitates to be in any way dogmatic. The virus has frequently been considered to be at the threshold of life, and although its macromolecular structure might add some credence to this concept the total knowledge of virus morphology and biochemistry has only served to emphasize the vast gulf that exists between the virus and the simplest of prokaryotic cells, a gulf far greater than that between the pro- and eukaryotic cells.

Statements about Viruses

(1) Viruses are obligate intracellular parasites which can infect both prokaryotic and eukaryotic cells but, surprisingly, perhaps the most morphologically complex are the bacteriophages which infect bacteria and similar viruses of blue-green algae.

(2) Viruses are generally, but not always, small (diameters 100–3000 Å), the largest being about the size of a small bacterium.

(3) The shapes of viruses vary substantially. They may be acyclic, tubular, helical structures as is common in many plant viruses such as tobacco mosaic virus; polyhedral as in the tipula irridescent virus, the icosahedron being perhaps the most common form; morphologically complex with head, body and tail through which nucleic acid is extruded as in some bacteriophages including T4 phage and algal viruses; and complex viruses including the myxoviruses and polyhedroses (insect viruses) which are enclosed in an envelope containing material other than nucleic acid and protein. These last are among the more interesting of known viruses since they seem to form a sort of natural link between the simple viruses and the prokaryotic cell. However, it is now reasonably well established that the vaguely described lipoprotein enveloping membrane in such particles is derived from the host cell during exocytosis of the virus. The envelope has the serological properties of the host and this undoubtedly is important for permitted viral penetration.

(4) Viruses all contain nucleic acid and usually protein too. A few "viruses" are reported to consist of RNA alone (e.g. potato spindle tuber virus with double-stranded RNA, and tobacco necrosis virus with single-stranded RNA). In these cases resistance to nucleases is perhaps achieved by specialized coiling and supercoiling of the RNA molecules. Viruses only contain one type of nucleic acid, either DNA or RNA, but apparently never both together. The nucleic acid may be

single- or double-stranded and cyclic (phage) or acyclic (T4 phage). In some viruses (e.g. Tobacco mosaic, T4 phage) there is a single molecule of nucleic acid, but in many others (reoviruses, influenza viruses and Rous sarcoma viruses) more than one molecule of nucleic acid is present. Sometimes only one of the nucleic acid components appears to be active (BMV plant virus), but in others (reoviruses, etc.) the complex of the mixed nucleic acids are required for infectivity.

(5) With a few exceptions all viruses contain protein sub-units which in various ways coat the nucleic acid moeity to produce a complex in which the nucleic acid is presumably protected from attack by environmental nucleases. There are normally many more protein molecules than nucleic acid molecules; thus in TMV virus each RNA molecule (molecular weight $2{\cdot}05 \times 10^6$) is associated with 2130 protein molecules which are arranged around the helix so that three nucleotide units are bound per protein molecule, probably via two arginine and one lysine residue. Many of the simple viruses including TMV contain only a single type of protein but others contain several types. These include poliomyelitis virus, mouse encephalitis virus, etc., which contain possibly 14 different proteins.

(6) Some viruses contain more than one type of particle. These are the coviruses which are associated with plants and include the cowpea mosaic virus. This contains two components, 95s and 115s, containing 23 and 32% RNA respectively, with differing base sequences. Only the combination of the two units results in infectivity.

(7) All viruses are dependent on the host cell for the machinery required for replication. Although the virus may possess a complex code in its RNA or DNA genophore (Fig. 7.29) sufficient to provide not only for structural protein but also for specific enzymes such as viral lysozymes which may be required to dissolve cell membranes, especially during exocytosis, it does not possess any means of providing its own energy sources or any source of the various monomeric precursors required for polymer reproduction. It is here that the great gulf between the cell and the virus is most obvious. Many viruses, especially those transmitted by insects, can of course infect more than one type of cell.

(8) No viruses are known which specifically attack mitochondria, chloroplasts or similar cell organelles, although the membranes of these bodies seem capable of allowing passage of molecules as large as the viral genophore. In some ways this is a little surprising and a reasonable speculation might be that such viruses will ultimately be discovered.

(9) Entry to the host cells has to be gained by more or less physical

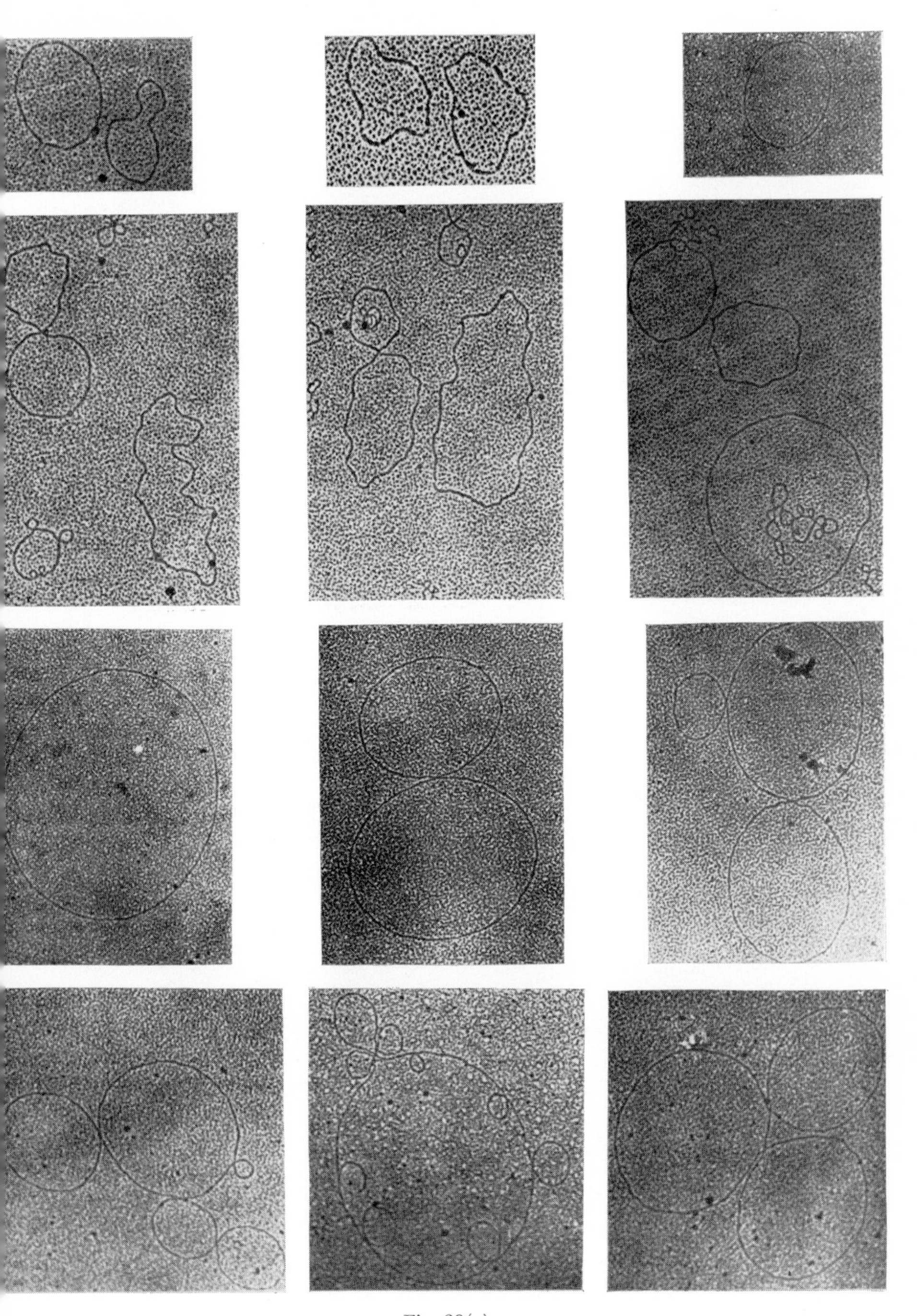

Fig. 29(a)

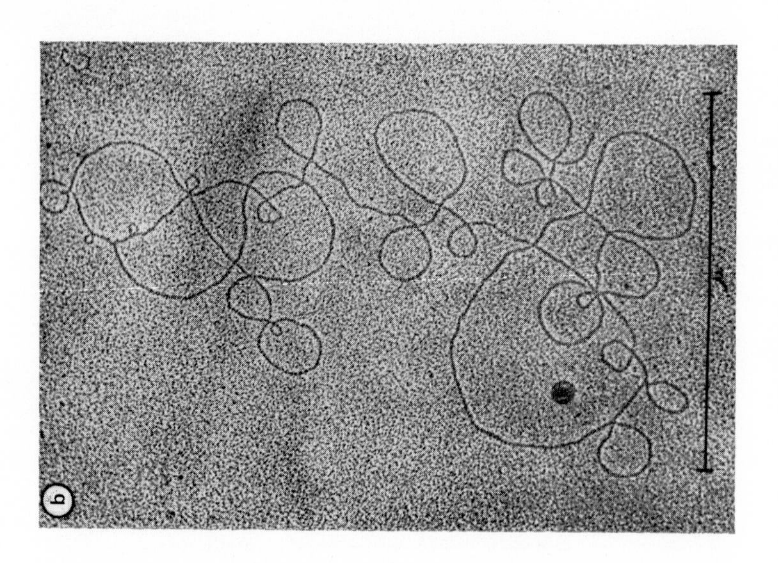

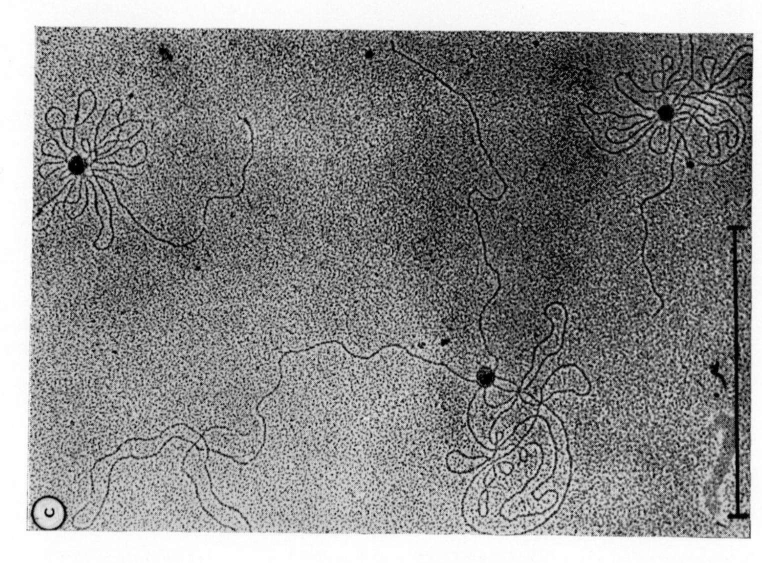

Fig. 7.29. Electron micrographs of SV40 monomeric and complex DNA. (a), (b) and (c) show the monomeric, circular and catenated SV40 DNA after repeated sedimentation through alkaline sucrose gradients. Photographs by kind permission of A. J. Levine (1971). *Virology* **44,** 480.

means (Fig. 7.30), whereas exit frequently engages the use of lysozymes which are coded in the genophore and produced during viral replication in the host.

(10) Although in the case of many bacteriophages only the viral

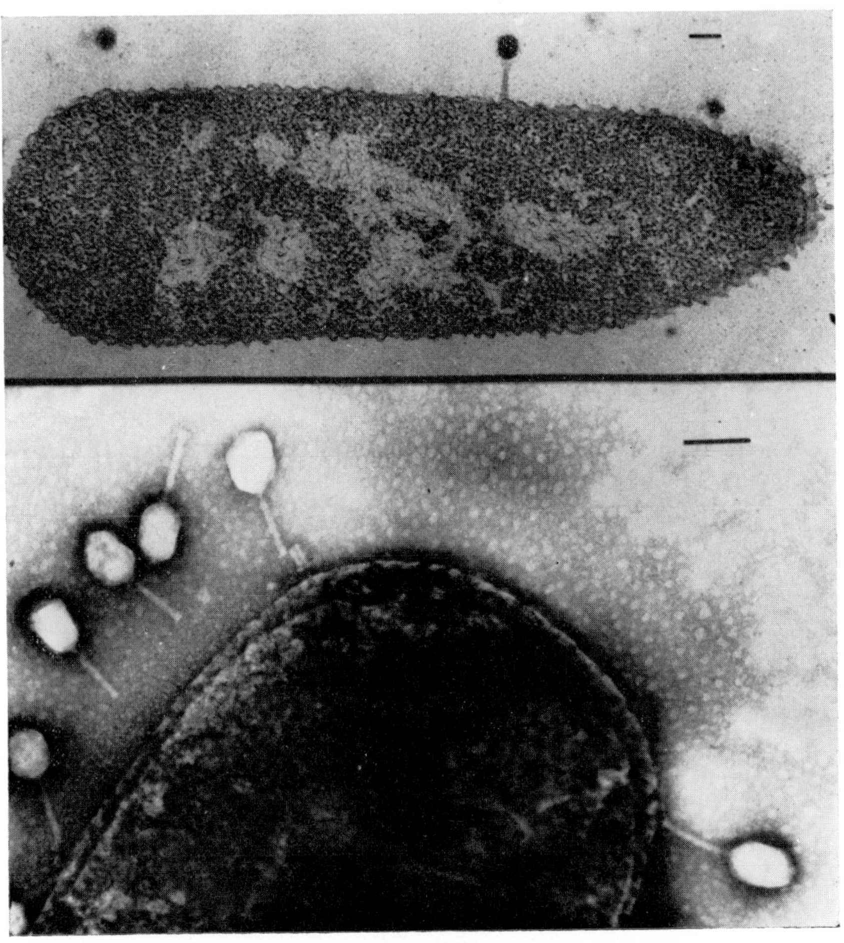

Fig. 7.30(a)

Fig. 7.30. Entry of viruses into host cells. (a), (b) and (c) showing the entry of T4 $11^- : 12^-$ bacteriophage into *Escherichia coli* cells. (d) T4 phages detached from *E. coli* cells, showing they have empty head units and contracted sheaths. Photographs by kind permission of L. D. Simon (1970). *Virology*, **41,** 77. (Scale mark = 0·1 μ.)

nucleic acid enters the cell, in the case of most animal or plant viruses both nucleic acid and protein are enveloped by the host. In the cytoplasm the protein coat is removed, presumably by cytoplasmic pro-

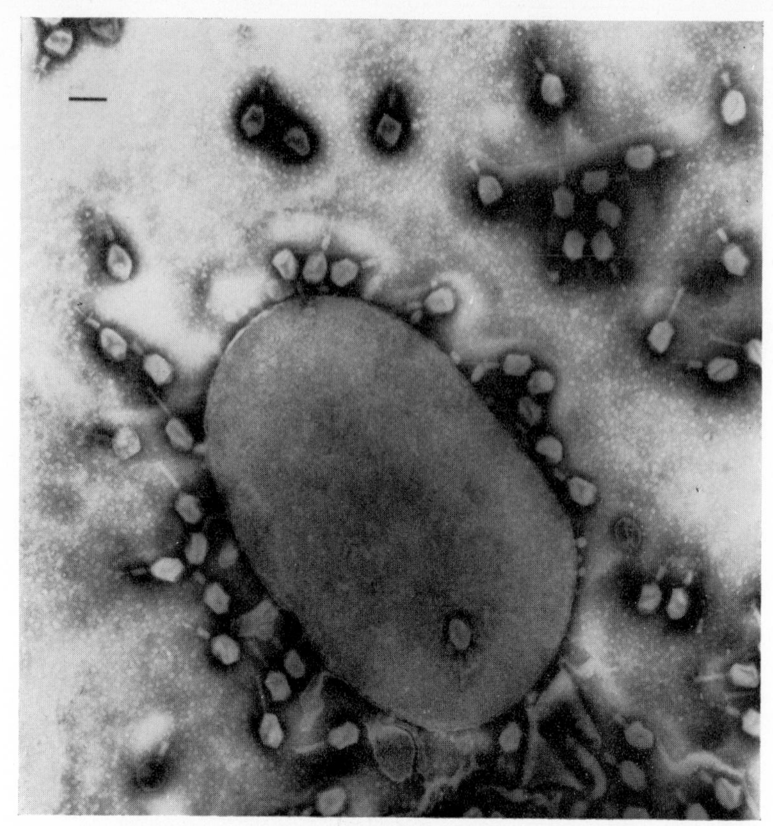

Fig. 7.30(b)

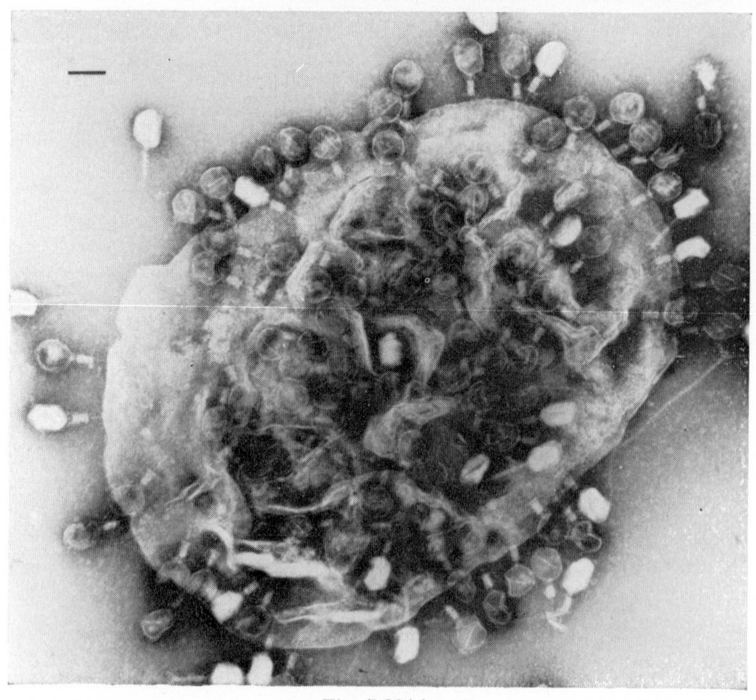

Fig. 7.30(c)

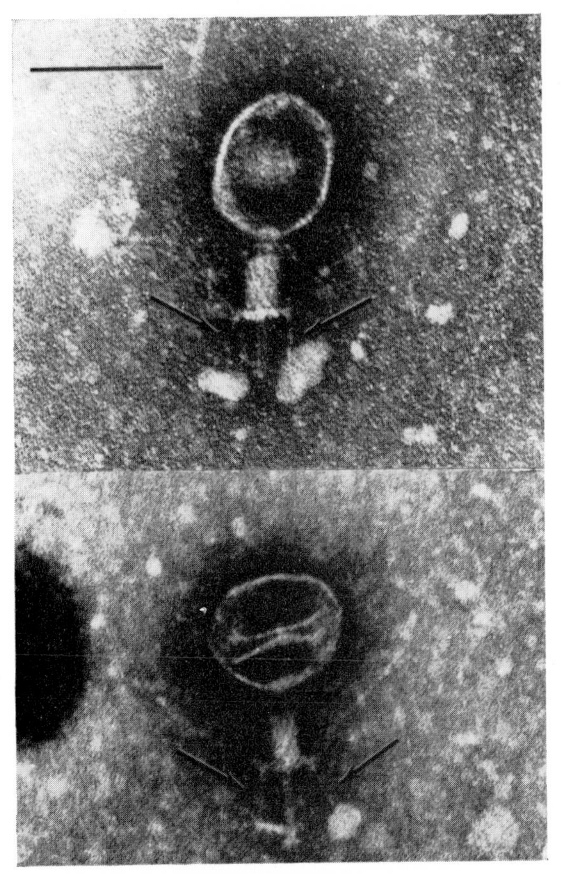

Fig. 7.30(d)

teinases, and the remaining nucleic acid can then: (a) if RNA operate as an mRNA and be translated by the host ribosomes to give rise to both sub-unit capsid proteins and enzymes (polymerases) required for viral RNA replication; or (b) if the nucleic acid is DNA, then this is transcribed in the host by a DNA-dependent RNA polymerase to form complementary mRNA. The various portions of the virus accumulate in the host and will aggregate to form a complete virus in a spontaneous manner.

Edgar and Wood subjected *E. coli* to treatment with 40 different mutants of T4 bacteriophage. Examination of the results showed that none of the extracts was infectious and that the mutants produced only partial virus particles, e.g. heads and not tails, or tails but no heads. However, when the extracts were mixed so that all functions were present infective whole virus was produced. Similar recombination occurred by separation of heads from tails, then mixing and incubating. In these cases no enzymes were required to persuade the viral components to interact with one another to form the complete particle. The general mechanism of viral nucleic acid and protein production are apparently much the same as for the host nucleic acid replication and protein biosynthesis. However, an early stage in viral nucleoprotein synthesis involves shutting off the host analogous reactions in some as

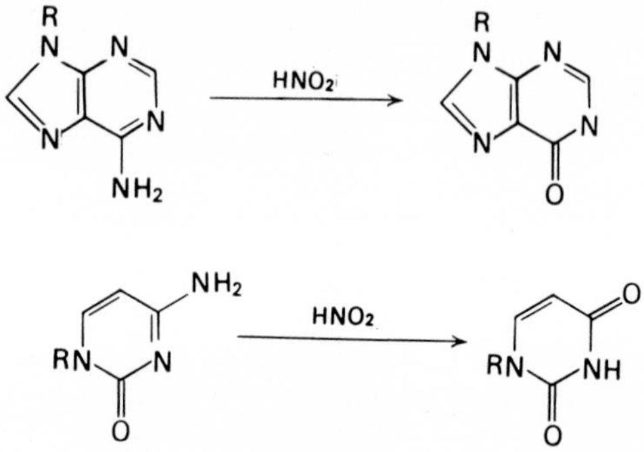

Fig. 7.31

yet unknown manner. Also a build-up of viral nucleic acid appears to precede protein formation.

(11) Mutation of viruses may occur spontaneously or can be induced in various ways using chemical or non-chemical methods such as UV radiation. Chemical agents include a wide variety of substances such as alkylating agents (including formaldehyde), acylating agents,

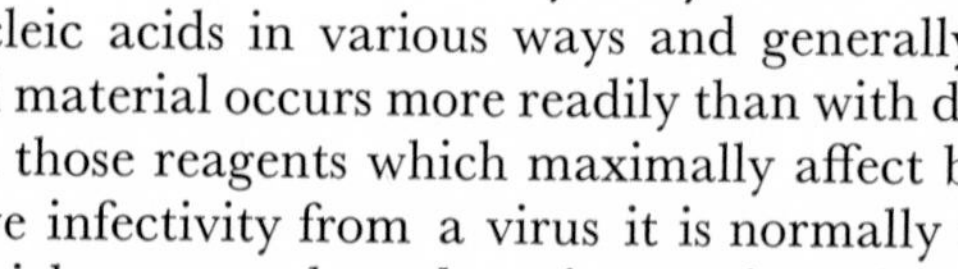

Fig. 7.32

compounds such as nitrous acid and hydroxylamine. These compounds react with nucleic acids in various ways and generally reaction with single-stranded material occurs more readily than with double stranded, especially with those reagents which maximally affect base pairing. In order to remove infectivity from a virus it is normally required to use compounds which prevent base–base interactions, but if mutants are required then reagents which interfere with the tautomeric equilibria of one or more bases should be sought.

For example, nitrous acid converts adenosine to inosine or cytidine to uridine and in each case the base proton equilibria is altered as is the coding and the result is mutagenic. On the other hand, alkylating agents may affect base–base interaction, as in Fig. 7.32, and lead to inactivation of the virus.

Spontaneous mutations probably arise in similar ways from the effect of toxic mutagenic cell metabolites or exogenous materials of various types. Certain dyestuffs, including acridine derivatives such as proflavine, may also induce mutations by intercalation in the helical structures with concomitant changes in the code. A small fraction of total spontaneous mutations is also derived by deletion of DNA segments in a manner not yet duplicated by artificial means, leading to removal of one or more genes or parts of genes.

SUGGESTED FURTHER READING Chapter 7

Books

Charles, H. P. and Knight, B. D. (Ed.) (1970). "Organization and Control in Prokaryotes and Eukaryotes". Cambridge University Press.

Fraenkal-Conrat, H. (1969). "The Chemistry and Biology of Viruses". Academic Press, New York.

Frankel, O. H. and Bennett, E. (Eds) (1970). "Genetic Resources in Plants". Blackwell Scientific Publications, Oxford.

Goodwin, T. W. (Ed.) (1966). "Biochemistry of Chloroplasts". Academic Press, London.

Gunsalus, I. C. and Stanier R. Y. (Eds) (1964). "The Bacteria: A Treatise on Structure and Function". Academic Press, New York.

Harbone, J. B. (Ed.) (1970). "Phytochemical Phylogeny". Academic Press, London.

Lehninger, A. (1964). "The Mitochondrion". Benjamin, Philadelphia.

Lehninger, A. (1970). "Biochemistry". Worth, New York.

Margulis, L. (1970). "Origin of Eukaryotic Cells". Yale University Press.

Miller, P. L. (Ed.) (1970). "Control of Organelle Development". Cambridge University Press.

Articles

Allsopp, A. (1960). Phylogenetic Relationships of the Prokaryotes and the Origin of the Eukaryotic Cell. *New Phytol.* **68,** 591–612.

Echlin, P. (1970). The Photosynthesis Apparatus in Prokaryotes and Eukaryotes. *In* "Organization and Control in Prokaryotes and Eukaryotes" (Eds H. P. Charles and B. D. Knight), pp. 221–48. Cambridge University Press.

Echlin, P. and Morris, I. (1965). The Relationship Between Blue-Green Algae and Bacteria. *Biol. Rev.* **40,** 143–187.

Erhan, S. (1968). Model for DNA polymerase. *Nature, Lond.* **219,** 160–162.

Granik, S. and Gibor, A. (1967). The DNA of Chloroplasts, Mitochondria and Centrioles. *In Advances in Nucleic Acid Research*, 143–186.

Jacob, F. and Monod, J. (1961). *J. mol. Biol.* **3,** 318.

Klug, A. (1968). Rosalind Franklin and the Discovery of the Structure of DNA. *Nature, Lond.* **219,** 808–844.

Knippers, R. (1970). DNA Polymerase II. *Nature, Lond.* **228,** 1050–1053.

Manton, I. (1967). Further Observations on the Fine Structure of Chrysochomelina Chiton with Special Reference to the Naptonema "Peculiar" Golgi Structure and Scale Production. *J. cell. Sci.* **2,** 265–272.

Margulis, L. (1968). Evolutionary Criteria in Thallophytes: A Radical Alternative, *Science*, **161,** 1020–1022.

Meselson, M. and Stahl, F. W. (1958). *Proc. natn. Acad. Sci.* **44,** 671.

Morowitz, H. J. (1967). Biological Self-Replicating Systems. *Progress in Theoretical Biology* **1,** 35–58.

Pelc, S. R. (1968). Turnover of DNA and Function. *Nature, Lond.* **219,** 162–163.

Polanyi, M. (1967). Life Transcending Physics and Chemistry. *Chem. Engng News*. August 21st, 55–56.

Perutz, M. F., Muirhead, H., Cox, J. M. and Goaman, L. C. G. (1968). Three-Dimensional Fourier Synthesis of Horse Oxyhaemoglobin at 2.8Å Resolution. The Atomic Model. *Nature, Lond.* **219,** 131–139.

Watson, J. D., and Crick, F. H. C. (1953). A Structure for Deoxyribose Nucleic Acid. *Nature, Lond.* **171,** 737–738.

Wilkins, M. H. F., Stokes, A. R. and Wilson, H. R. (1953). Molecular Structure of Deoxypentose Nucleic Acids. *Nature, Lond.* **171,** 738–740.

Wood, W. B., Edgar, R. S., King, J., Lielausis, I. and Henninger, M. (1968). *Fed. Proc.* **27,** 1160.

8

Abiotic Synthesis and Chemical Evolution

INTRODUCTION

Until the seventeenth century it is generally agreed that the concept of spontaneous generation of living matter was held universally inasmuch as any serious thought was given to matters concerning the origin of living creatures. It apparently seemed quite natural to most that plants and animals should arise in a spontaneous fashion from seemingly inanimate matter and confirmatory examples abounded in nature. Thus flies arose from rotting garbage, crocodiles from the mud of river deltas and maggots from rotting meat. There were in addition other more precise examples of metamorphosis in which caterpillars gave rise to butterflies and tadpoles to frogs which served in a sense to confirm the primary assumptions. Many of these assumptions arose less, one suspects, from the apparently inherent stupidity of savants of the day than from the almost complete lack of serious study of the subject which in any case was much shrouded in superstition, alchemy and witchcraft, and in which too close an interest might have been considered unhealthy.

Like so many major changes in scientific direction the major change in this field arose from the invention of a new technique—the microscope—which completely revolutionized man's concepts of nature. Although the first compound microscope was invented (by Janssen) in 1590 and an improved version made by Scheiner in 1628 from a design of Kepler's, the first simple microscope containing a single relatively high-power lens was produced by van Leeuwenhoek about 1680 and the higher power of this instrument provided the significant breakthrough into the micro world.

It was during this Renaissance period from about the mid-seventeenth century that Harvey showed that animals arose from eggs, Redi (late seventeenth century) proved that maggots were derived from eggs which had been laid by flies and Spallanzani in the early eighteenth century discovered that spermatozoa were required for the fertilization of eggs. Finally as a culmination of the microscopic work which had been carried out during the preceding century Pasteur in the 1850's finally demonstrated that fermentation of nutrient broths is due to infection by aerial microbes and their subsequent rapid growth, and not to any *de novo* life production from inanimate matter. In a sense modern biology began from that period and one of the basic tenets of the subject requires that living matter can only arise from living matter of like nature. This biological "law" of course does not in any sense prejudice the possibility of any abiotic formation of the very first living cell, which would clearly have to be a special case.

In recent years, thanks in many ways to the discovery and application of the various new techniques of chromatography, especially paper and ion-exchange chromatography, and applications of the techniques of X-ray crystallography and electron microscopy to chemical structure determination, there have been enormous strides made in our understanding of the chemistry and biochemistry of cellular function and especially of the chemistry of the genetic processes whereby duplication of living matter occurs from like material. The subject of molecular biology has arisen to combine the physical and biological sciences into a most powerful instrument for the study of living processes. Molecular biology sees living matter increasingly as a physico-chemical system whose components can be precisely defined in physical and chemical terms and whose functions can similarly be expressed in terms of the components. The success and results of this approach have led equally to a new and widening interest in the primary origins of living matter. This new interest is not merely confined to speculative sorties but seeks by experiment and logic to reach useful decisions about the origin of life in the universe. It is a relatively new study which none the less has now reached the end of what can be regarded as a primary stage and arrived at the conclusion that life arose on this planet not from like matter, nor by divine intervention, but by the chance formation and interaction of a wide variety of organic molecules, leading finally to a self-sustaining and duplicating physico-chemical system, the progenitor of all living matter on the planet.

Early theories of the abiotic synthesis of organic chemicals (Oparin) and of the origin of the solar system by condensation of primeval dust cloud (Urey) led in each case in a quite unique sense to a common concept of the possible nature of the primeval atmosphere existing on earth at the time of its formation. The outstanding feature of this hypothetical atmosphere was that it had reducing properties and contained hydrogen, in direct contrast to the oxygen-containing atmosphere that we know today. The presence of hydrogen is essential if complex organic compounds are to be formed from simple molecules since in an oxidizing environment any tendency for more complex structures to form will be accompanied by corresponding rapid oxidative degradation of such structures to CO_2, H_2O, etc. The reducing atmosphere or at the very least a neutral atmosphere is therefore necessary if organic compounds of any degree of complexity are to form in quantity.

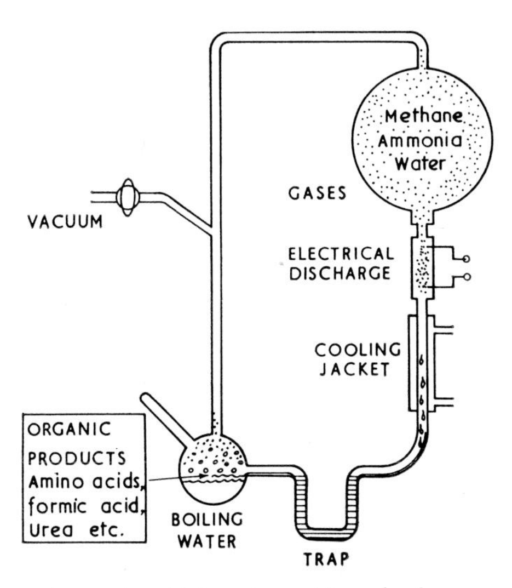

Fig. 8.01. Miller's experiment, in which amino acids and other organic compounds were synthesized from a mixture of gases

In 1953 Miller announced the results of some preliminary experiments which were designed to test the validity of these particular theories (Fig. 8.01). He brought together in a sealed system a mixture of water (steam), CO_2, NH_3, hydrogen and subjected this to electrical discharges over a period of three to six days. The experiments were conducted in such a manner as to exclude biological contamination. After cooling the mixture the resulting aqueous solution obtained was examined by the relatively new techniques of paper chromatography which of course were not available to workers in the period before the Second World War. He found that the aqueous solution contained very small amounts of a mixture of amino acids (including glycine, alanine and glutamic acid), sugars, urea and other organic chemicals, many of which were of normally natural occurrence or part of naturally-

Table 8.01. Compounds produced from sparking a mixture of ammonia, hydrogen, methane and water similar to the experiments carried out by S. L. Miller (1953)

Compound	
Formic acid	Decreasing yields of compounds ↓
Glycine	
Glycolic acid	
Alanine	
Lactic acid	
β-Alanine	
Acetic acid	
Propionic acid	
Iminodiacetic acid	
α-Amino-*n*-butyric acid	
α-Hydroxybutyric acid	
Sarcosine	
Succinic acid	
Urea	
Iminoacetic-propionic acid	
N-Methyl urea	
N-Methylalanine	
Glutamic acid	
Aspartic acid	
α-Aminoisobutyric acid	

occurring polymers such as proteins (Table 8.01). In later experiments Miller was able to show that the amino acids were formed by a Strecker reaction which involves preliminary formation of an aldehyde condensation of this with hydrogen cyanide and ammonia and subsequent ready hydrolysis of the amino nitrile formed with dilute aqueous ammonia as follows:

$$CO_2,H_2O,CH_4,H_2 \longrightarrow RCHO \xrightarrow{HCN} R.CH(OH)CN$$

$$R.CH(OH)CN \xrightarrow{NH_3} R.CH(NH_2)CN$$

$$R.CH(NH_2)CN \xrightarrow{aqu.\ NH_3} R.CH(NH_2)CO_2H$$

Fig. 8.02

The derived aldehydes of course offer a ready source of a wide variety of other types of organic compounds, especially carbohydrates. In fact it had been shown many years prior to Miller's experiments that formaldehyde and sodium hydroxide in aqueous solution readily produced a complex mixture of sugars. Many other reactions of this type were carried out during the 1920's and 1930's and include the classical synthesis of tropinone by Robinson:

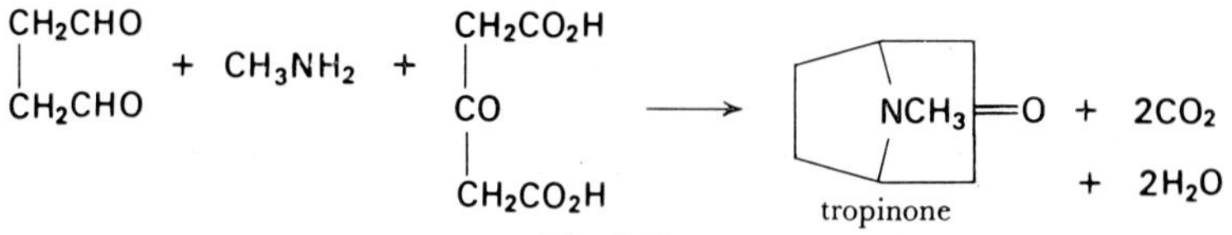

$$\begin{matrix} CH_2CHO \\ | \\ CH_2CHO \end{matrix} + CH_3NH_2 + \begin{matrix} CH_2CO_2H \\ | \\ CO \\ | \\ CH_2CO_2H \end{matrix} \longrightarrow \text{tropinone } (NCH_3, =O) + 2CO_2 + 2H_2O$$

Fig. 8.03

and from this followed a spate of chemical reactions which were claimed to operate under so-called physiological conditions. The precise conditions used may not always have been truly acceptable to a biological species, but these experiments have always convinced organic chemists at least that there is no great problem in elaborating complicated organic molecules from simple precursors. Indeed the average organic chemist normally endeavours to escape from this annoying and all-too-ready tendency of organic molecules to form such diverse numbers and types of compounds. The Miller type of experiment has been repeated by many other workers using a variety of simple compounds, variation of the energy source from radiation to shock tubes and adjustments to the synthetic atmosphere, and there is no question that the basic monomer requirements (α-amino acids, carbohydrates, purines and pyrimidines) required for the abiotic formation of proteins and nucleic acids and polysaccharides may be readily satisfied by conditions used in these types of experiment.

In addition to the radiation experiments many related experiments have been carried out with various mixtures of suitable small molecules in aqueous solution which have led to the detection, usually of course in very small amounts, of a wide variety of α-amino acids, many of which figure in the list of twenty found in most naturally occurring proteins. These include reaction of formaldehyde with hydrogen cyanide at room temperature, formaldehyde and hydroxylamine, N-carboxy amino acid anhydrides and a trace of water, aminonitriles and activation by dicyandiamide over kaolin at 100°, and excess heat treatment (1300°K) of a mixture of methane, ammonia and water, and various combinations of these types of experiment.

In the original experiments Miller was unable to detect the presence of purines or pyrimidines, but subsequently Ponnamperuma produced a small (0·01%) yield of adenine by irradiation of the mixture of gases used by Miller with a 4·5 MeV electron beam. In addition to the radiation type experiments which have been modified and duplicated several times with production of numerous various small amounts of amino acids, etc., other simpler experiments have also readily produced purines, carbohydrates and pyrimidines and by suitable combinations of these it has been possible to obtain nucleosides and nucleotide derivatives.

Oro in 1960 showed that adenine was produced by heating a concentrated solution of ammonium cyanide, and yields of up to 0·1% could be obtained when the ammonium cyanide solution was heated at 90° over 1 day or 70° over 25 days. Of special interest was the fact that aminoimidazole aglycone intermediates, analogous to the nucleotides produced during the biosynthesis of purine nucleotides, are involved in the mechanism of purine synthesis from cyanides. Thus 4(5)-amino-imidazole-5(4)-carbonitrile and the related amidine and amide were isolated together with formamidine from reaction mixtures of the types outlined above. The conversion of aminoimidazole carboxylic acid derivatives to purines by reaction with formates, formamidines and related substances was well known from the earlier work of Cook, Shaw and others and operates in the conversion of AICAR to IMP in the biosynthetic pathway (Chapter 6).

Various mechanisms have been proposed to explain adenine synthesis in these reactions. Oro suggested a sequence of reactions which commenced with the formation of a trimer of HCN (amino malononitrile) converted by ammonia to aminomalondiamidine and subsequent cyclization of this with formamidine (also produced from HCN and NH_3) to adenine as follows:

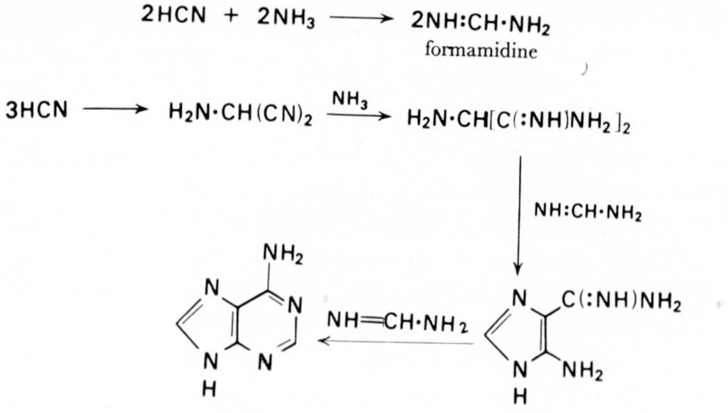

$$2HCN + 2NH_3 \longrightarrow \underset{\text{formamidine}}{2NH{:}CH{\cdot}NH_2}$$

$$3HCN \longrightarrow H_2N{\cdot}CH(CN)_2 \xrightarrow{NH_3} H_2N{\cdot}CH[C({:}NH)NH_2]_2$$

$$\downarrow NH{:}CH{\cdot}NH_2$$

adenine (NH₂-substituted purine, N–H) $\xleftarrow{NH{=}CH{\cdot}NH_2}$ imidazole bearing $C({:}NH)NH_2$ and NH_2 (N–H)

Fig. 8.04

In this scheme five equivalents of HCN produce each molecule of adenine. In a similar manner other purines may be produced from appropriate intermediate aminoimidazoles and formic or carbonic acid derivatives by well-known cyclization reactions. Thus hypoxanthine would arise from the carboxamide and formamidine, guanine from the amide and guanidine and xanthine by heating with urea:

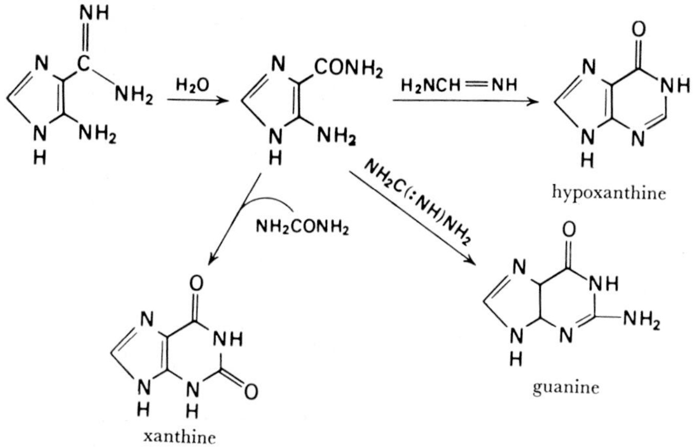

Fig. 8.05

Indeed Lowe found hypoxanthine in addition to adenine in heated solutions of ammonium cyanide and also showed that the solution obtained contained at least 50 other ultraviolet absorbing compounds.

Orgel and co-workers have to some extent more recently clarified the aminoimidazole reaction sequence involved in dilute aqueous solutions of ammonium cyanide. Thus an early important intermediate was found to be the hydrogen cyanide tetramer (diaminomaleonitrile) produced by reaction of aminomalononitrile with ammonia in 50% yield. The maleonitrile derivative readily reacts with formamidine to give adenine in addition to the aminoimidazole amidine or carboxamide and the step is not inhibited by cyanide ion. However, when the maleonitrile derivative was exposed to ultraviolet light the aminoimidazole carbonitrile was produced in 80% yield. At the same time photolytic destruction of the aminonitrile also occurred but was inhibited by cyanide. Similarly ultraviolet radiation of ammonium cyanide solutions produced both the maleonitrile and the aminoimidazole nitrile directly. The mechanism of the photoinduced reaction has been shown to involve an initial isomerization of the maleonitrile to the related fumaro derivative, which had also been found in small amounts in the ammonium cyanide solutions. The conversion of this substance to an aminoimidazole is suggested to involve an azetine derivative (Fig. 8.06).

The maleonitrile production was shown to be enhanced by freezing and yields of 10% were obtained after 11 months at −20°.

The synthetic aminoimidazole nitrile was also found to react with aqueous ammonium cyanide at 30–100° to give adenine and aminoimidazole carboxamide and by extrapolation to −20° adenine is presumed to be formed in 10% yield within 1–10 years.

Orgel has used these results to indicate how adenine might have been produced in a lake which alternately froze and thawed. Thus if in summer a 10^{-5} molar solution of cyanide was acquired, during the

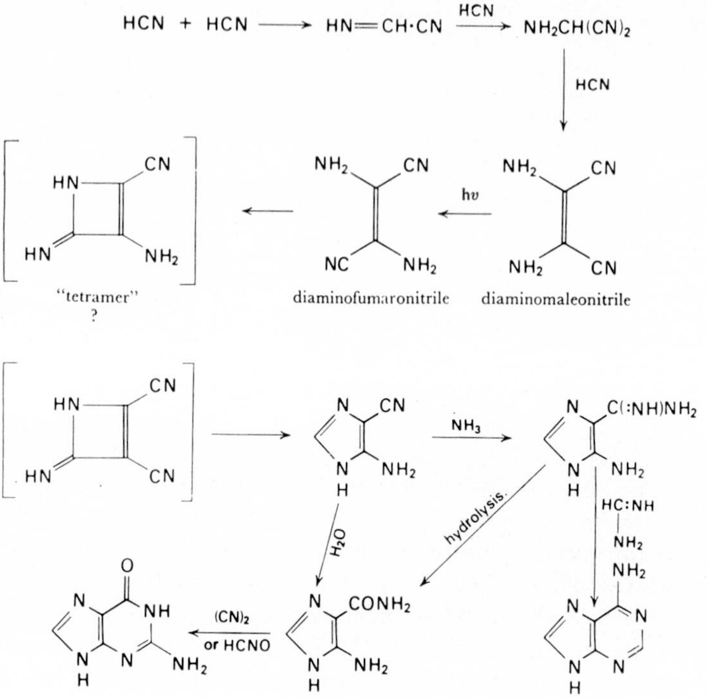

$$HCN + HCN \longrightarrow HN{=}CH \cdot CN \xrightarrow{HCN} NH_2CH(CN)_2$$

HCN

CN
HN
HN
NH_2
"tetramer"
?

NH_2 CN
NC NH_2
diaminofumaronitrile

$h\nu$

NH_2 CN
NH_2 CN
diaminomaleonitrile

CN
HN
HN
CN

N CN
N NH_2
H

NH_3

N C(:NH)NH_2
N NH_2
H

hydrolysis

H_2O

HC:NH
NH_2

NH_2
N N
N N
H

O
N NH
N N NH_2
H

$(CN)_2$ or HCNO

N $CONH_2$
N NH_2
H

Fig. 8.06

winter a 10% conversion to organic material would occur and material of molar concentration might then accumulate every million years.

Guanine is also formed in a similar manner by reaction of synthetic

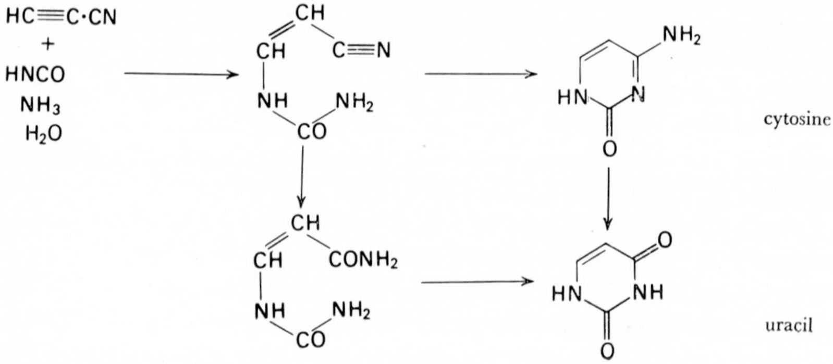

HC≡C·CN
+
HNCO
NH_3
H_2O

CH
CH C≡N
NH NH_2
CO

NH_2
HN N
O

cytosine

CH
CH $CONH_2$
NH NH_2
CO

O
HN NH
O

uracil

Fig. 8.07

aminoimidazole carboxamide with cyanogen or cyanate. Cyanogen in fact may be produced under various so-called prebiotic conditions from cyanide and cyanate is a normal hydrolysis product of cyanide.

The pyrimidines, uracil and cytosine have also been produced under abiotic conditions by exposing mixtures of cyanoacetylene, cyanate and cyanogen to electric discharges.

A well-known preparation of uracil involves reaction of malic acid with urea in the presence of sulphuric acid:

$$HO_2CCH_2CHOHCO_2H \xrightarrow{H^+} OHC.CH_2CO_2H + \text{urea} \longrightarrow \text{uracil}$$

Fig. 8.08

Shaw and co-workers have similarly prepared uracil derivatives including glycosides by the similar reaction scheme.

$$HC \equiv C.CO_2H \longrightarrow HC \equiv C.CO.NHCO_2Et \xrightarrow{NH_3} \text{uracil}$$

Fig. 8.09

Thus small amounts of adenine, guanine and cytosine have also been detected among the products formed in Fischer–Tropsch type synthesis from mixtures of CO, H_2 and NH_3 at temperatures from 200–950° with (or without) Fe-Ni, alumina or silica catalysts.

Nucleosides do not appear to have been produced directly under prebiotic conditions from small molecular precursors. However, there are several syntheses involving preformed units which have been produced in the prebiotic process. The formation of carbohydrates from formaldehyde and alkali dates back to the work of Butlerow in 1861, and in 1933 Orthner and Gerisch showed that pentoses were obtained from mixtures of glyceraldehyde and dihydroxyacetone in a formally analogous system to that operating in the biosynthesis of these types of substances. The nucleic acid sugars, ribose and deoxyribose, have been produced in a similar fashion. Thus Oro and co-workers obtained 2-deoxyribose from glyceraldehyde and acetaldehyde, and both ribose and deoxyribose were produced by Ponnamperuma in 1962 by electron radiation of a methane–ammonia–water mixture and also by ultraviolet irradiation of dilute weakly acid solutions of formaldehyde or of hydrogen cyanide. Various pentoses, including ribose and deoxyribose, were also produced from an aqueous formaldehyde solution by refluxing over kaolinite, and disaccharides and glucose-6-phosphate were obtained from mixtures of glucose, dicyandiamide and phosphoric acid with or without ultraviolet irradiation.

When mixtures of adenine, ribose and phosphate are subjected to ultraviolet irradiation adenosine is formed and deoxyadenosine similarly results from deoxyribose, adenine, HCN and ultraviolet radiation.

Nucleosides are readily converted into phosphorylated derivatives by a wide variety of inorganic phosphates. Thus when uridine was heated with various simple phosphates, including sodium potassium or calcium mono-, di- and tri-hydrogen phosphates, or related pyrophosphates at 160° for times from 0·1 to 336 h uridine 2′-, 3′- and 5′-phosphates were formed in total yields of up to 29%. Not unexpectedly the best yields were obtained using pyrophosphates, and with longer times, and the phosphorylation with orthophosphates is probably due to initial conversion of phosphate to pyrophosphate at the particular temperature used. The phosphorylation also occurred with these materials in the presence of limited amounts of water and a close correlation was found between orthophosphates which gave poor yields of ribonucleotides and those which in separate experiments gave low yields of inorganic polyphosphates.

More efficient phosphorylation would, of course, exist at lower temperatures using a more active phosphoric acid derivative. A presumably requisite prebiotic activation may be produced in several ways including, for example, the reaction of phosphoric acid with cyanamide or dicyandiamide, both of which are produced by irradiation of methane, ammonia, water mixtures or with cyanogen or its hydrolysis products cyanoformamide, cyanate and also cyano-acetylene.

$$(1)\quad HCNO + H_3PO_4 \rightarrow NH_2CO{\cdot}OPO(OH)_2 \xrightarrow{H_3PO_4} (HO)_2\text{—}\overset{O}{\overset{\|}{P}}\text{—}O\text{—}\overset{O}{\overset{\|}{P}}\text{—}(OH)_2 + NH_2CO_2H$$

$$(2)\quad NH_2CN \rightarrow NH_2{\cdot}C(:NH){\cdot}NHCN \rightarrow NH_2C(:NH)NH_2$$

$$NH_2CN \xrightarrow{H_3PO_4} NH_2C(:NH)O{\cdot}P(O)(OH)_2$$

$$(3)\quad (CN)_2 \xrightarrow{H_2O} NH_2C(:NH)OP(O)(OH)_2$$

$$(4)\quad NC{\cdot}C{\equiv}CH \rightarrow NC{\cdot}CH{=}CHOP(O)(OH)_2$$

Fig. 8.10

Of particular interest in the formation of purines by the above abiotic processes is that they all appear to proceed by way of intermediates, especially aminoimidazoles which parallel to some extent the sequence of reactions which operate in the apparently universal pathways of *de novo* biosynthesis of these compounds from small molecules (see Chapter 6). Indeed Shaw and co-workers and others have demonstrated that almost all the *de novo* biosynthetic pathway reactions may be duplicated under relatively mild non-enzymic conditions. Thus the aminoimidazole (AIR—Chapter 6) is converted in up to 60% yield into carboxy-AIR by warming at 60° for about 15 min with potassium hydrogen carbonate. Other reactions in this sequence which may be similarly duplicated include:

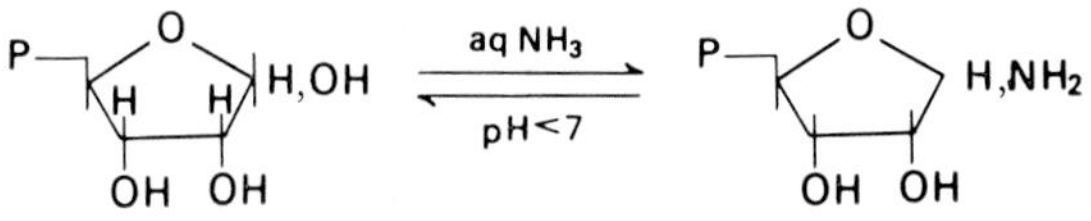

Fig. 8.11

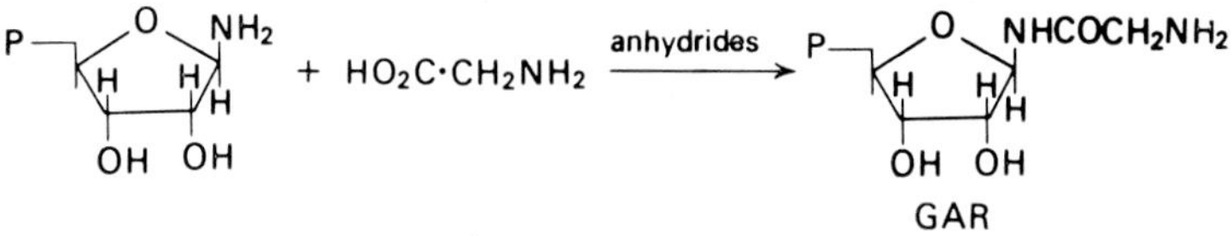

Fig. 8.12

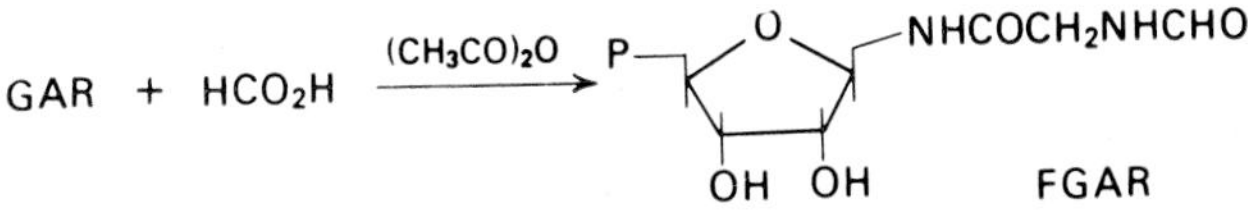

Fig. 8.13

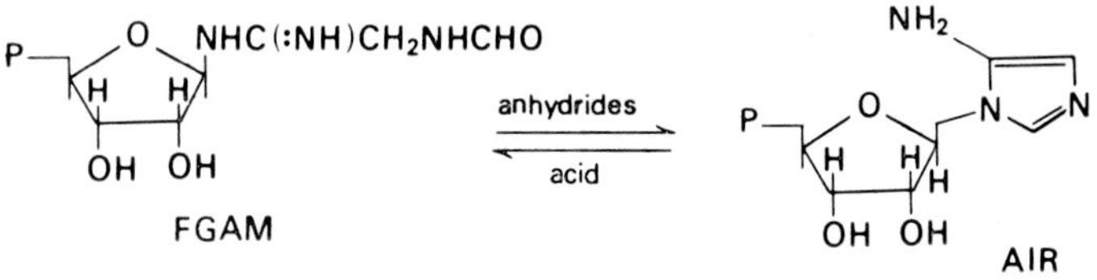

Fig. 8.14

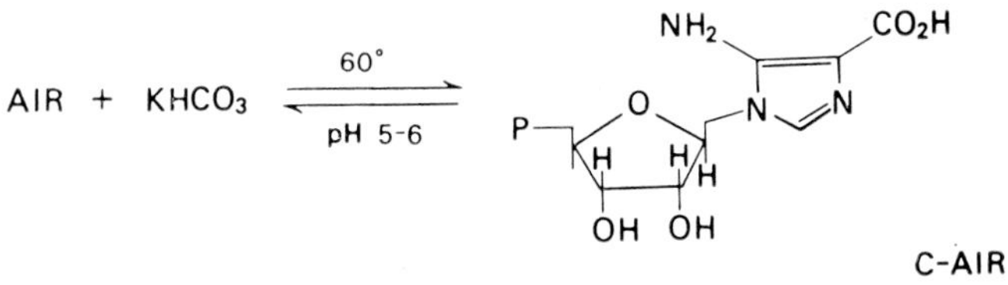

Fig. 8.15

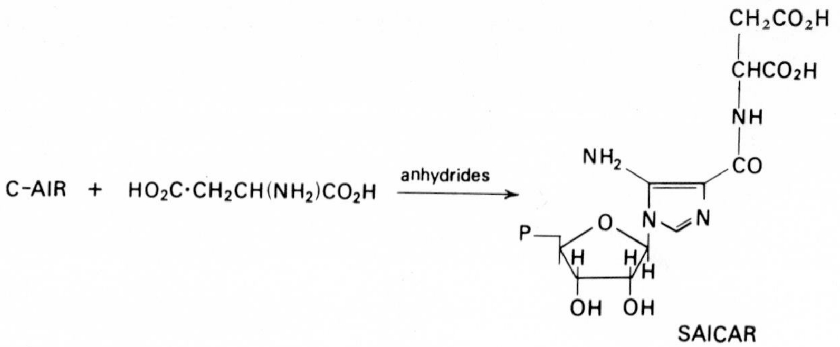

C-AIR + $HO_2C \cdot CH_2CH(NH_2)CO_2H$ $\xrightarrow{\text{anhydrides}}$ SAICAR

Fig. 8.16

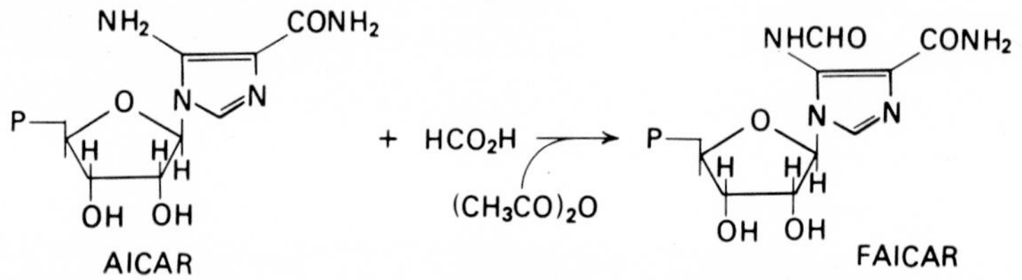

AICAR + HCO_2H $\xrightarrow{(CH_3CO)_2O}$ FAICAR

Fig. 8.17

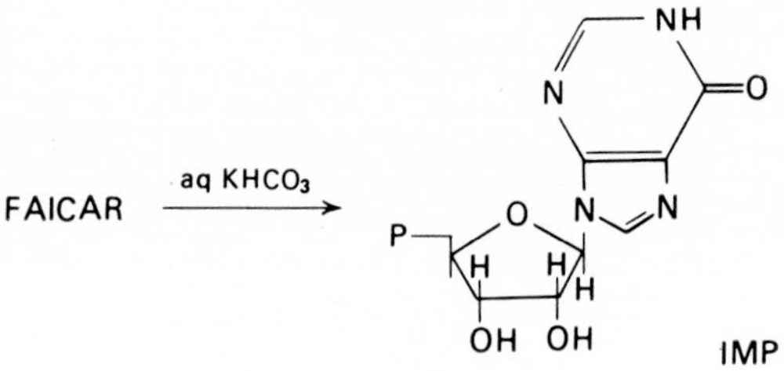

FAICAR $\xrightarrow{\text{aq } KHCO_3}$ IMP

Fig. 8.18

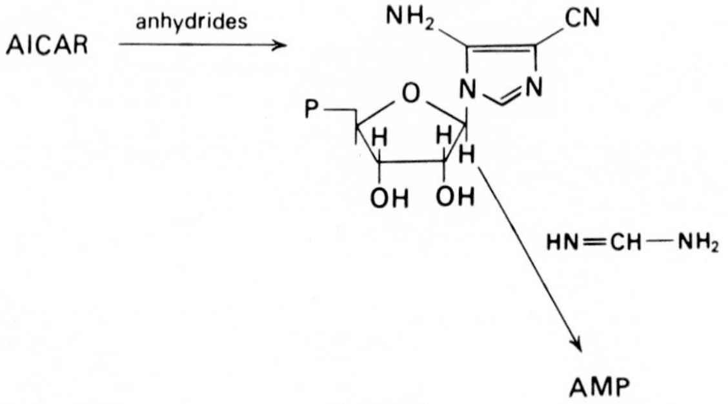

AICAR $\xrightarrow{\text{anhydrides}}$ (CN compound) $\xrightarrow{HN{=}CH{-}NH_2}$ AMP

Fig. 8.19

All these reactions are very much of the abiotic type and in many ways much more pertinent to the general theory than many of the previously mentioned reactions. In addition the phosphate group may have a special role as for example in the intramolecular nucleophilic catalysis of the carboxylation–decarboxylation between AIR and C-AIR which has been suggested from kinetic studies of this reaction sequence.

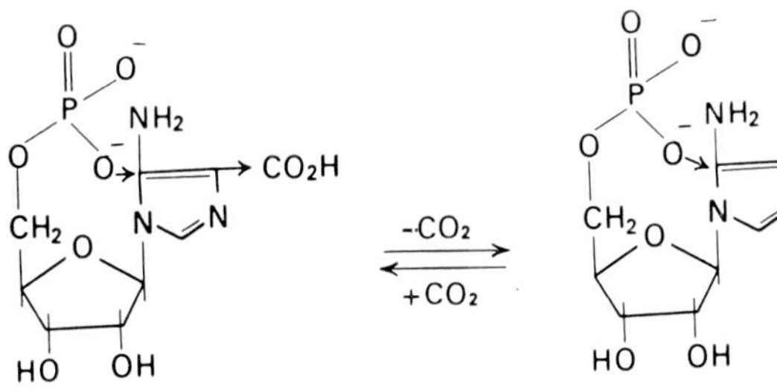

Fig. 8.20

It is interesting to note, however, that abiotic routes to pyrimidines have rarely paralleled their normal biosynthetic pathway which involves initial formation of N-carbamoyl aspartate which cyclizes to a dihydrouridine *in vivo*, whereas *in vitro* hydantoin acetic acid is formed.

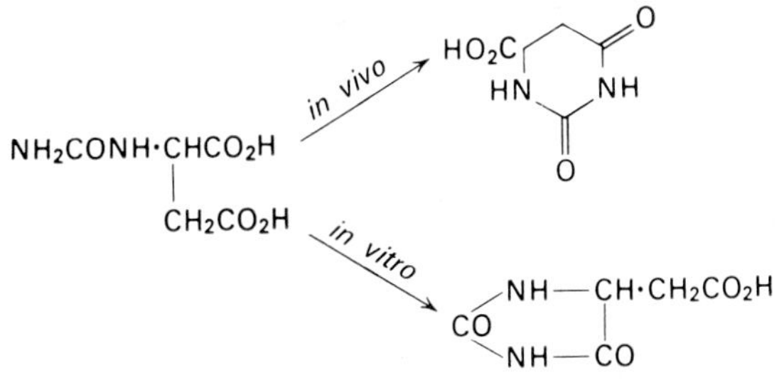

Fig. 8.21

It is possible, of course, that these types of dihydropyrimidine intermediates have not been detected because they lack ultraviolet absorbing properties and are generally difficult to detect when present in small quantities.

ABIOTIC FORMATION OF POLYMERS

Poly-amino acids

There can be little doubt that by adopting the appropriate reaction conditions simple molecules such as alkanes, ammonia, aldehydes, ketones, etc., can be a source, albeit generally in minute yield, of all the simple monomer building units required for the elaboration of the polymers, polysaccharides, proteins, and nucleic acids which are so essential a part of the life process (Chapter 6). Abiotic polypeptide-like materials have been produced by many workers by subjecting α-amino acids to a variety of conditions including heat treatment and condensation with numerous types of anhydrides. Fox in particular was able to obtain relatively high molecular weight (10,000) materials, so-called proteinoids, which possessed polypeptide properties, by heating mixtures of amino acids at 170° or by heating for longer times at lower temperatures with inorganic polyphosphates. The proteinoids have some properties of polypeptides; thus they give amino acids on acid hydrolysis, although they also contain other molecular units. In addition their composition, not unexpectedly, is influenced by the specific mixture of amino acids used in their preparation, large amounts of the acidic amino acids such as aspartic acid and glutamic acid not unnaturally give essentially acidic proteinoids, and basic amino acids including lysine give basic proteinoids. Various workers have also shown that many of these polymeric materials have various types of catalytic activities which to some extent parallel those of certain enzymes. Thus claims have been made for catalysis of a wide variety of reactions including hydrolysis of esters such as *p*-nitrophenyl acetate and ATP, decarboxylation of pyruvic acid and oxaloacetic acid, and oxidation of glutamic acid to α-ketoglutaric acid. When the proteinoids are formed in the presence of haem, peroxidase and catalase like activities are reputed to appear in the products. It must be emphasized of course that these catalytic activities in many cases only slightly exceed those possessed by some of the simple amino acids from which the polymers are derived. There have been various other claims for the direct formation of what are usually described as small "polypeptide-like" molecules by direct radiation of various materials including hydrogen cyanide, but in the main, apart from the proteinoids, progress to discrete reasonably sized proteins has been disappointing.

Polynucleotides

There has been little or no worthwhile progress towards an abiotic synthesis of a nucleic acid molecule and indeed it is difficult to envisage the synthesis of a macromolecule with a molecular weight of a million or so, by the sort of crude stewing together of monomer units that has dominated this field. If in fact nucleotide mono-, di- or triphosphates are heated, with or without anhydrides, certainly some polymerization occurs, although it leads to materials of very small chain length containing 2′-, 5′-phosphodiester linkages. Many claims in this field for the production not only of nucleic acid-like substances but also of polysaccharides by heating monomer units with substances such as polyphosphoric acid and ethyl metaphosphate have not been substantiated, and to some extent this field has fallen slightly into disrepute.

Of some special interest, however, is the observation that polymerization of AMP in the presence of single-stranded poly-U (as a template) with a water soluble carbodi-imide results in the formation of a complementary poly-A which possesses, however, the 2′-5′-linkage.

PRIMEVAL SOUPS AND CHEMICAL EVOLUTION

There is little doubt that many organic chemicals have formed in nature from simple precursors by abiotic processes. Indeed such chemical reactions are quite certainly proceeding today in biological sediments, where it must be emphasized not all the reactions are necessarily under enzyme control. It may well be that some of the reactions outlined in the preceding notes may also have occurred to some extent. Certainly the so-called abiotic experiments have successfully illustrated that the sort of organic monomer precursors of the basic macro-molecules, proteins, polysaccharides and nucleic acids required for the formulation of a living system can be elaborated from even simpler molecules such as hydrogen cyanide, formaldehyde, carbon dioxide and so on. These results have led to what has now almost become a total belief, namely that all these things really happened and that living matter had its primary origins in these materials. The theory, which in general form is briefly outlined below, has been termed the chemial evolutionary theory. The theory assumes that:

(1) The original primeval atmosphere of this planet Earth contained hydrogen and was therefore a reducing atmosphere.

(2) The temperature of the planet's surface was relatively low (100°?) at an early stage of its formation; low enough, that is, to enable the survival at least of substances such as proteins which of course are normally rapidly denatured by heat.

(3) In these mild reducing conditions simple molecules such as ammonia or nitrogen, methane, water, and of course hydrogen, were subjected to various forms of energizing radiation including various electric discharges derived from natural phenomena such as lightning, radioactivity, meteorite impact (Fig. 8.22) and ultraviolet radiation which was probably more intense than it is now. Under these conditions a host of simple organic chemicals were formed by the various mechan-

isms outlined earlier but especially α-amino acids, sugars and the nucleic acid bases.

(4) All these chemical substances accumulated in the oceans. They accumulated because (a) there was no life which might remove them by metabolic processes and (b) we still had a reducing environment, so that these molecules maintain their relative complexity and do not break down, as they well might in the present environment by an oxidative mechanism.

(5) With the passage of time the gross amount of organic chemicals increase to form what has been described as a primeval soup or broth. That is the oceans contain ever-increasing amounts of organic matter, large amounts of which will of course be nitrogenous.

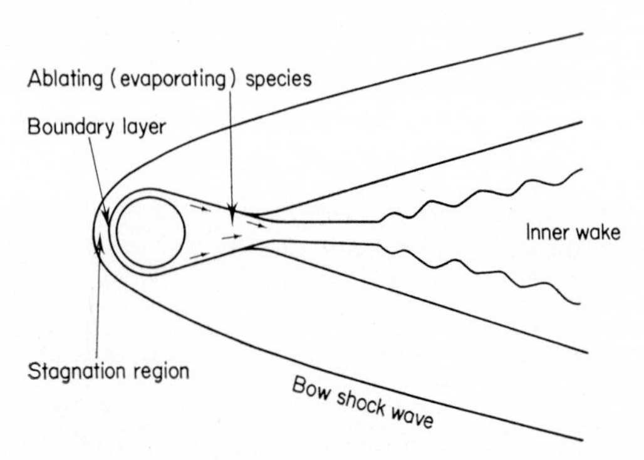

Fig. 8.22. Physical forces produced by impact of a sphere (e.g. meteorite)

(6) Given enough time it is considered that the simple chemicals which compose the soup will condense together so that amino acids give rise to proteins, monosaccharides to polysaccharides, nucleic acid bases and sugars and inorganic phosphates, to nucleotides and hence to polynucleotides. All the assymetric molecules produced by the abiotic processes will be racemic, and consequently at this or perhaps a somewhat later stage there would have to be a separation of the various optical enantiomers from each other, since we know that all proteins today with very few minor exceptions contain only L-amino acids and the nucleic acids only D-ribose or D-deoxyribose. Indeed it is only possible to produce a stable α-helical protein structure from either pure L- or pure D-amino acids. In general, theories regarding the formation of pure chiral molecules fall into two parts according to whether resolution of enantiomorphs preceded or followed the production of the first living system. Although there have been many attempts to produce optical activity from racemic mixtures by numerous devices, including reactions

in the presence of circularly polarized light or one-handed naturally occurring crystals such as quartz, they have mainly been disappointing and such activities as have been claimed have usually been very small and near the limit of accuracy of the particular polarimeter used. There have been other similar concepts involving separation of diastereoisomers, crystallization of one chiral molecule under special conditions, but they are all most unconvincing. In particular they almost embarrassingly fail to acknowledge that all the conditions supposedly present in our primeval environment are almost ideal for racemization processes, the very reverse of the desired phenomenon. Accordingly much attention has been given to the possibility that resolution of racemic compounds occurred in the post-life period by chance selection of specific macromolecular complexes such as helices, and the development of some type of autocatalysis to account for the continuing production of resolved material. These arguments, although superficially more attractive, are in many ways even less convincing partly because the need to explain resolution has to a large extent disappeared once the living system has been created, and partly because the choice of L-amino acids and D-ribose is assumed in the first place to have been a purely chance phenomenon. We can see at this stage the beginning of a major weakness in the whole chemical evolutionary theory, namely that chance has to be invoked not once in the abiotic sequence but many times.

(7) The next stage is a little less clear, but the general assumption is that the nucleic acids and proteins came together in profitable unison to form a particle (nucleoprotein) perhaps somewhat analogous to a virus. This further became associated with other proteins which could act in association with suitable substrates as energy-providing enzymes, and with a membrane which would bind the whole thing together into a stable self-duplicating system. This would be the first primeval cell. The subsequent evolution of such a cell could then proceed, perhaps initially by mutations then by symbiosis with other simple macromolecular units to form a more complex cell and ultimately the pro- and eukaryotic cell with which we are familiar. The subsequent development of both types of cells would then proceed by the more or less accepted steps of biological evolution. There are, of course, several variations on this theme, but the final result is much the same and the mechanisms only vary in detail. There is little doubt that ready acceptance of theories of this type has been much encouraged by the remarkable developments in molecular biology which have occurred over the past two decades. In particular the discovery of the precise nature of the genetic processes, the relationships between nucleic acids and protein biosynthesis, and the fact that many viruses, which in a sense have some

features of a living system, consist of nucleic acid and protein alone. Indeed analysis of the life types extant on the planet reveals a sort of evolutionary pathway from the simple nucleoprotein viruses through more complex viruses which may possess lipid or lipo-protein membranes through simple prokaryotic cells such as the minute psittacosis bacterium; the somewhat larger rickettsia and more familiar bacteria; algae; and eventually to the complex eukaryotic cell system and its subsequent differentiation and integration to tissues and organizations of great beauty and complexity. It is a relatively easy step, therefore, to extrapolate backwards from the viral particle to its macromolecular components and ultimately to the simple monomer units themselves.

In this particular sequence of part chemical and part biological evolution the greatest mental hurdle involves the transition from a nucleoprotein with no doubt (like a virus) a potential reproductive capacity to the simplest of prokaryotic cells which will contain all the requisite apparatus for cell division bound by a membrane into a tight efficient functioning system. There have been many attempts to duplicate *in vitro* the formation of this type of system and these have involved especially the encirclement of macromolecular (proteins, nucleic acids) complexes with membranes or boundaries of various types.

COACERVATES AND MICROSPHERES

When aqueous solutions of certain polymers are concentrated they frequently separate spontaneously into two phases, one of which will have a high polymer concentration and the other a low concentration. This phenomenon is known as coacervation and the droplets which frequently form and constitute one of the phases are known as coacervate droplets or coacervates. This phenomenon has been used by Oparin in particular to suggest a possible mechanism whereby the first primitive cells (called protobionts) might have formed. This theory envisages that droplets of this type trap both macromolecules such as proteins and also perhaps small molecules such as α-amino acids and monosaccharides. If the trapped protein has catalytic activity the droplet could produce a metabolic unit taking in, say, glucose and excreting a glucose metabolite. Oparin and his co-workers have prepared coacervates from aqueous solutions of polysaccharides, polypeptides and nucleic acids with volumes ranging from 10^{-8} to 10^{-6} cm^3 and with polymer concentrations ranging from 5 to 50%. The droplets contain a boundary membrane, and in an interesting series of experiments it was shown that glucose-1-phosphate could be converted into a starch which appeared in the droplets when they contained the enzyme glycogen phosphorylase.

Similarly if the droplets also contained the enzyme maltase, then the starch produced was broken down into maltose which was excreted into the aqueous medium.

More than 200 different coacervate systems have now been prepared with diameters ranging from 0·5 to 640 μm. They have been shown to be capable of absorbing enzymes and so acting as centres for many types of enzyme-catalysed reactions including hydrolysis, oxido-reduction and synthesis. Most coacervates are, however, quite unstable and eventually settle out of the aqueous environment into thin films. Some droplets claimed to be stable over periods of up to two years have, however, been prepared from polysaccharides or DNA and histones with an included enzyme polyphenol oxidase which catalyses the oxidation of tyrosine and pyrocatechol. The stability of the drops was ascribed to the quinones produced during the enzymic reaction.

Fox and his co-workers have produced an alternative source of possible primordial membranes which they term microspheres. They allowed hot aqueous extracts of their synthetic proteinoids (produced by heating amino acid mixtures) to cool slowly and uniform spherical droplets separate with diameters about 2·0 μm. These droplets appear to be stable between pH 3 and 7 and also when seen at appropriate pH values possess (generally two) membranes. The membranes have semi-permeable properties and will contract or expand when placed in solutions of differing ionic strength. They also have some small catalytic activity on the rate of hydrolysis of ATP, although the kinetic results must be interpreted with care. The microspheres also undergo splitting in a manner reminiscent of cell division, when set aside with solutions of magnesium chloride, and if allowed to stand in contact with their proteinoid precursors they tend to form buds of new microspheres, presumably by some initial accretion process of new polypeptide material. They form an interesting model of a possible primordial cell membrane but contain no lipid or nucleic acid.

There are of course many other possible ways in which a hypothetical membrane might have formed in a primitive organism. They include direct formation of membranes from a collapsed lipid protein monolayer. Examples of such systems have been demonstrated by Calvin and others and they have the advantage of being composed of the sort of materials that characterize some membranes of some simple cells and have analogies with some more complex viruses and rickettsial organisms. There is of course no reason why lipid droplets stabilized in suspensions or emulsions by substances such as proteins and nucleic acids should not also form suitable model systems. Chemical reactions in such systems especially have been reasonably clearly defined. The reaction kinetics vary according to the droplet size. Initially they presumably would not contain a major aqueous phase (this could in some circumstances be an advantage), but there seems to be no strong objection to one appearing by some simple physicochemical variation of the aqueous environment and it is perhaps surprising that so little work has been carried out with this system which in many ways has been exhaustively studied for other purposes.

In addition to the pursuit of a discrete boundary membrane some attention has been paid especially by Bernal and co-workers to the possibility of orientation of molecules and macromolecules after adsorption on suitable natural inorganic materials (clays, etc.) in a manner which would lead ultimately to their organization into a chemically evolved chemosynthetic unit. Such ordered units, however, would still ultimately require the protection of a boundary membrane before they could escape the somewhat limited environment.

THE FINAL ASSEMBLAGE

In the preceding sections we have seen how it is possible to create by numerous ingenious *in vitro* model experiments most if not all of the chemical parts which are minimal basic requirements of the simplest bounded living system. These experiments make many assumptions about the physico-chemical nature of the primeval planet's environment and also claim abiotic synthesis for what has in fact been produced and designed by highly intelligent and very much biotic man in an attempt to confirm ideas to which he was largely committed in the first place. These points aside, there still remains the problem of how and especially in what specific sequence the various parts of our ultimate abiotic system came to be finally assembled.

Oparin has suggested that the first cells which he calls protobionts may have arisen from proteins alone and that the genetic material developed at a later period. These particular concepts were first mooted at a time when the chemistry, biochemistry and functions of the nucleic acids were little understood. More recently others have emphasized the prime importance of nucleic acids and with increasing knowledge of the molecular genetic processes this particular suggestion has perhaps most supporters.

There are many advantages in first producing a protein. There becomes immediately available enzymes capable of augmenting various types of organic chemical reactions. In particular they offer an immediate route to large molecular weight nucleic acids by polymerization of mononucleotides and, by degradation of monosaccharides, a continuous source of energy. On the other hand, the broad mechanism of protein biosynthesis in all types of cells is known to be roughly similar and to involve essentially transcription of a base triplet code in a genetic nucleic acid through a mRNA with the aid of small tRNA molecules, so that the protein ultimately produced by the cell is directly related to the nucleic acid structure. In this sense therefore the nucleic acid dominates the protein synthesis. These points tend to favour primary production of nucleic acids. At the same time nucleic acids offer the only route to cell division and replication. A major problem here, however, is the extreme difficulty of producing an enormous macromolecular nucleic acid by abiotic methods. There are certainly no known ways of getting anywhere near the enormous chain lengths required by even the simplest living system by such methods. In our opinion if one accepts these theories at all, perhaps the most convincing system would involve the initial production of protein capable of catalysing the polymerization of mononucleotides to nucleic acids. These nucleic acids would exist in numerous sizes, some very large, some very small, and there seems to be no reason why the smaller units should not by base pairing line up along the larger units in much the same way as the tRNA molecules line along the mRNA molecules in the ribosomes. This could be the beginning of a type of protein synthesis which would mechanistically be directly related to current protein biosynthetic processes.

The proteins produced by such a mechanism would then presumably by chance have to possess catalytic properties which would serve to conserve and extend both the nucleic acid replicating process, the protein synthesis and cell membrane formation, and operate an energy producing metabolic sequence of reactions at the same time.

Further discussion, critical analysis and summary of some of these concepts appears in the final chapter of this book.

SUGGESTED FURTHER READING Chapter 8

Books

Bernal, J. D. (1967). "Origin of Life". World Publishing Co., Ohio.

Buvet, R. and Ponnamperuma, C. (Eds). "Molecular Evolution I: Chemical Evolution and the Origin of Life". North-Holland, Amsterdam.

Calvin, M. (1969). "Chemical Evolution". Oxford University Press.

Fox, S. W. (Ed.) (1965). "The Origins of Prebiological Systems". Academic Press, New York.

Haldane, J. B. S. (1929). "Science and Human Life", p. 149. Harper Bros. New York.

Kimball, A. P. and Oro, J. (1971). "Prebiotic and Biochemical Evolution". North-Holland, Amsterdam.

Margulis, L. (Ed.) (1970). "The Origins of Life". Gordon and Breach, London.

Oparin, A. I. (1924). "Proiskhozdenie zhizny". Izd. Moskovski Rabochii.

Oparin, A. I. (1936). "The Origin of Life" (translation of Russian edition by S. Margulis). MacMillan, London.

Oparin, A. I. (1968). "Genesis and Evolutionary Development of Life". Academic Press, New York.

Ponnamperuma, C. (Ed.) (1972). "Exobiology". North-Holland, Amsterdam.

Rutten, M. G. (1971). "The Origin of Life by Natural Causes". Elsevier, Amsterdam.

Articles

Fox, S. W. (1965). Experiments Suggesting Evolution to Protein. *In* "Evolving Genes and Proteins" (Eds V. Bryson and H. J. Vogel), 359–370. Academic Press, New York.

Fox, S. W. (1968). Spontaneous Generation, the Origin of Life and Self Assembly. *Current Modern Biol.* **2,** 235–240.

Fox, S. W. (1972). "Molecular Evolution to the First Cell". Plenary Lecture at IUPAC International Symposium on Chemistry in Evolution and Systematics, Strasbourg, France, July 3–8, 1972.

Haldane, J. B. S. (1929). "The Origin of Life". *Rationalist Annual.*

Lemmon, R. M. (1969). "Chemical Evolution". *Chem. Rev.* **70,** 95–109.

Miller, S. L. (1953). A Production of Amino Acids Under Possible Primitive Earth Conditions. *Science, N.Y.* **117,** 528–529.

Miller, S. L. (1959). Formation of Organic Compounds on the Primitive Earth. *In* "The Origin of Life on Earth" (Ed. A. I. Oparin), 123–135. Pergamon Press, Oxford.

Orgel, L. E. (1968). Evolution of the Genetic Apparatus. *J. mol. Biol.* **38,** 381–393.

Oró, J. and Yang, C. C. (1971). Synthesis of Adenine, Guanine, Cytosine, and Other Nitrogen Organic Compounds by a Fischer-Tropsch Process". *In* "Molecular Evolution" (Eds R. Buvet and C. Ponnamperuma), 152–167. North-Holland, Amsterdam.

Oró, J., Ibanez, J. and Kimball, A. P. (1971). The Effects of Imidazole, Cyanamide and Polyornithine on the Condensation of Nucleotides in Aqueous Systems. *In* "Molecular Evolution" (Eds R. Buvet and C. Ponnamperuma), 171–179. North-Holland, Amsterdam.

Ponnamperuma, C. and Gabel, N. W. (1968). Current Status of Chemical Studies on the Origin of Life. *Space Life Sci.* **1,** 64–96.

Sagan, C. (1972). Interstellar Organic Chemistry. *Nature, Lond.* **238,** 77–80.

Shaw, G. *et al.* (1954–1972). Series of Papers "Purines, Pyrimidines and Imidazoles". *In J. Chem. Soc.*

Sylvester-Bradley, P. C. (1970). Environmental Parameters for the Origin of Life. *Proc. geol. Assoc.* **82,** (1), 87–136.
West, M. A. and Ponnamperuma, C. (1970). Chemical Evolution and the Origin of Life: A Comprehensive Bibliography. *Space Life Sci.* **2,** 225–295.

9

The Precambrian Fossil Record. I

INTRODUCTION

Rocks are formed in many ways. Some by great catastrophic processes—the sort of events which we see in extant volcanoes, spewing out their great lava streams of molten rock, later to solidify. These are igneous rocks. There are, however, other types of rocks. They are formed in a more subtle, gradual manner by the slow inexorable deposition over many millions of years of small particles of mineral matter, at the bottom of seas and lakes. These are the sediments. They ultimately become compressed, they harden, they slowly become rocks. During the sedimentation process not only are mineral particles deposited but also any other particles which may be around and which can survive the rigours of the environment and the process of the sedimentation itself. Such include organic particles which must in particular survive the massive hydrolytic forces occasioned by the vast amounts of water to which they will be subjected over great periods of time. They must also survive the initial attack of micro-organisms. Few organic materials possess such survival properties. Almost the only materials known which do have such properties which are retained over enormous lengths of time are the pollen or related spores: at least, not the whole pollen or spore but the extremely resistant (to the sort of processes outlined above) outer coats of such organisms (Fig. 9.01). They survive because they are composed of the substance sporopollenin which has been shown (Brooks and Shaw, 1968) to be a class of compounds produced by the oxidative polymerization of carotenoids and carotenoid esters.

They may survive, however, in different forms. On the one hand they may retain almost fully the morphology of the original pollen or spore.

On the other hand they may be partly broken by natural, chemical or physico-chemical abrasive forces into fragments which may nevertheless be recognizable as parts of a spore coat, or into an amorphous structureless material. In the latter case, and generally this type of material, termed kerogen, represents the major amount of all sedimented organic

Fig. 9.01. (a) *Lilium henryi* pollen grain × 350; (b) *L. henryi* pollen surface × 1700; (c) *Lilium longiflorum* pollen grain × 350; (d) *L. longiflorum* pollen surface × 1700.

matter, the main hope of assigning a specific biological origin to the material rests in chemical analysis and subsequently relating the particular sporopollenin structure obtained to the type of microorganism from which it was originally derived. Although progress in this direction is necessarily slow there is considerable hope that it may be possible to establish taxonomic relationships of this sort between the chemistry and the morphology of the living system.

In Chapter 8 we outlined various theories concerned with the possible abiogenic synthesis of chemical substances. These have been claimed to represent models of reaction sequences which took part in a *de novo* production of the living system. Many workers in this field, however, have not always been at pains to correlate their *in vitro* results with the evidence of the geological record or to examine the type of biological material which was in existence at the time when the events outlined in the theories are claimed to be occurring. Thus it is by no means certain that the earliest of detected micro-fossil organisms were prokary-

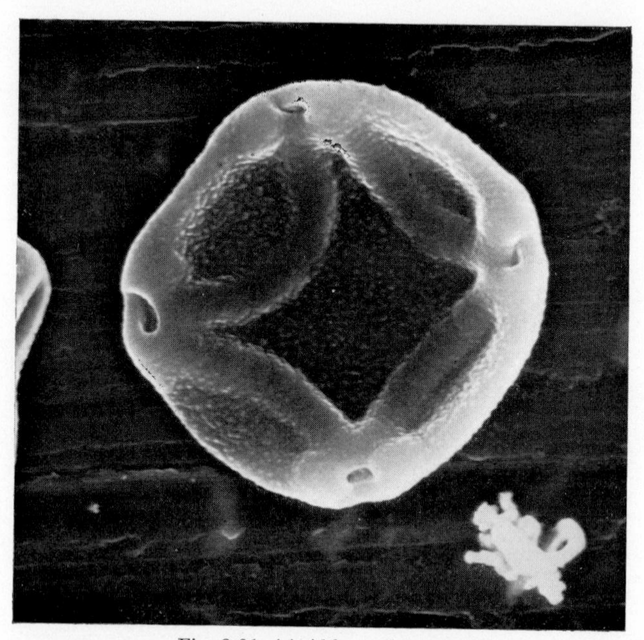

Fig. 9.01. (e) Alder pollen grain

otic cells as is frequently assumed; indeed the evidence equally favours the very early occurrence of eukaryotic cells (Chapters 7 and 12). There are, of course, many other records in the earth, not always of a biological nature, which are nevertheless of equal importance in assessing past environments and events which have been of great importance to the living system's survival, evolution and structure.

The presence of fossil corals, for example, or shells from marine organisms in limestones indicates that they were deposited on what was once a sea floor. In many areas of the world, old lava flows reflect the eruptions of ancient volcanoes, and record areas where volcanic vents were once active. Areas of rock-salt found in many Middle East deposits indicate the former existence of inland seas that ultimately evaporated. Coal seams, the compressed and chemically altered remains of large accumulations of vegetative plant material, provide evidence for the presence of widespread swamps, large forest lands and extremes of luxuriant vegetation some 300 million years ago. Smoothed and striated rock surfaces associated with beds of boulder clay indicate the passage of glaciers or ice-sheets in the area. Even the nature of the weather in geological times can sometimes be determined from rock structures. A brief rain shower falling on a smooth surface of fine-grained sediments covers the surface with tiny imprints known as "rain prints" which can be observed in geological samples. Suncracks found in ancient mud flats of tidal reaches or flood plains resulted when mud dried up and shrank, causing cracks to appear, and these subsequently filled with sand and the sediment records the "fossil weather" of that particular era. Observations of this nature indicate that wind, rain, sunshine and even gravity have always been the same as they are today. But the distribution and change of climate over the earth's surface has varied in an astonishing manner from area to area at different times.

Although there is an impressive amount of geological data now available, to give us an insight into much of the earth's history, it is only during the last few years that a search has been made throughout the whole geological column for residues of biological matter and an attempt made to determine when living systems first occurred on earth and to follow the subsequent development of these systems from the evidence of the rocks.

THE GEOLOGICAL COLUMN

Geological history is normally represented by the time scales of the geological column (Figs 9.02 and 9.03). For stratigraphical purposes the scales are based on geophysical (lithostratigraphy), geochemical or biological data (biostratigraphy). The age of rocks is determined by measurement of radioactive decay processes taking place within them (see Chapter 2). Chronological scales are used for the sequential collecting of events in time, rather than as a direct measure of the time interval since the event. Such chronological events can result from changes in solar or cosmic radiation, changes in the atmosphere, climatic changes, geological changes including volcanic activity and reworking processes on the earth's surface and also magnetic reversals. Events taking place within the earth's crust may also be studied by observations of changes in the composition of rocks during the various geological events. Geological events can also be related to biological events which mirror the origin and the development of organisms throughout the different periods of the earth's history.

Evolutionary changes do not appear to proceed at a constant rate; they are not fully understood and cannot be used as a measurement of

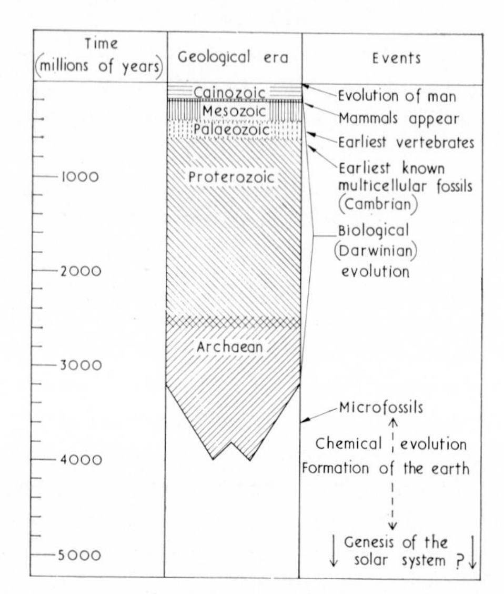

Fig. 9.02. The Geological time scale (after Calvin, 1969)

Eras: Cainozoic, Mesozoic, Palaeozoic

Duration of earth time

Oldest rocks dated 3900 million years.

Origin of earth 4000-5000 million years.

Periods	Epochs	Millions of years: Duration	Millions of years: Before the present
Cainozoic	Pleistocene	2	
	Pliocene	5	
	Miocene	19	
	Oligocene	11	
	Eocene	16	
	Palaeocene	12	65
Cretaceous		70	
Jurassic		55	
Triassic		35	225
Permian		55	
Carboniferous		65	
Devonian		50	
Silurian		35	
Ordovician		70	
Cambrian		70	570

Pre-Cambrian time

Fig. 9.03. The Phanerozoic time scale (after Calvin, 1969)

time. Biostratigraphical methods are used largely to classify Phanerozoic rocks into a time scale of Eras (long intervals) and Periods (shorter intervals) (Fig. 9.03).

During the nineteenth century, geologists were able to extend the geological (stratigraphical) column from the bottom of the Mesozoic to the base of the Palaeozoic by careful study of the fossil fauna of certain restricted areas, such as Central Wales (W. Smith, 1815). The fossil types were used as a basis for subdivision of the various parts of the stratigraphical column. In turn fauna types occurring in the restricted areas were also subsequently found in other rocks and this led to an extension of the classification to other areas. Further studies with both older and younger rocks led to a better understanding of the changes that had taken place in the earth's crust and produced information about the advance and decay of the Phanerozoic seas and mountains which became incorporated in the geological record. Thus many fundamental geological theories and ideas which arise from the nineteenth century pioneering geology of Smith and Cuvier laid the foundation of the geological record of the Phanerozoic.

Unfortunately geologists were unable to apply similar methods to a classification of Precambrian rocks and it has proved very difficult to correlate events responsible for noted Precambrian phenomena and to place them in their correct position within the geological record. These limitations to a complete geological classification are gradually being overcome and recently Russian palaeontologists have been able to subdivide various ancient geological eras, especially the Late Precambrian (Table 9.01). Many terms previously used to classify the divisions of the Precambrian era are also used for the Cenozoic, Mesozoic and Palaeozoic Eras and terms such as Proterozoic (meaning early life), Archeozoic (archaic life), Agnotozoic (unknown life), Azoic (no life), Eozoic (dawn of life), Cryptozoic (hidden life) and even Hadeoic have been suggested and used in various combinations to describe different parts of the Precambrian.

Understanding of the historical geology of the Precambrian has been much extended and transformed in the last decade as a result of various (and more accurate) radiometric age-determinations which have assisted the correlation between fossiliferous and unfossiliferous rocks (Holmes and others, 1966). These relatively new techniques (see Chapter 2) have lengthened by more than 3000 million years the span of geo-time now accessible and have provided new tools to study those changes in the earth's crust whose effects are hardly discernible within the time span of the Phanerozoic. The accurate dating methods, together with associated advances in the study of events taking place in igneous,

Table 9.01. Some divisions of Precambrian time

Range of Limits × 10^8 years	1	2	3	4	5	6	7	8
950	Upper Proterozoic	Upper Proterozoic	Grenville	Riphaean III	Upper Proterozoic	Epi Proterozoic	Upper Proterozoic	Upper Precambrian
↓ 1150	Riphaean 1100	950 ± 90	1100	1000–1100	1150–1200	1000–1100		1000–1100
1400	Lower Proterozoic	Middle Proterozoic	Svecofennid	Riphaean II	Lower Proterozoic	Upper Proterozoic		Middle Precambrian
↓ 1600				1550–1600		1400–1500	1600 ± 50	
1700				Riphaean I		Middle Proterozoic	Middle Proterozoic	
↓ 2100		1700 ± 25	1900	1900–2000	1900	1800–2100	1900 ± 100	1800–1900
2500	Archaean	Lower Proterozoic	Shamvaian	Lower Proterozoic	Archaean	Lower Proterozoic	Lower Proterozoic	Lower Precambrian
↓ 2800	2700	2500 ± 150	2700	2500–2700	2700	1600–2800	2600 ± 100	
↓ 3600	Katarchaen		Kola	Archaen	Katarchaean			

[1] Vinogradov and Tugrinov, 1961. [2] Stockwell, 1963. [3] Sutton, 1964. [4] Obrutchev, 1964.
[5] Tectonic Map of Europe, 1964. [6] Salop, 1964. [7] USSR Committee Absolute Age 12th Congress.
[8] International Geological Map.

Table 9.02. Precambrian time scale

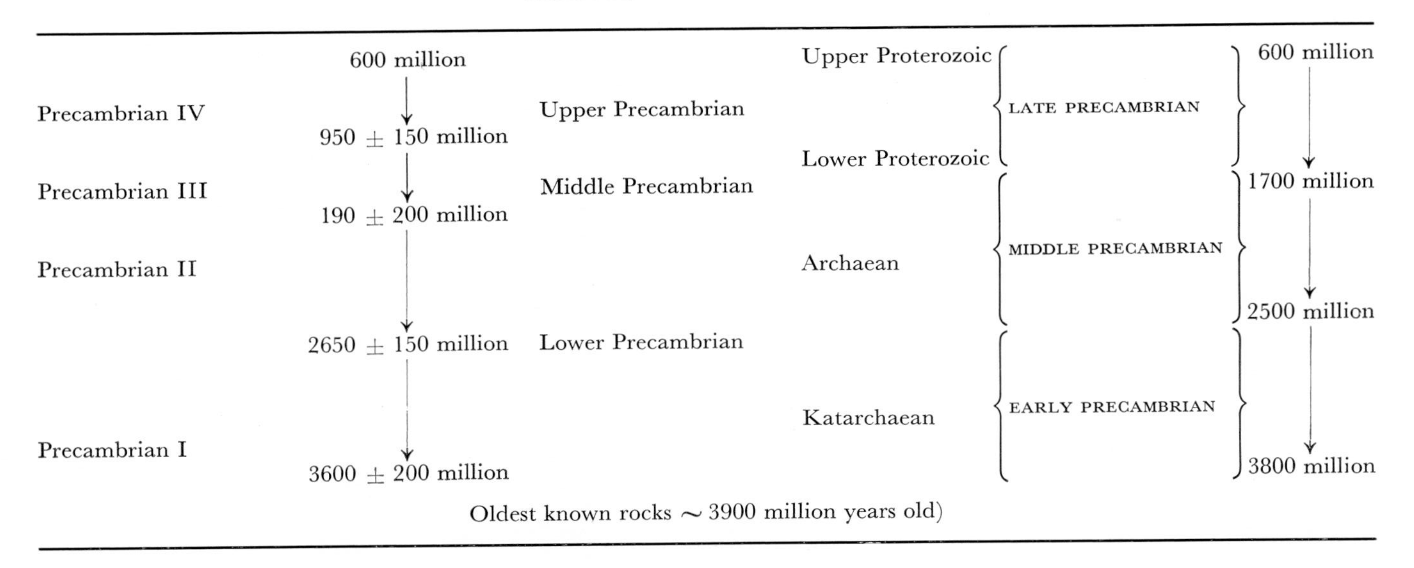

	600 million		Upper Proterozoic	LATE PRECAMBRIAN	600 million
Precambrian IV	↓	Upper Precambrian			↓
	950 ± 150 million		Lower Proterozoic		
Precambrian III	↓	Middle Precambrian			1700 million
	190 ± 200 million		Archaean	MIDDLE PRECAMBRIAN	↓
Precambrian II	↓				2500 million
	2650 ± 150 million	Lower Precambrian			↓
	↓		Katarchaean	EARLY PRECAMBRIAN	
Precambrian I					
	3600 ± 200 million				3800 million

Oldest known rocks ~ 3900 million years old)

metamorphic and related sedimentary rocks, have placed Precambrian geology on a new footing.

There have been two main suggestions for the subdivision of Precambrian time. The first is an arbitrary division into periods of equal length covering the whole of the Precambrian, and the second, and most favoured, uses geological events as time markers. There are many geological events which can be used within the Precambrian and this has resulted in lack of general agreement on the specific choice of chronological boundaries. We propose to use an amended scheme presented by Professor J. Sutton (1965), which in turn is a modification of the recommendations of the Commission which prepared the International Geological Map (Table 9.02).

EVIDENCE FOR LIVING SYSTEMS IN THE PRECAMBRIAN

The last period of 600 million years has been well documented by palaeontologists and other investigators and has led to an improved understanding of the nature and evolution of various biological systems. In recent years there has been a markedly increased interest in Precambrian rocks and their organic contents.

The Precambrian or Archaeozoic Era by definition spans the time (Table 9.02) between the origin of the earth and the beginning of the Cambrian Era and lasted from about 4500 to 600 million years ago. It accounts for approximately seven-eighths of geological time and includes at least 80% (probably the most important period) of the time required for the presumed origin, development and evolution of living systems (see Chapter 4).

The studies of Precambrian organisms and organic matter were pioneered by C. D. Walcott in 1883. Walcott was then the acknowledged leader in the early search for evidence of biological fossils in the Precambrian and was the first to suggest the probable algal origin of Precambrian laminated stromatolites. Both Walcott's work and that of Gruner (1923–1925), who claimed the discovery of filamentous microfossils in Precambrian Cherts, aroused scepticism among many palaeontologists. During the past decade, however, detailed examination of Precambrian rocks has conclusively demonstrated the presence of micro-organism remains and all available data shows that the earlier suggestions of Walcott (1883) and Gruner (1923) which associated cherts, stromatolites and micro-organisms within the Precambrian have been fully confirmed and their general theses now form the basis of many current palaeontological investigations.

ORGANIC MATTER IN SEDIMENTARY ROCKS

Sediments are the main rock types in which organic matter is deposited and chemists are frequently surprised to discover that more than 95% of the organic material present in the earth's crust is in the form of insoluble organic matter, generally called kerogen. The term kerogen is derived from the Greek word *kerós*, meaning oil- or wax-forming, and was originally used to describe such materials (e.g. shales), but more recently the term has been applied in a more general manner to include all the insoluble organic matter which occurs in sedimentary non-reservoir rocks.

Estimates (Hunt, 1962) which had been collected from 200 forma-

Table 9.03. Estimates of the amount of organic matter present in rock specimens (Hunt, 1962)

Sediment	% Organic Matter
Shales	2·10
Carbonates	0·29
Sandstones	0·05

Table 9.04. Estimate of the amount of organic material in the earth's crust

Nature	Amount in tons
Crude oil	$84{\cdot}6 \times 10^9$
Coal	$42{\cdot}3 \times 10^{11}$
Lignite	$12{\cdot}9 \times 10^{11}$
Peat	$1{\cdot}2 \times 10^{11}$
Insoluble organic matter—"Kerogen"	$16{\cdot}9 \times 10^{14}$

tions in 60 major sedimentary basins showed that shales contained by far the highest proportions of organic matter (Table 9.03). From these results and from data concerning the total volume of known sediments and the amounts of shale, carbonate and sandstone, it has been calculated that the total amount of organic matter entrapped in the earth's sediments is about $3{\cdot}8 \times 10^{15}$ metric tons, with by far the majority present as insoluble matter. These insoluble organic deposits are almost 500 times greater than the known coal deposits and about 20,000 times the amount of known petroleum reserves (Table 9.04).

NATURE OF SEDIMENTARY ORGANIC MATTER AND THE ORIGIN OF PETROLEUM

There is now little doubt that the primitive source of natural oil and gas is the organic matter which is formed by decomposition of vegetable (and possibly animal) deposits in subaqueous sediments. The high concentrations of isoprenoid hydrocarbons present in natural petroleum point strongly to its biological origin. Recent work on petroleum constituents has identified relatively high concentrations of pentacyclic triterpanes and steranes, which strongly suggests that these geolipids are derived from the important plant lipids, the pentacyclic triterpenes and sterols probably by hydrogen exchange reactions. Although there is general agreement about the biological origin of petroleum, the nature of the source material and the chemistry of its formation is still debated. Many different naturally occurring substances (lipids, carbohydrates, lignin, proteins, etc.) have been suggested as possible progenitors of petroleum. However, it seems most likely that the insoluble organic matter (kerogen) is the main source material contributing to the genesis of petroleum, and in part to the formation of hydrocarbon gas, although there are known associations of gas fields with coal measures.

Recent work by Brooks and Shaw has produced evidence that sporopollenin derived from terrestrial and marine plants may contribute in many cases significantly to the insoluble organic matter present in sediments. The products obtained from thermally degraded sporopollenin, at various temperatures and different periods of time, include not only aliphatic hydrocarbons, but also aromatic compounds, fatty acids and high molecular weight bitumen-like and residual materials. It has also been shown that a spectrum of products similar to those present in many fossil fuels and source rocks are formed by the diagenesis of sporopollenin present in both fresh and fossil higher and lower plant spores, Tasmanites, dinoflagellates and related insoluble sedimentary organic matter (see later).

OCCURRENCE AND DISTRIBUTION OF PETROLEUM

According to the latest estimates from authoritative sources, the world's proven crude oil reserves were approximately 84,567,350,000 tons at the beginning of 1971. This compares with 47,686,600,000 tons at the beginning of 1965 and 40,787,800,000 tons at the same time in 1961. Estimates of crude oil reserves have therefore more than doubled in the last 10 years, even though accumulative world oil production during this time has reached 16,810,995,000 tons.

The greater portion of the world's proven crude oil reserves (56·5%) are located in the Middle East (Fig. 9.04), which is normally defined for most practical purposes by oilmen as the countries surrounding the Persian (Arabian) Gulf. The Gulf countries are estimated to possess 47,816,730,000 tons of crude oil reserves, or nearly double the known 1961 reserves. Other major areas have all increased their proven crude oil reserves during the last decade: the U.S.A. now possesses 6,100,000,000 tons of reserves and the Sino-Soviet block have approximately trebled their reserves in this period to 13,698,600,000 tons. With the large oil discoveries in Libya, Algeria and Nigeria, the proven reserves of Africa have increased nine-fold to 9,600,580,000 tons.

These figures show that a very large proportion of the known crude oil reserves occur in a relatively few areas which appear to have acquired special geological characteristics. Warman (1971) has summarized these in general terms. The primary requirements are for marine sedimentary basin rocks of Cretaceous and Tertiary age laid down on continental shelves or platforms (Fig. 9.04), rather than in mobile geosynclinal troughs with their attendant rapid rates of deposition and strong deformations. The reasons for these conclusions are that relatively extended periods of undisturbed deposition are required, plus an appropriate degree of deformation to give large structural traps, and sufficient inclinations to allow migration into traps to produce crude oil reservoirs.

Although oil accumulations are generally of Cretaceous/Tertiary age, there are numerous exceptions to this and oil has been found in much older formations. Relatively large reserves have been discovered in Pre-Jurassic reservoirs in the U.S.A., Canada and in the Soviet Union, where there exist large unique areas of Palaeozoic platform sediments. Sediments of Palaeozoic age on continental margins have generally had too turbulent a history to have retained many large crude oil accumulations, whereas similar spreads of relatively undisturbed Palaeozoic sediments do not have a very widespread occurrence on other main continental plates. Until recently, estimates of crude oil (and gas) potential of ancient platforms considered only Palaeozoic deposits, and the unmetamorphosed sedimentary formations of earlier stages in the earth's development were virtually ignored by oil geologists,

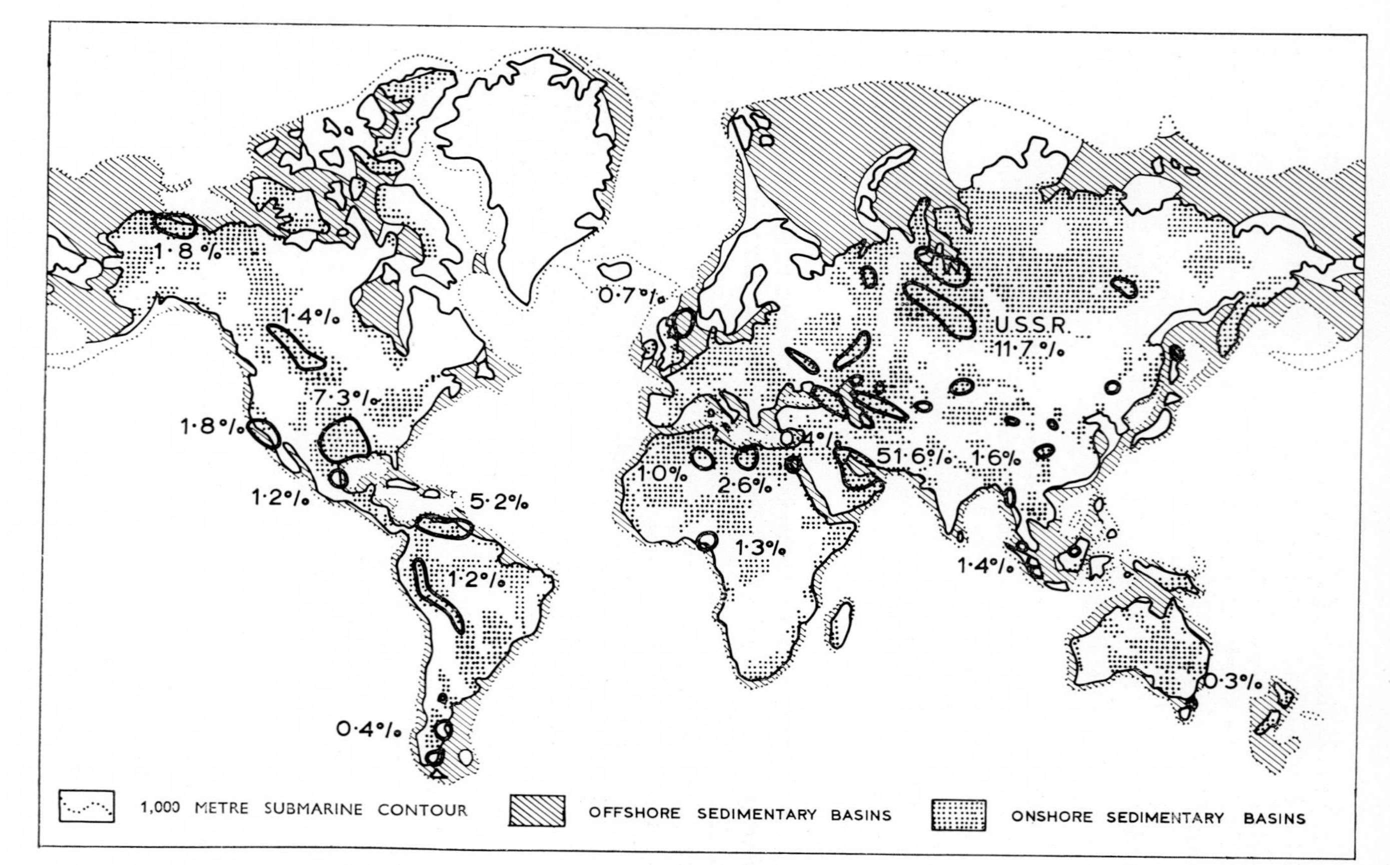

Fig. 9.04. The world's sedimentary basins (estimated crude oil reserves—Warman 1971)

because they believed that Cambrian rocks, and even more so Precambrian rocks, did not contain sufficient quantities of organic matter to allow the formation of significant amounts of hydrocarbons (Fig. 9.05). It was also generally supposed that unaltered or only slightly altered Precambrian deposits do not occur widely, so that the organic matter they may contain can have played almost no significant part in the formation of crude oil. Recent geological and geochemical studies have radically changed these assumptions. At the beginning of the late Proterozoic Era ($\sim 160 \times 10^9$ years ago) the tectonic development of the earth's crust was no different in principle from its development in the Phanerozoic. By this time, the major platforms were already in existence: the Laurentian, Fennosarmatian, Angaran and Sinian in the Northern Hemisphere, and the Gondwana, ancestral Brazilian and others in the Southern Hemisphere, and these early platforms were surrounded by geosynclinal systems and the ancestral oceans. On these platforms were formed sedimentary mantles that have been well preserved. The continental formations are represented by red clastic rocks and by platform basalts, whilst the marine formations consist of sandstones, siltstones, shales and argillites, dolomites and limestones. The greatest thickness of Precambrian rocks occur in the sandstone–shale formations and the thickness of the sedimentary mantle in some places can be more than 5–7 km.

Precambrian sedimentary formations have developed in all the old platforms (Fig. 9.06), and on the Russian and Siberian platforms they are present in especially large volumes. These platform mantles are generally considered to have formed under conditions of uniform and continuous subsidence over very long periods of time. Geochronological determinations have established that the south-eastern part of the Siberian platform, represented by the Maya and Uy series, were formed almost without interruption over a period of about 500 million years. Similar conditions of formation have been established for the eastern part of the Russian platform.

Oil and gas exploration has begun on these ancient platforms and current observations reveal numerous direct indications of the oil and gas potential of Precambrian rocks.

Precambrian deposits that are known to contain oil and gas deposits are shown in Fig. 9.06. Within the Soviet Union, thick bodies of Late Precambrian rocks are included in the structure of several major oil and gas basins. The largest of these are the Central Russian and Volga–Urals basins within the Russian platform and the Angara–Tunguska and Lena–Vilyuy basins of the Siberian platforms. The central Russian basin is filled with a layer of up to 6 km of Upper Proterozoic rocks,

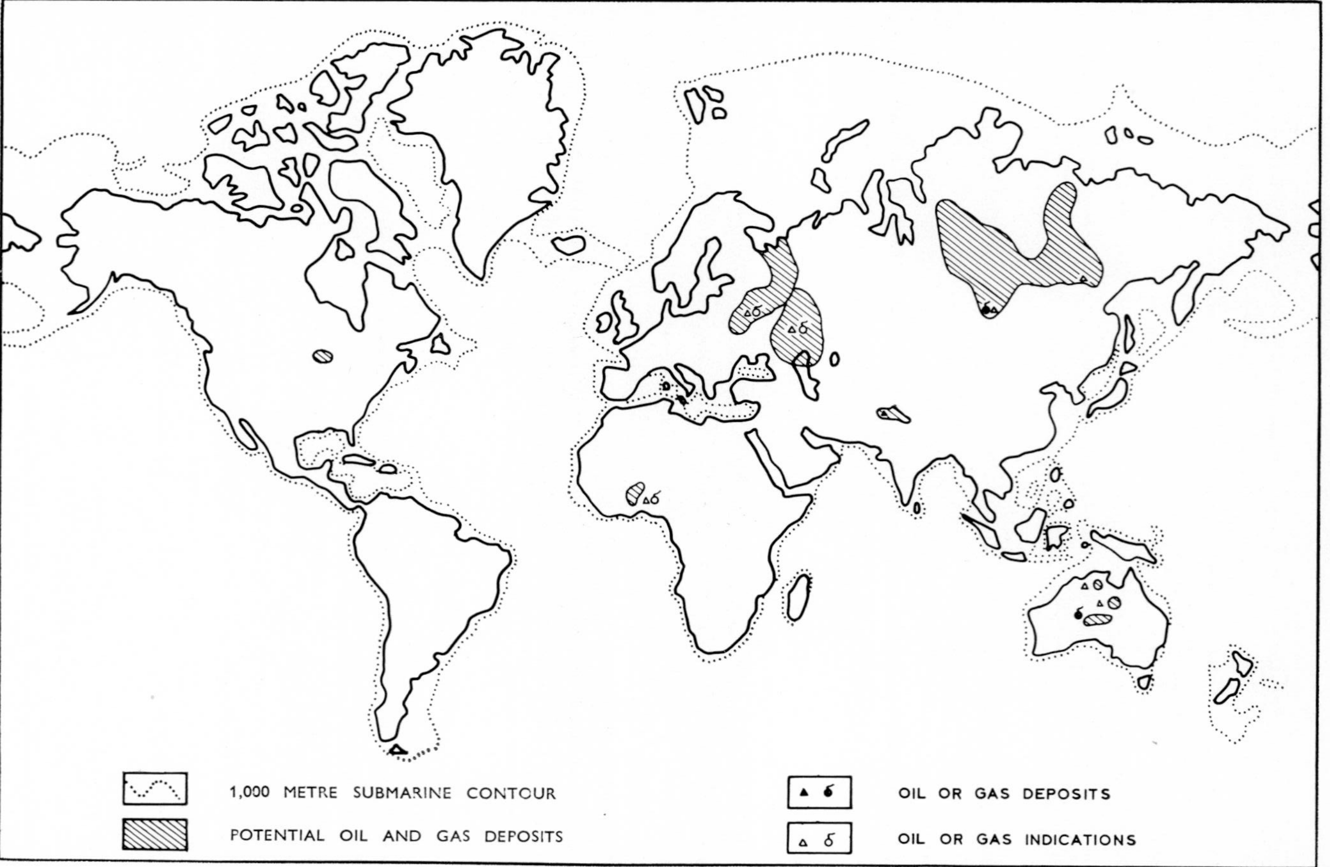

Fig. 9.05. General map showing the oil and gas potential of the world's Precambrian deposits

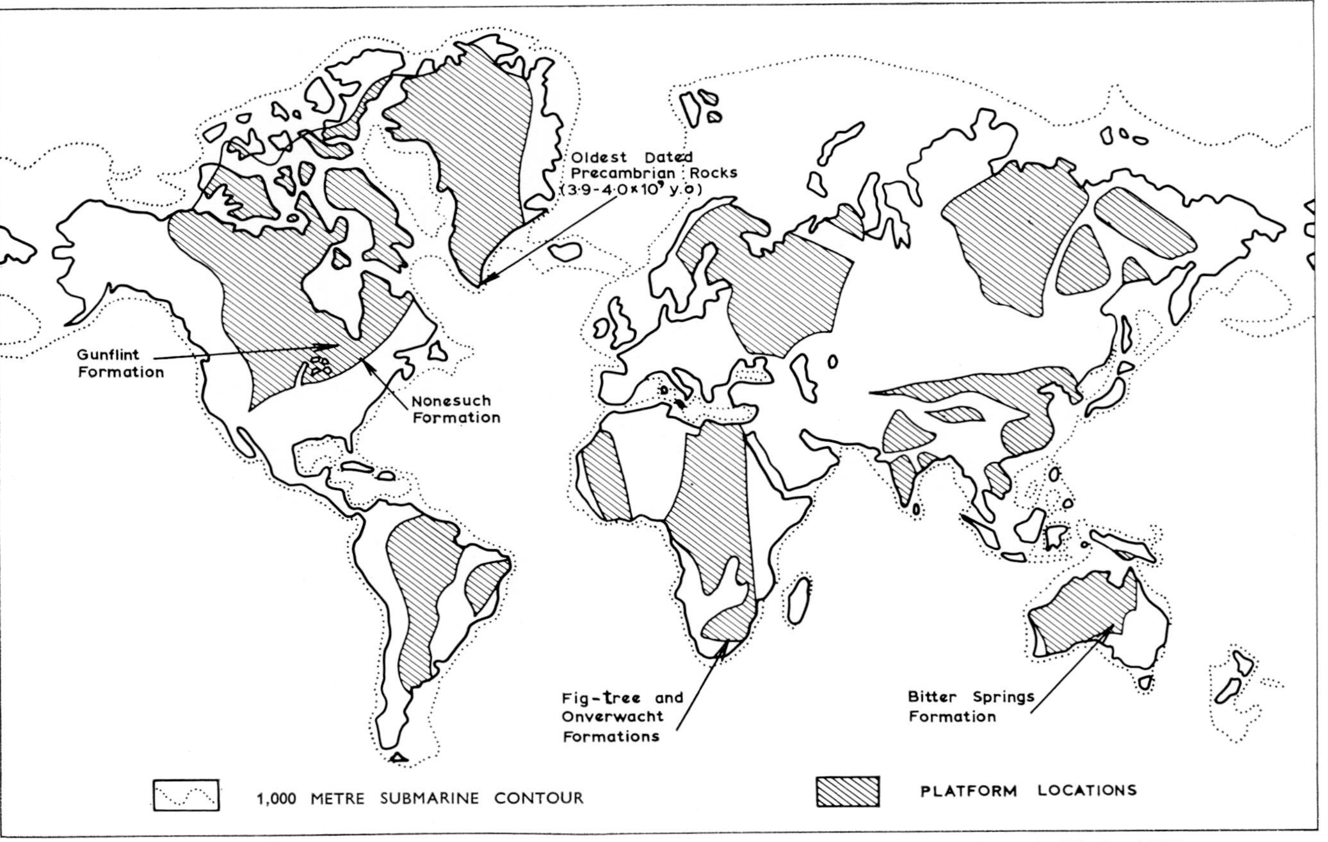

Fig. 9.06. Map showing the major redistribution of old platforms (after A. A. Bogdanov and V. Ye. Khain, 1968)

which are covered by a relatively thin mantle of younger deposits. Underlying these formations is a Precambrian complex—the Wendian Riphean which contains black bituminous shales up to 20 m thick. These sequences, which are characterized by a high content of organic matter (up to 30%), are known as the "Wendian Domanik" Formation. The total thickness of this Wendian Riphean section through the mantle of the Russian platform is estimated as 2–4 km.

Ancient platform formations are widespread on the China–Korean and Hindustan platforms and unmetamorphosed Late Precambrian ("Late-Middle Riphean") Formations of probable marine origin and have thicknesses up to 7–8 km of sandstones and shales in the Sinian deposits in the Yen Shan basin which contain crude oil. Geological structures of the Upper part of the Precambrian of territory in Iran shows great similarity to corresponding sections in West Pakistan. Upper Precambrian unmetamorphosed deposits are quite extensively developed on the African continent, where they form the lower part of the sedimentary mantle in the Taudeni, Congo and Volta synclines. There is little data on oil occurrences in the Late Precambrian deposits of the African platform, apparently because they have yet been little studied, although comparison with known oil occurrences in similar geological structures indicates the possible oil and gas potential.

Thick unmetamorphosed sediments occur in many basins of the Australian platform including the Kimberley, Daly River, the Georgina, the Saint Vincent and the Pirie–Torrens basins. Gas flows have been found from the formations in addition to the presence of asphalts and oil smears and solid bituminous matter in the rocks.

Although ancient formations of considerable thickness are well developed on the platforms of North and South America and Antarctica, there is very little indication of their having any oil potential. The Precambrian Nonsuch Shale Formation ($\sim 1{\cdot}0 \times 10^9$ years old), one of several cupriferrous deposits of northern Michigan, U.S.A., appears to be the oldest formation in the American continent in which crude oil occurs. Any possibility of petroleum migration from younger rocks is negated by consideration of the great thickness of coarse-grained, permeable sediments devoid of organic matter directly overlying the Nonsuch formation, and by the nearly complete absence of crude oil in the 180 m of essentially similar silts and sands directly overlying the basal oil-bearing cupriferrous zone.

In the Precambrian deposits therefore there are vast distributions of unmetamorphosed sedimentary formations which contain relatively large amounts of organic matter including hydrocarbons, and the deposits can and should be regarded as potential material for oil exploration.

MICRO-ORGANISM AND CHEMICAL FOSSILS IN PRECAMBRIAN ROCKS

Criteria used to establish evidence for the presence of former living systems in the Precambrian include:

(1) The presence of morphologically intact micro-organisms.
(2) The presence of soluble and readily extractable organic compounds which have chemical structures and occur in ratios characteristic of compounds one might expect to have arisen from a biological system.
(3) The presence of amorphous insoluble organic matter, which is usually by far the most important quantitative organic constituent of the sediments, and its relationship to materials of known occurrence in living systems.

Precambrian Microfossils

Micro-organisms in Precambrian rocks have been identified as microfossils in sediments as far apart as Northern Ontario, Southern California, Southern Africa, Central and Southern Australia and the Soviet Union. Well-preserved organic micro-organisms, thought to be the remains of algae, bacteria and possibly aquatic fungi, isolated as both spheroidal and filamentous structures, have been described in the Bitter Springs Formation of Central Australia ($\sim 1{\cdot}0 \times 10^9$ years old); the Skillogalee Dolomite of Southern Australia ($\sim 1{\cdot}0 \times 10^9$); the Belt Series of Montana, U.S.A. ($\sim 1{\cdot}1 \times 10^9$ years old); the Beck Spring Dolomite of Southern California ($\sim 1{\cdot}2 \times 10^9$ years old); the Gunflint Iron Formation of Ontario ($1{\cdot}6 \times 10^9$ years old), the Pokegama Quartzite of India ($\sim 1{\cdot}6 \times 10^9$ years old), the Soudan Iron Formation of Minnesota, U.S.A. ($\sim 2{\cdot}7 \times 10^9$ years old) and the Fig-Tree ($3{\cdot}1 \times 10^9$ years old) and the Onverwacht ($3{\cdot}4$–$3{\cdot}7 \times 10^9$ years old) of the Swaziland Group in Southern Africa (see Chapter 10). In addition to these organic micro-organisms, evidence of Precambrian biological activity is evident from stromatolitic structures.

Stromatolites are finely laminated "organo-sediments" mainly composed of calcium carbonate, and result from the accretion of detrital and precipitated minerals on successive layers of micro-organism (mainly filamentous blue-green algae) communities and are laid down in sheet-like mats. As early as 1858, W. E. Logan from a study of *Eozoon canadense* suggested that stromatolite-like deposits were evidence for Precambrian life, and since then many studies and correlations have been carried out on these types of structures. Recently Schöpf (1971) and co-workers have studied the oldest known stromatolitic structures (Figs 9.07 and 9.08) which occur in limestones of the Bulawayan Group of Rhodesia ($\sim 2{\cdot}6 \times 10^9$ years old). They used optical microscope and stable carbon isotope methods and demonstrated that the Early Precambrian stromatolites appear to place a minimum age of

Fig. 9.07. Biogenicity and significance of the oldest known stromatolites. J. W. Schöpf, D. Z. Oehler, R. J. Horodyski and K. A. Kvenvolden (1971). *J. Paleontol.* **45,** 477–485

$2{\cdot}6 \times 10^9$ years on the time of origin of cyanophycean algae which are filamentous and form integrated biological communities possessing an oxygen-producing photosynthesis biochemistry. Identification of well-authenticated stromatolitic structures of presumed blue-green algal origin in limestones, dolomites and cherts from sediments more than 2×10^9 years old in the Witwatersrand System of Southern Africa have also been reported. Cloud (1968), however, questions the reports of stromatolitic structures found in the Fig-Tree Sediments and also comments that the reports of similar structures in the Bulawayan rocks of Rhodesia are neither completely convincing morphologically nor

Fig. 9.08. Biogenicity and significance of the oldest known stromatolites. J. W. Schöpf, D. Z. Oehler, R. J. Horodyski and K. A. Kvenvolden (1971). *J. Paleontol.* **45,** 477–485

unequivocally dated. However, the oldest known calcareous stromatolites in the Bulawayan Group ($2 \cdot 6 \times 10^9$ years old) are now generally considered to have had a biological origin and further evidence for this has been obtained from carbon isotope studies which support an algal origin for these limestones.

Perhaps the best preserved of all Precambrian fossils have been discovered in the black cherts from the Bitter Springs Formation of Northern Territory, Australia. These bedded carbonaceous cherts of the Late Precambrian occur in the Ross River area of Central Australia and contain varied assemblages of exceptionally well-preserved spheroidal and filamentous plant microfossils. The fossiliferous cherts, which occur about 4200 ft beneath members of the uppermost Precambrian Ediacaran fauna, underlie sediments having a Rb/Sr age of 820 million years and they are possibly younger than 1440 million years. It is estimated that the age of the Bitter Springs microflora is approximately 1000 million years. The micro-organisms (Figs 9.09 and 9.10), which are three-dimensionally preserved in a matrix of primary chalcedonic chert, consist primarily of algal biocoenses which probably grew as laminar sheets or mats in the apparently marine environment of the Amadeus Basin, and evidently contributed to the formation of widespread algal stromatolites. Schöpf has identified the dominant micro-organisms of the Bitter Springs cherts as filamentous and coccoid blue-green algae. Of the nineteen species of blue-green algae observed in the microflora, fourteen have been referred to modern algal families. These studies show, therefore, that blue-green algae were highly diversified in the Late Precambrian, and that certain species have experienced little or no evolution, at least in terms of organismal surface morphology, since Bitter Springs time. It is also interesting to note that the blue-green algae (*Myxococcoides minor*) (Fig. 9.10) organically preserved in the chert and isolated by Schöpf from acid-resistant residues shows a granular, reticulate surface texture including similar dimples or depressions and of very similar dimensions to the micro-organisms isolated from the very early Precambrian Onverwacht Series Chert of Swaziland (see Chapter 10). The organisms provide an excellent example of the limited change that has taken place in evolution of living systems during the vast periods of time between these two eras ($\sim 2 \cdot 5 \times 10^9$ years).

Small assemblages of organically and structurally preserved plant micro-organisms have been identified in the silicified dolomites and black cherts from the Late Precambrian Skillogalee Dolomite of the Adelaide Geosyncline of South Australia, which is dated at between 900 and 1050 million years old. The micro-organisms referred to as blue-green algae (*Archaeonema longicellularis* Schöpf) are considered to be shallow water assemblages preserved *in situ* near the sediment–water interface. Although this study described only a small number of micro-organism assemblages geological evidence has correlated the Skillogalee

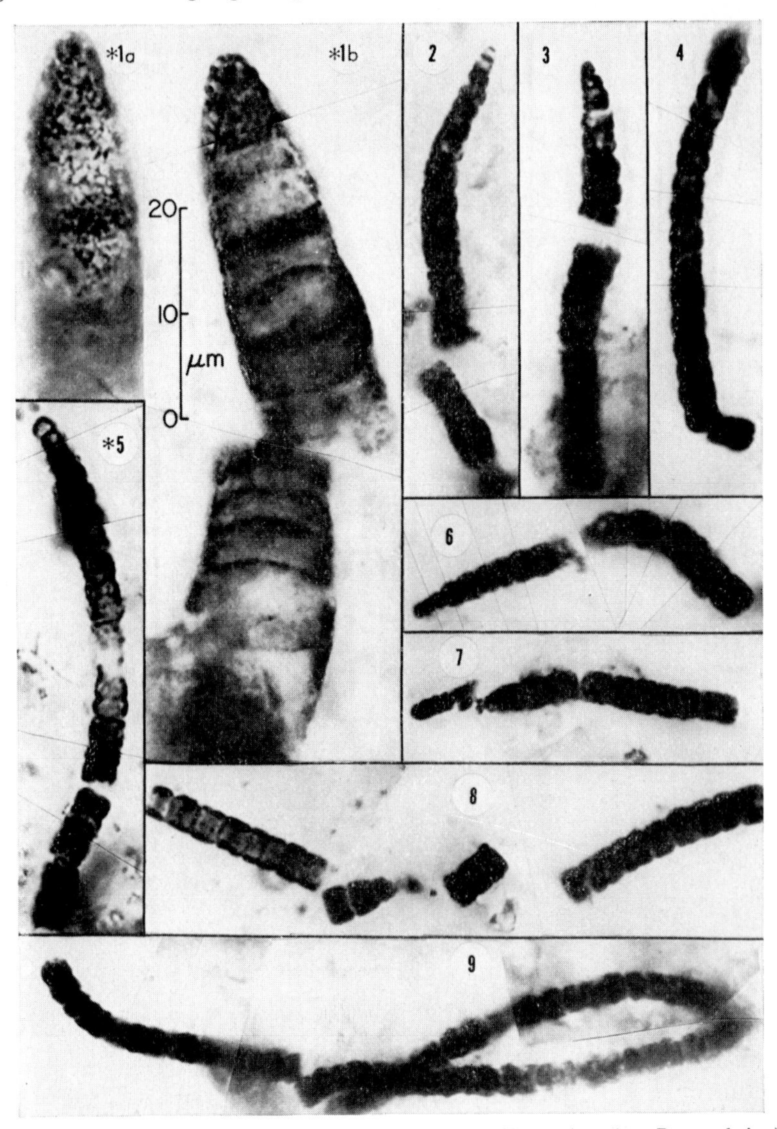

Fig. 909a. New micro-organisms from the Bitter Springs Formation (late Precambrian) of the north-central Amadeus Basin, Australia. J. W. Schopf and J. M. Blacic (1971). *J. Paleontol.* **45,** 925–960

Dolomite with the Bitter Springs Formation of the Amadeus Basin of Central Australia, and the three types of micro-organism present in Skillogalee Dolomite are also present in the slightly older Bitter Springs Formation. Schöpf and Barghoorn point out that the occurrence of

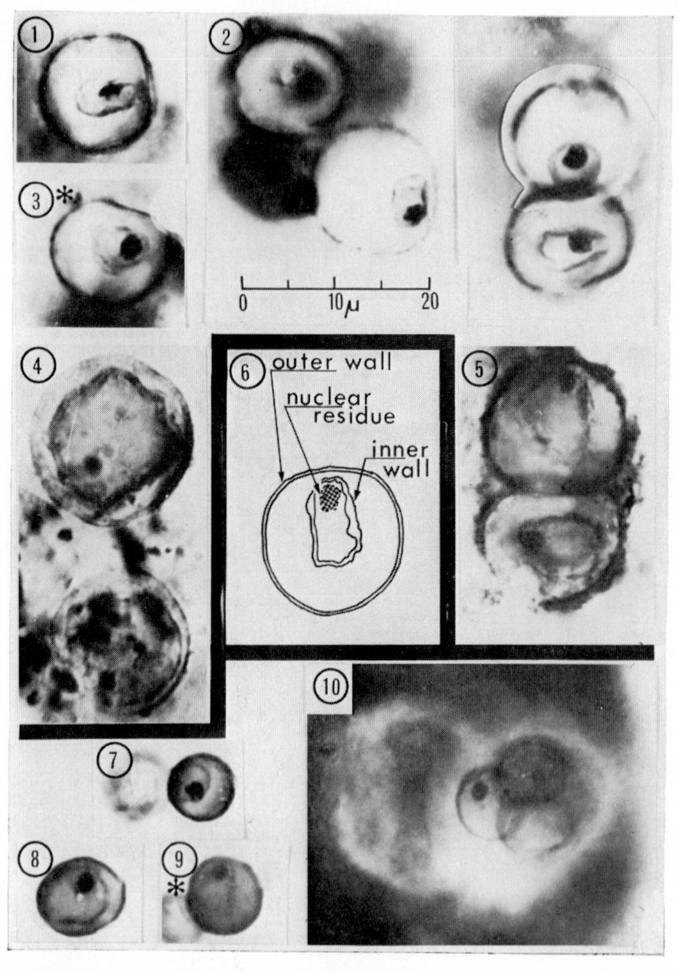

Fig. 9.09b. Microflora of the Bitter Springs Formation, late Precambrian, central Australia. J. W. Schöpf (1968). *J. Paleontol.* **42,** 651–688

structurally and organically preserved algae in the Skillogalee and Bitter Springs Formations and the numerous reports of plant microfossils in other sediments of similar age establishes that photosynthetic organisms were widely distributed during the Late Precambrian. These observations, together with other geochemical and palaeontological evidence, seem to disprove the often held hypothesis that postulates a low oxygen content for the earth's atmosphere during Late Precambrian times.

Perhaps the most studied Precambrian sediments are the black cherts of the Gunflint Formation near Lake Superior which is approxi-

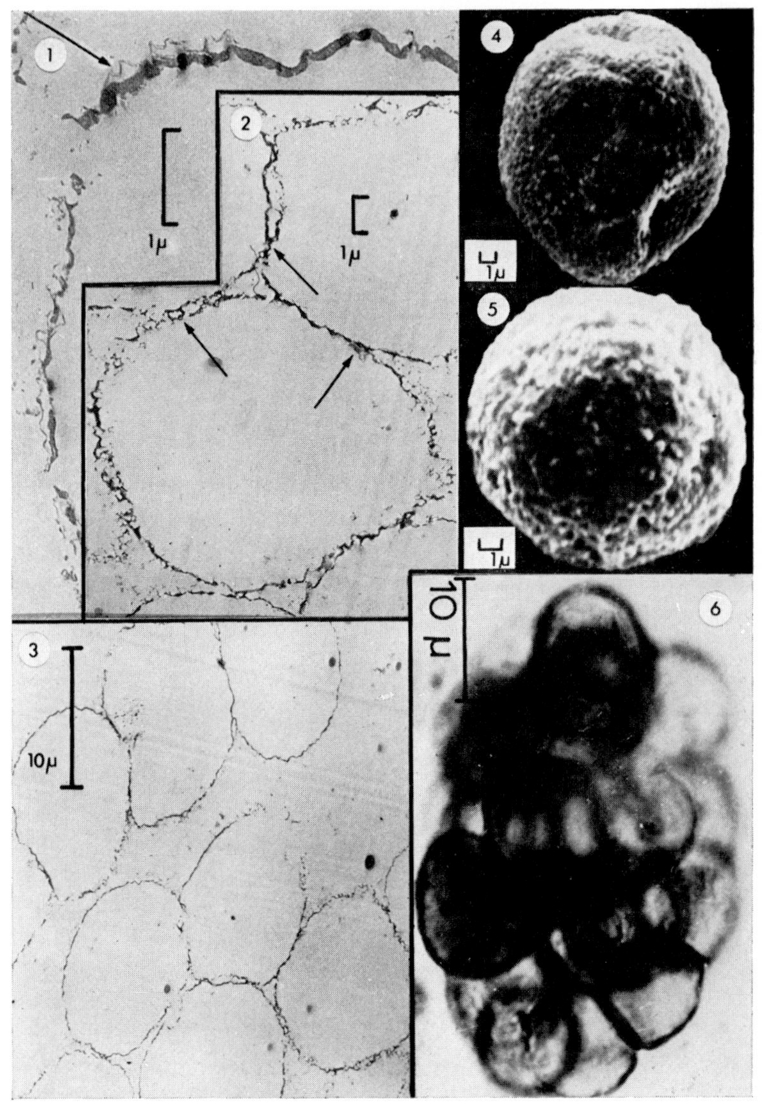

Fig. 9.10. Electron microscopy of organically preserved Precambrian micro-organisms. J. W. Schöpf (1970). *J. Paleontol.* **44,** 1–6

mately 2000 million years old; it is composed of a relatively thick sequence of shales, limestones and cherts, occurring in a latitudinally trending trough along the northern edge of Lake Superior. In the sequence the lowest member of the formation contains the fossiliferous chert horizon which varies in thickness from a few inches to about two feet and is generally stromatolitic throughout its lateral extent of some 150 miles parallel to the sedimentary sequence.

In the eastern part of the formation near Schreiber, Ontario, there are a series of unmetamorphosed cherts which contain well-preserved micro-organisms. In addition to the eastern sections, micro-organisms are found throughout most of the formation even in the more altered sections to the west. Various palaeobiologists have included the Gunflint Chert in their list of "classic horizons" in palaeontology because of the preservation and evolutionary significance of the contained microfossil assemblage. The Gunflint microfossils consist primarily of microscopic algae (Fig. 9.11) which probably formed in laminar sheets or mats at or near the sediment–water interface. The most commonly occurring species in the twelve recognized microscopic plant assemblages are unicellular spheroids and filaments. Barghoorn and Tyler (1965) reported that the ellipsoidal to spheroidal unicellular organisms (called *Huroniospora* spp.) had size distributions from less than 1 μm to about 16 μm diameter, and varied in surface textural detail from psilate to reticulate. This wide size range and different surface detail suggested that the spheroids were of diverse biological affinities and probably related to coccoid blue-green algae and the organic filaments represented a heterogeneous mixture of apparently prokaryotic micro-organisms. The results have been confirmed by electron microscopic examination of the Gunflint assemblages, and from morphological comparisons with extant micro-organisms, it is suggested that these microfossils are related to modern iron-bacteria, such as *Sphaerotilus* and *Siderococcus*.

Schöpf has commented that the occurrence of iron-bacteria in the Gunflint may indicate that the iron-bearing facies of the formation were deposited, at least partially, by biological activity and this could account for much of the so-called reduced iron beds and dispose of the hypothetical reducing atmosphere which is usually invoked to explain these phenomena.

Similarities between the detailed morphology of the organic microfossils with extant blue-green algae, the occurrence of chemicals of known biological origin within the formation, and the $^{12}C/^{13}C$ ratios measured on the organisms and surrounding chert (see later) all serve to indicate the photosynthetic nature of the Gunflint micro-organisms. The

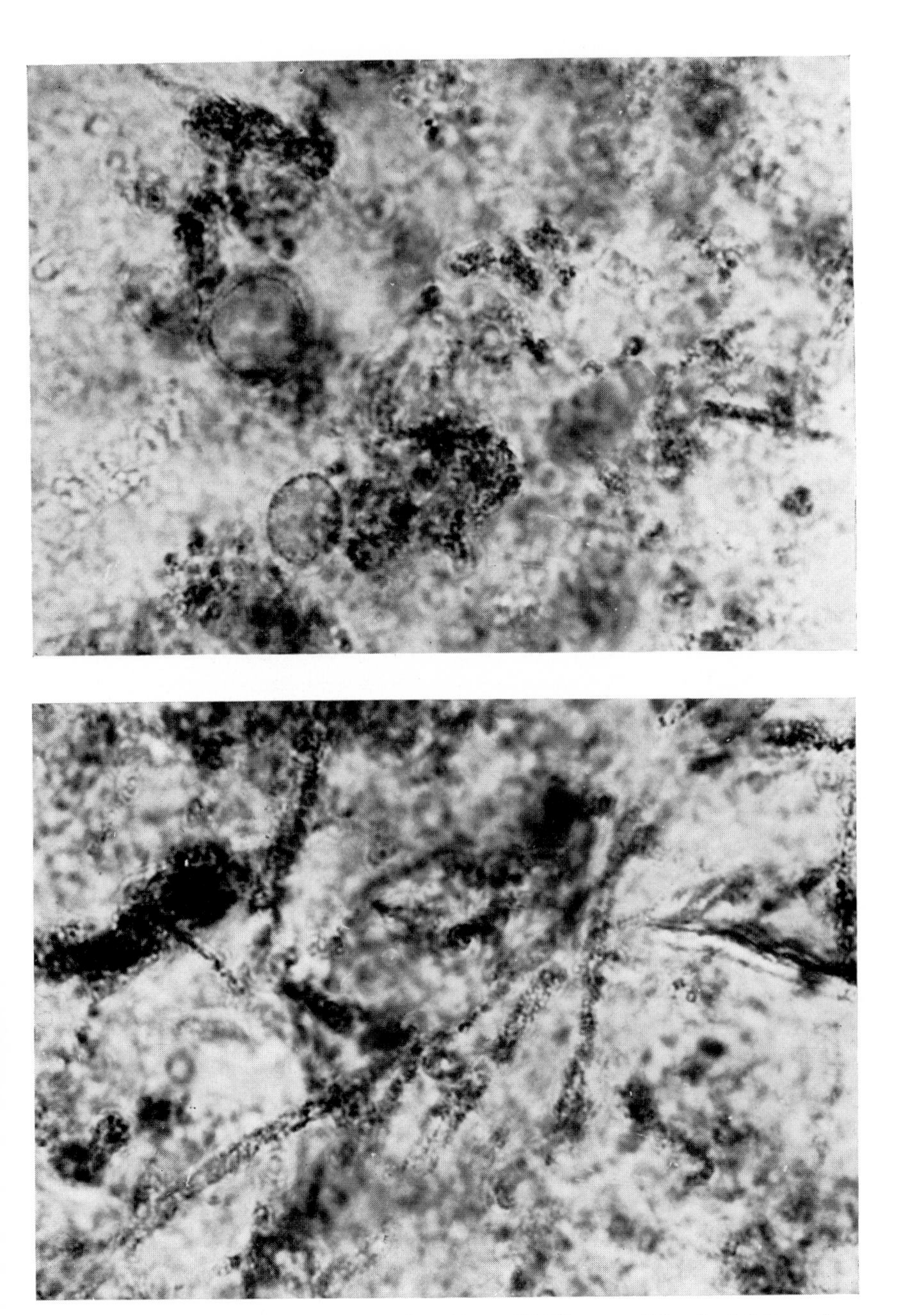

Fig. 9.11. (a) "Huroniospora" from Gunflint Chert. (b) Filaments "Gunflintia" from Gunflint Chert. Photographs by M. D. Muir 1972

presence of photosynthetic micro-organisms in the Gunflint formation ($\sim 2{\cdot}0 \times 10^9$ years old) shows that the environment must then have been oxygenated. Also, since the organisms have been identified as planktonic algae and not just benthic stromatolites, it has been inferred that an ultraviolet absorbing ozone layer (to absorb the high influx of solar radiation) and an oxygen atmosphere, was established at least 2000 million years ago.

The nature of the morphologically intact material which include spheroids and filamentous material varies considerably in rocks of increasing age; the micro-fossil material becomes increasingly less differentiated until in Precambrian sediments such as the Gunflint Formation, virtually the only recognizable bodies are intact walls of micro-organisms. These results reflect the greater stability of the micro-organism wall material and the lesser stability of other materials which make up the greater part of plants.

Plants are mainly composed of cellulose and other polysaccharides and lignin which are readily subject to attack by a wide range of micro-organisms; they are also highly susceptible to chemical hydrolysis and under the normal conditions of sedimentation there will be no lack of water for this purpose. The rate of hydrolysis of such materials will vary with temperature but especially with the pH of the aqueous phase. At low pH values the cellulose component will survive for only a brief period; in alkaline or neutral conditions it will be more stable but would not be expected to survive the enormous time spans into the Precambrian in any but the most minute quantity. White has clearly outlined (Fig. 9.12) how initially cellulose, then lignin are the first substances to undergo decomposition in aqueous media, whereas the outer walls of spores and related micro-organisms have maximum stability under these conditions.

In addition to the preserved micro-fossil (spore) walls in Precambrian sediments, filamentous materials have also been described and this raises the question as to whether these filaments, which survive acetolysis, are also composed of the same chemical material (sporopollenin) as the resistant spore wall material. One perhaps normally thinks of material of this type in terms of a cellulose or polysaccharide structure. It would be unusual if cellulosic filaments are, however, to become suddenly resistant to acetolysis and so to survive into the Precambrian, since they are relatively so easily hydrolysed. Carbohydrates are also less stable than most organic matter, and taking all sources of information about the stability of sugars into consideration, Degans suggests that the bulk of the carbohydrates are eliminated biochemically in the early stages of diagenesis.

It is often suggested that laminar sheets of microscopic planktonic algae, at or near to sediment–water interfaces, formed mound-shaped biothermal masses in shallow waters. These micro-organisms become entrapped and embedded in the colloidal silica of the sedimentation processes, which permeated the algal matrix and filled the cells with silica, leaving the organic cell wall intact. During compaction of the sediment, the amorphous silica recrystallized to form a very fine-grained

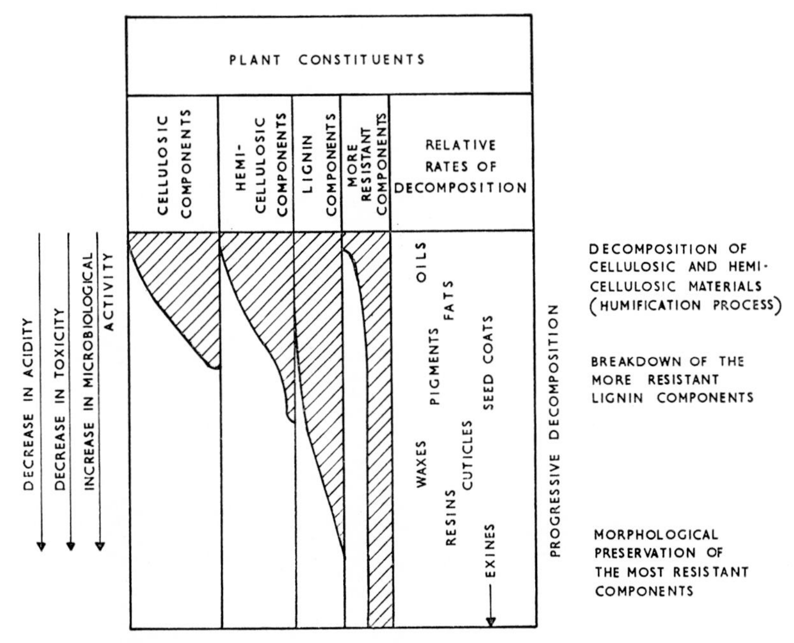

Fig. 9.12. Progressive decomposition of plant components under normal sedimentary conditions (after D. White (1933). *Econ. Geol.* **28,** 556.

quartz which assists in the three-dimensional preservation of the micro-organism.

An alternative explanation suggested by Brooks and Shaw for the survival of acetolysis-resistant filamentous materials in sediments is derived by examination of a process taking place in modern plants. The process of pollen- and spore-resistant exine formation (sporopollenin deposition) involves deposition on a cellulose template of small, sporopollenin-containing (or producing) globules (orbicules); coating of the sporogenous tissue also occurs, in the same manner. This has been observed in many plants and provides the particular tissues with the same resistant coating as the pollen and spores, and makes them inert to microbiological and non-oxidizing chemical attack. Thus the filamentous materials observed in Precambrian sediments may well be preserved together with the "spores" because they have the same

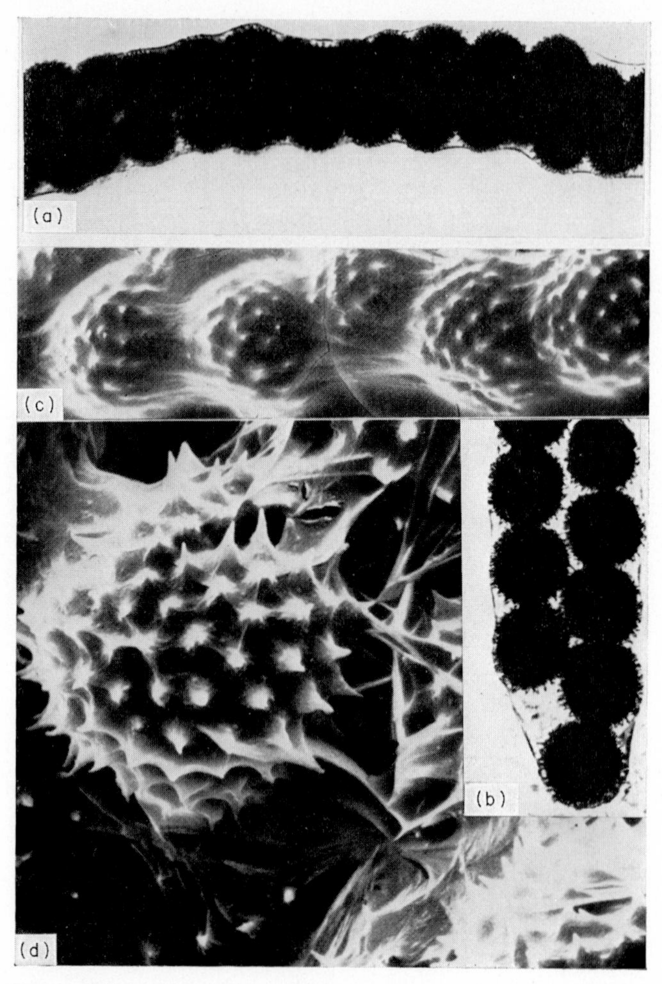

Fig. 9.13. Spores of *Tagetes patula* invested in an extra-tapetal membrane. (a) and (b) are optical micrographs; (c) and (d) are scanning electron micrographs. (By kind permission of J. Heslop-Harrison, 1968.)

resistant coat of sporopollenin. A comparison of photographs (Figs 9.13 and 9.14) of the resistant membrane which invests tapetal and sporogenous tissue and the extra-tapetal membrane which contains the spores in certain Compositae, with photographs of the structurally

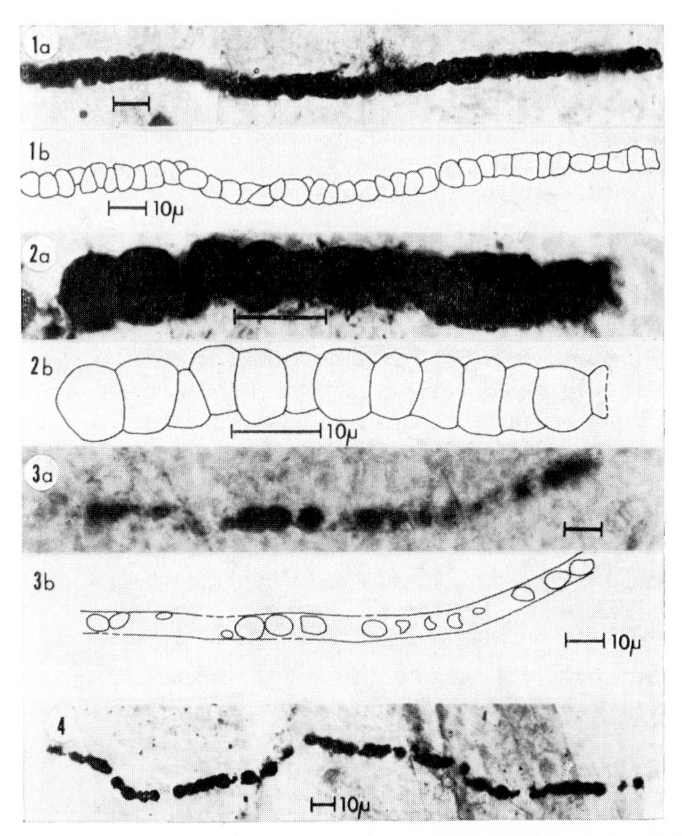

Fig. 9.14. Possible organic sheaths on algal fossils from the late Precambrian of California. (By kind permission of J. W. Schöpf (Gutstaft and Schöpf, 1969).)

and organically preserved micro-organisms of the Precambrian Bitter Springs Formation and Beck Spring Dolomite of Southern California, where the preserved micro-organisms are encompassed in a resistant sheath, show many similarities between the modern and fossil "spores" and their resistant tissues and filaments.

Recent advances in Precambrian palaeobotany have provided a reasonable guide to geological events and to the origin and development of primitive biological systems. Results show that living systems originated prior to the period 3·4–3·7 $\times$ 10^9 years ago (see Chapter 10) and younger Precambrian rocks provide conclusive evidence for the presence of living systems throughout the geological column up to the Cambrian period. The Early Precambrian rocks (3·7–2·5 $\times$ 10^9 years ago) contain fossil micro-organisms which have generally been classified as primitive, unicellular prokaryotic cells with little morphological complexity and as such are considered to have had limited biological diversity. The micro-organisms, which are perhaps too readily equated with bacteria or modern blue-green algae, were apparently capable of photosynthesis, which means that free oxygen was being added to the earth's environment at a very early state ($\sim 3{\cdot}7 \times 10^9$ years ago) and throughout most of geological time.

The Middle Precambrian period ($2{\cdot}5$–$1{\cdot}7 \times 10^9$ years ago) had much stromatolitic algal production and communities were formed in shallow lakes and coastal environments on quite large scales through the earth's surface, whilst planktonic algae appear to have been more abundant in isolated local surroundings. The Middle Precambrian biota consisted mainly of prokaryotic type micro-organisms and chemosynthetic bacteria. Schöpf records the presence of other species, of less certain biological definition, and it is yet to be established whether these organisms were anucleate, eukaryotic or maybe some intermediate stage in cellular organization and development. All these micro-organisms are considered to have been capable of photosynthesis, so the Middle Precambrian atmosphere would contain significant proportions of oxygen.

In the Late Precambrian ($1{\cdot}7$–$0{\cdot}6 \times 10^9$ years ago) there is a marked increase in the diversity of quantity and the microflora present in the rocks. Late Precambrian rocks contain both prokaryotic and eukaryotic micro-organisms, with some showing morphological similarities to extant taxa.

From these observations of Precambrian biota, palaeobotanists have postulated that during the Early and Middle Precambrian slowly evolving prokaryotic micro-organisms existed which were supplanted by advanced mitotic algae about $1{\cdot}5 \times 10^9$ years ago, which resulted in sexual organisms with an alternation of generations. Around the Middle–Late Precambrian boundary came the appearance of eukaryotic cells, which presumably led to the high level of biological organization of biotas near the beginning of the late Precambrian.

Organic Chemicals Present in Precambrian Rocks

In recent years there has been much interest in the isolation and identification of "biological markers" from Precambrian rocks and attempts to relate such compounds as have been obtained to living systems and use them to decipher the time and mode of formation of the earliest forms of life and the assumed transition from "non-living" to "living" chemical systems.

The so-called biological markers are geochemically stable organic compounds, usually lipids, which possess skeletal features and occur in ratios which are characteristic of compounds from known biological sources and can therefore be related to living organisms. Many simple biological markers can be synthesized abiogenically and extreme care must be taken in studies using molecular criteria for hydrocarbon genesis to distinguish between the chemicals produced by non-biological means and those produced by living systems.

Optical activity is commonly associated with molecules of biological origin (Chapter 7) and the biological fixation of carbon in photosynthesis leads to the concentration of ^{12}C at the expense of ^{13}C in photosynthetic organisms. These properties associated with biochemicals are used as parameters to assist in defining the origin of soluble organic chemicals which may or may not have originated from living systems.

Stable Carbon Isotope Studies and Evidence for Photosynthetic Processes

Photosynthesis (Chapter 5) is very characteristic of many living systems, including plants, algae, diatoms, mosses, ferns and higher plants, and also certain micro-organisms such as protozoa (e.g. *Euglena*), and a number of different bacteria (e.g. the purple sulphur bacteria, *Thiorhodaceae*; the purple non-sulphur bacteria, *Athiorhodaceae* and the green sulphur bacteria *Chlorobacteriaceae*). Thus photosynthetic processes not only provide many living systems with the biochemicals necessary for growth and development, but have recently become an important tool in the armoury of organic geochemistry as a parameter to identify geolipids which have been produced via photosynthesizing organisms and can thus be used to distinguish biologically produced chemicals from those which could have resulted from non-biological (abiogenic) reactions.

Organic carbon from plant material exhibits a range of $^{13}C/^{12}C$ ratios, since the biosynthetic pathways in photosynthesis tend to discriminate against ^{13}C and selectively absorb $^{12}CO_2$ rather than $^{13}CO_2$. In effect the plant behaves as an "isotope separator" concentrating the ^{12}C isotope in the plant, and in this manner photosynthesis exerts a major control either directly or indirectly on the distribution of the stable carbon isotopes within the plant. Marine, non-marine and terrestrial organisms show different stable carbon isotope values (expressed as $\delta^{13}C$ in ppm). Plants extract carbon from two main sources, atmospheric CO_2 and molecular carbonate and bicarbonatn from their aqueous environment. The differences in isotope compositioe of the organisms will be influenced largely by the $\delta^{13}C$ value of the carbon source used for photosynthesis.

Absolute determinations of ^{12}C and ^{13}C are not usually measured since natural ^{13}C has a low abundance. Mass spectral analysis of the ratio of ^{12}C to ^{13}C in carbon dioxide produced by complete combustion of the plant material or geological sample gives an indication of the original environment of the material and provides information about the cycle of carbon in Nature. Comparison of this ^{12}C to ^{13}C ratio to that of carbon dioxide produced from a standard carbonate sample (usually PDB-1, a Belemnite from the Peedee Formation, South Carolina) produces the $\delta^{13}C$ values

$$\delta^{13}C\ ^{0}/_{00} = 1000 \times \left(\frac{R}{R_s} - 1\right)$$

where R = $^{13}C/^{12}C$ ratio in sample
R_s = $^{13}C/^{12}C$ ratio in PDB standard carbonate.

Nier and Gulbrandsen (1939) first observed that carbon present in biological systems had a lower $\delta^{13}C$ value than that found in atmospheric carbon dioxide or carbonate minerals. Park and Epstein (1960) showed that the enzyme catalysed CO_2 fixation by ribulose diphosphate (see Chapter 5) used $^{12}CO_2$ 1·017 times faster than $^{13}CO_2$. Graham's Law of Diffusion states that for gas molecules the velocities of isotopic species are proportional to the inverse square root of their molecular weights. Thus for carbon dioxide:

$$\frac{\text{Velocity } (^{12}C\ ^{16}O\ ^{16}O)}{\text{Velocity } (^{13}C\ ^{16}O\ ^{16}O)} = \left(\frac{45}{44}\right)^{1/2} = 1{\cdot}011$$

which implies that collisions of $^{13}CO_2$ of molecular weight 44 with a photosynthesizing portion of a plant will be 1·11% more frequent than those of $^{12}CO_2$ of molecular weight 45. Solution of carbon dioxide in water results in a slight concentration of the heavier ^{13}C in the dissolved ions, so land plants are expected to contain substantially less ^{13}C relative to marine plants because the latter utilize dissolved and not gaseous carbon dioxide during photosynthesis. The differences in $\delta^{13}C$ values in plants can therefore be largely attributed to the fact that during photosynthesis marine organisms utilize carbonate and bicarbonate from the ocean, the non-marine organisms use carbonate from fresh water, whereas land plants use the isotopically lighter carbon dioxide in the atmosphere (Fig. 9.15) and this is mirrored in the different $\delta^{13}C$ values for organic matter present in plants from the various environments (Fig. 9.16).

Stable carbon isotope values have been measured for the organic matter in Precambrian sediments. Hoering (1962) measured the $\delta^{13}C$ values in organic carbon and associated carbonates in Precambrian stromatolites from the Belt Series, Montana; Iron River Formation, Michigan; Bulawayan Series, Rhodesia and Transvaal Series, South

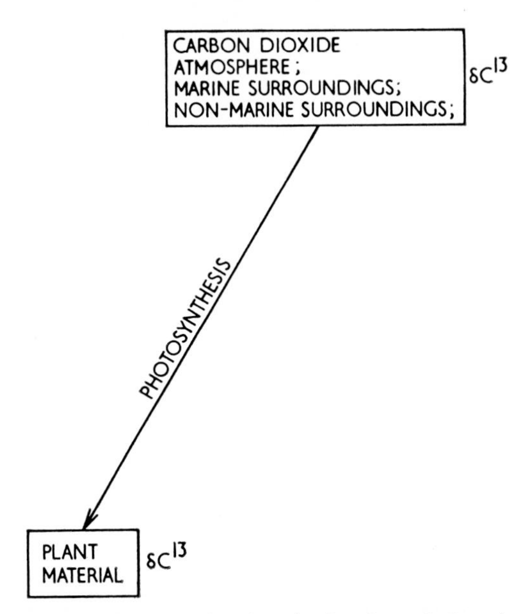

Fig. 9.15. Stable carbon isotope fractionation by plants during photosynthesis

Africa; and these stromatolites were interpreted to result from algal mat sedimentation. The $\delta^{13}C$ values indicated a depletion of the ^{13}C isotope in the organic carbon similar to that observed for sediments of younger age, where the organic carbon is known to be of biological origin. These stable carbon isotope results, in conjunction with morphological and other geochemical studies, strongly support the inferences

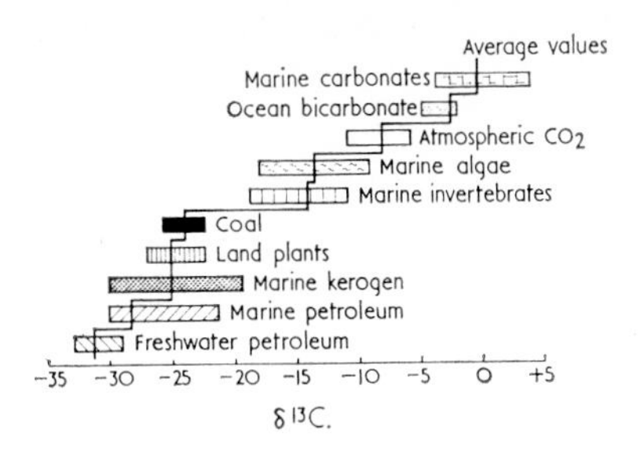

Fig. 9.16. Carbon-13 heavy isotope deficiencies (^{13}C) in various types of organic matter

that algal photosynthesis occurred in Precambrian times. However, there is not exact agreement between $\delta^{13}C$ values for organic matter in Lower Palaeozoic and Precambrian younger rocks and it appears that diagenesis and maturation may have altered the ratio of the carbon isotopes in the organic matter. The $\delta^{13}C$ value for the organic carbon of marine origin normally falls in the range -25 to -30 (Fig. 9.16). The organic carbon present in Precambrian sediments generally show lower $\delta^{13}C$ values than other sedimentary organic matter. Various reasons for this have been suggested:

(i) a high abundance of carbon dioxide in the Precambrian hydrosphere probably resulted in slightly more acidic aqueous media;
(ii) the $\delta^{13}C$ ratio of the bicarbonate ion of the aqueous media used in photosynthesis was different in Precambrian times than in more recent times;
(iii) during diagenesis (especially thermal alteration) of organic matter, degradation products (methane and ethane particularly) with a higher ^{13}C content are released leaving the organic residue enriched in ^{12}C;
(iv) lower temperature of the aqueous media in Precambrian times than in more recent and present-day conditions.

A lower pH, due to dissolved carbon dioxide, of Precambrian aqueous media and/or a change in temperature from present-day conditions would alter the characteristics of the calcium bicarbonate/carbonate equilibria and could therefore be a significant factor in the variation of $\delta^{13}C$ values for Precambrian organic matter. However, the factor affecting the $\delta^{13}C$ values of Precambrian organic matter is probably maximally heat. This is discussed in more detail in relation to the alteration of $\delta^{13}C$ values of the Fig-Tree and Onverwacht Series organic matter in the following Chapter 10.

Soluble Organic Compounds

n-Alkanes (Fig. 9.17a) occur in geological samples throughout the geological column. Although they have been identified in most Precambrian sedimentary rocks and their distributions and individual components analysed and related to possible biological precursors (mainly primitive algae), they are generally considered not to have sufficient structural details and characteristics to supply unequivocal evidence of having been produced by living systems. Alternatively hydrocarbons having an isoprenoid-type structure (Fig. 9.17b) are of much more geochemical significance, since they can be directly related to isoprenoid compounds present in extant organisms. Isoprenoid hydrocarbons are postulated to have arisen by decomposition of chlorophyll (Fig. 9.18a) into two fragments: the phytyl-chain which is

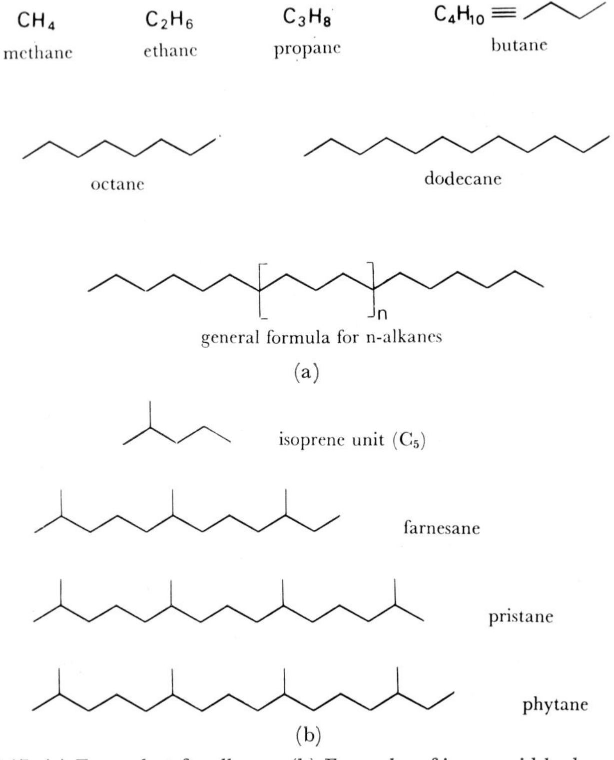

Fig. 9.17. (a) Examples of *n*-alkanes. (b) Examples of isoprenoid hydrocarbons

isoprenoid and a porphyrin (Fig. 9.18b). So when isoprenoids (phytane, pristane and farnesane) (Fig. 9.17b) and porphyrins are found together in sediments it is assumed that chlorophyll, characteristic of photosynthetic organisms, was once present. Alkanes, including the saturated isoprenoids phytane and pristane, have been identified in extracts from the Bitter Springs Formation ($\sim 1{\cdot}0 \times 10^9$ years old); the Nonsuch Shale ($\sim 1{\cdot}0 \times 10^9$ years old); the Paradise Creek Formation, Queensland, Australia ($\sim 1{\cdot}6 \times 10^9$ years old); the Soudan Iron Formation of Minnesota, U.S.A. ($\sim 2{\cdot}7 \times 10^9$ years old); and various other Precambrian sediments including the Fig-Tree and Onverwacht Series (see Chapter 10).

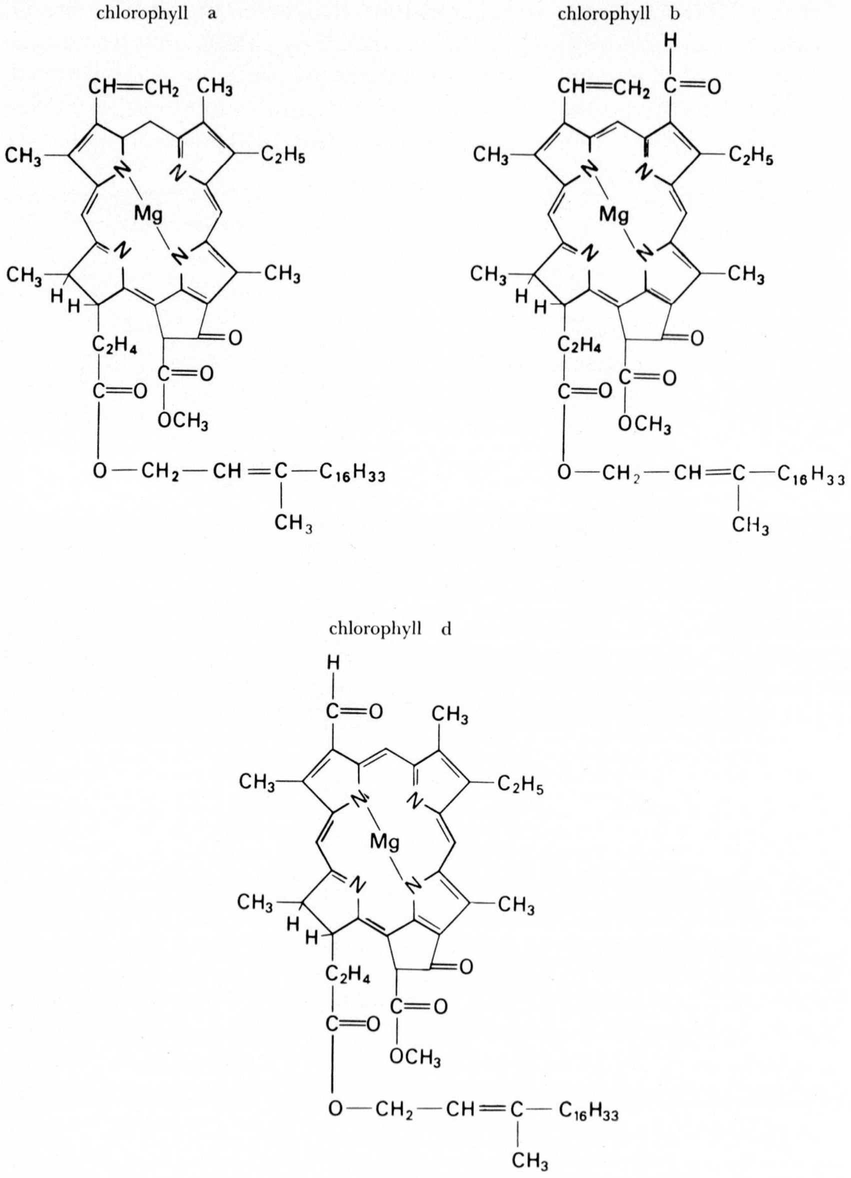

chlorophyll a

$CH{=}CH_2$ CH_3

CH_3 C_2H_5

N N

Mg

N N

CH_3 CH_3

H H

C_2H_4 =O

$C{=}O$

$C{=}O$

OCH_3

$O-CH_2-CH=C-C_{16}H_{33}$

CH_3

chlorophyll b

H

$CH{=}CH_2$ $C{=}O$

CH_3 C_2H_5

N N

Mg

N N

CH_3 CH_3

H H

C_2H_4 =O

$C{=}O$

$C{=}O$

OCH_3

$O-CH_2-CH=C-C_{16}H_{33}$

CH_3

chlorophyll d

H

$C{=}O$ CH_3

CH_3 C_2H_5

N N

Mg

N N

CH_3 CH_3

H H

C_2H_4 =O

$C{=}O$

$C{=}O$

OCH_3

$O-CH_2-CH=C-C_{16}H_{33}$

CH_3

(a)

Fig. 9.18. (a) Chlorophylls

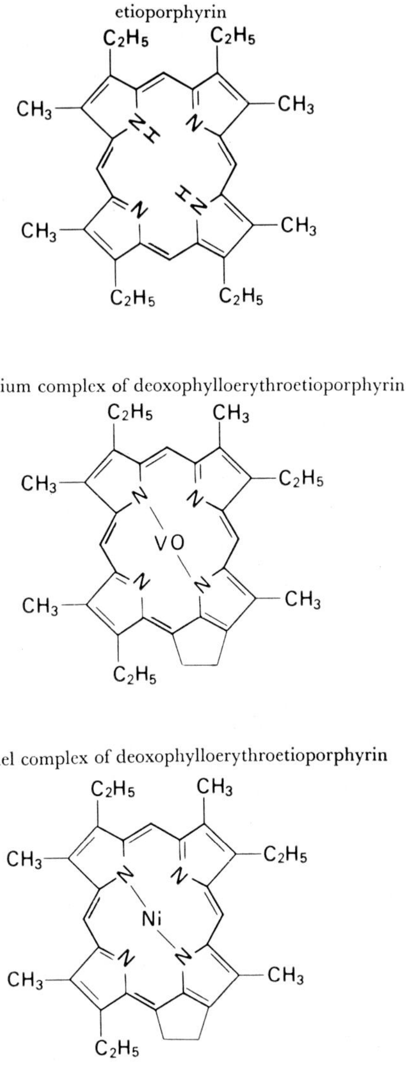

etioporphyrin

vanadium complex of deoxophylloerythroetioporphyrin

nickel complex of deoxophylloerythroetioporphyrin

(b)

Fig. 9.18. (b) Porphyrins

Amino acids, including serine which is unstable, have also been identified in various Precambrian sediments (including the Bitter Springs Formation, the Fig-Tree and Onverwacht Series sediments) and examination of their stereochemistry, using recently developed capillary column gas chromatography, showed that often the amino-acids were present as mixtures of D and L enantiomers. This is not in keeping with the observation (Chapter 5) that α-amino acids produced by biological systems are of the L-configuration. Geochemists have shown that over periods of time and exposure to heat, the α-amino acids will probably racemize and less stable amino acids such as serine would be destroyed. This suggests that some of the amino acids, at least which are present in the samples, may not be indigenous to the rocks but were introduced more recently. It is interesting to relate at this point that recent work with the Orgueil, Murchison and Mokoia carbonaceous chondrites (see Chapter 11) has shown that both D- and L-forms of α-amino acids are present in nanogram (10^{-9} g) quantities, and these have consequently been interpreted as being abiogenic in origin. But since the presence of D- and L-forms of amino acids in Precambrian sediments is readily explained by racemization, it appears inconsistent that the same reasoning is not applied to the meteorite studies.

The nitrogeneous organic matter, in addition to having some extra-terrestrial thermal history (Chapter 11) is usually extracted from the meteorite with boiling aqueous 6N hydrochloric acid for 20 h and later chromatographed on "Dowex 50" ion-exchange columns using ammonium hydroxide as an eluant. Comment is made in later chapters about the dangers involved in the use of ammonium compounds for the isolation of nanogram quantities of amino acids from rocks and meteorites; but this aside, some geochemists are still apparently using different criteria and logic for Precambrian as distinct from meteorite studies.

The soluble organic chemicals present in Precambrian sediments occur in very minor quantities, and although the distribution and occurrence of the various classes of compounds and of certain individual components may contribute to our understanding of the early forms of life, their significance must be treated with much reserve since soluble organic compounds are likely to have undergone movement in total or in part from their point of origin (e.g. petroleum accumulations) since they will have potential mobility in solvent or aqueous emulsion phases, and it is therefore not always possible to assign such chemicals to a particular geological sample.

Insoluble Organic Matter Present in Precambrian Rocks

The insoluble organic matter present in Precambrian sediments is usually amorphous and is known to occur throughout the Pecambrian as far back as the earliest periods 3700 million years ago and it accounts for as much as 99% of the total organic matter. Geochemical studies have shown that there are different types of insoluble organic matter found in rocks and that the nature of the material depends, amongst other things, on:

(i) the source of the organic matter;
(ii) the age of the sediment;
(iii) the degree of microbiological and chemical alteration or degradation;
(iv) the amount of metamorphosis;
(v) the correct definition of "insoluble", i.e. the organic solvents and chemical reagents in which the organic material is insoluble.

Factors affecting Organic Matter during Sedimentation

During sedimentation a mixture of fine-grained mineral particles and dispersed organic material is gradually deposited in an aqueous, often marine, environment. The organic matter present in aqueous surrounding can obviously vary in different areas and according to conditions, but a useful illustration is provided by the Caspian Sea. This is the largest isolated salt-water inland lake in the world and the various types of living matter which are to be found therein have been well documented (Table 9.05); it has also been calculated that the Caspian Sea contains 2% biological matter. There will of course be considerable variation in the amounts of bio-organic material found in various aqueous environments. Lakes and pools in Precambrian times are thought to have contained quite high concentrations of organic matter, similar to those of modern "volcanic pools" in North America and New Zealand (see Chapter 10). Although relatively large amounts of organic matter were apparently present during the sedimentation processes, rather less than 1% is normally found in Precambrian rocks. The organic matter will of course have undergone chemical and microbiological transformation and degradation whilst the sediment was being laid down and compressed.

Organic material in the most recent sediments will include the immediate decomposition products of much of the recent plant and animal kingdoms. This has been confirmed from examination of benthos and other lake-bottom deposits, and detailed studies of a Mud Lake ooze from Florida (about 1100 years old) shows it to be composed almost exclusively of blue-green algal cell walls and some pollen grains. Examination of sediments of increasing geological age indicates that the nature and properties of the organic matter tends to become more readily definable in terms of organic chemistry and to become increasingly stable. The insoluble organic matter present in Precambrian sediments is a chemically stable substance, and unlike much of the "insoluble" material present in younger sediments which can in

Table 9.05. Principal forms of life in the Caspian Sea

Type of organism	Weight of material (in millions of tons) Gross weight	%	Dry weight	%
Bacteria	2000	61·0	400	75·0
Phytoplankton	1000	30·5	100	18·7
Zooplankton	150	4·5	15	2·8
Zoobenthos	120	3·6	18	3·3
Phytobenthos	3	0·1	0·37	0·07
Fish	3	0·1	0·90	0·16
Total	3276	99·8	534·27	100·03

The Caspian Sea is 760 miles long, average 270 miles wide and average depth of 830 ft and contains $16{\cdot}1 \times 10^{12}$ tons of water.

∴ % Weight "Life forms" in Caspian Sea = 2·04%

∴ % Weight Organic Material in Caspian Sea = 0·33%

(after S. V. Bruevich)

fact be readily solubilized by various chemical reagents. As the sediments become older the more readily degraded organic compounds (cellulose, cutin, chitin, lignins, etc.) are degraded by microbiological and by chemical reactions. Chemical transformations, including reduction, oxidation, decarboxylation and especially hydrolysis, will readily degrade the large mass of organic matter to simple water-soluble compounds. Microbiological degradation of organic material will take place maximally in the early stages of sediment formation. In the uppermost layers of newly forming sediments it has been found that there is very strong microbiological activity which results in extensive alteration and degradation of the vast majority of the deposited organic matter. After these initial but frequently continuing reactions, the nature of the remaining insoluble resistant organic material gradually changes and assumes a constant chemical nature which represents a measure of the concentration of the most chemical and microbiological resistant components of the biological organisms.

The differential preservation of certain parts of the organic material at the expense of the degradation of the less resistant material during

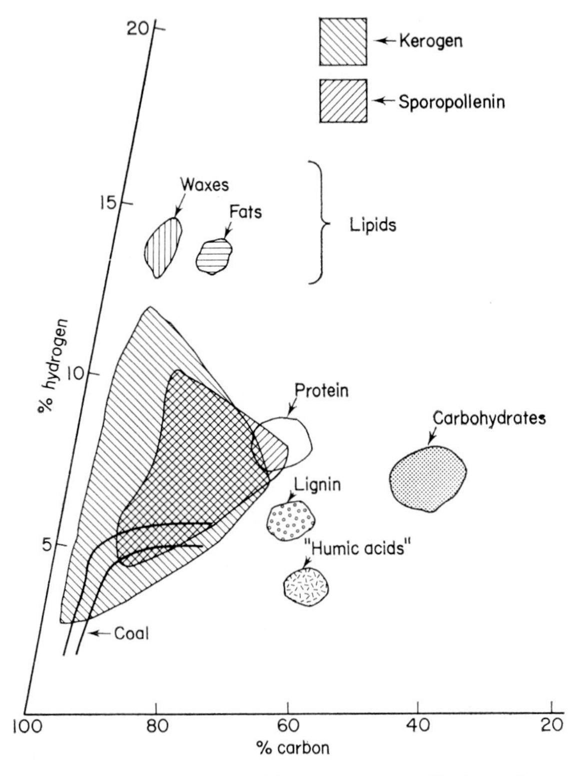

Fig. 9.19. Relationship of composition of kerogen, sporopollenin and natural products

sedimentation is well illustrated by the remarkable resistant materials called sporopollenins. Sporopollenins are chemical substances formed, for example, in plant anthers by the oxidative polymerization of carotenoids and carotenoid esters; they form the resistant protective outer wall (exine) of pollen grains and certain spores. When pollen or spores fail to reach their intended destination, they may soon become infertile (but not always) and the cytoplasm, genetic material and polysaccharide components are rapidly destroyed by chemical and microbiological agents leaving the intact sporopollenin containing wall. The study of fossil pollen and spore walls constitutes the science of palynology, derived from the Greek verb *palynein* (to scatter). The extreme resistance of sporopollenin to chemical, microbiological and other processes is shown by the preservation of fossil pollen in sediments 450 million years old and of spores in strata laid down in the earliest Precambrian era, up to 3700 million years ago (see Chapter 10). This unique resistance to decay makes pollen and related spores the most ubiquitous of plant fossils. Most sediments also contain insoluble organic matter

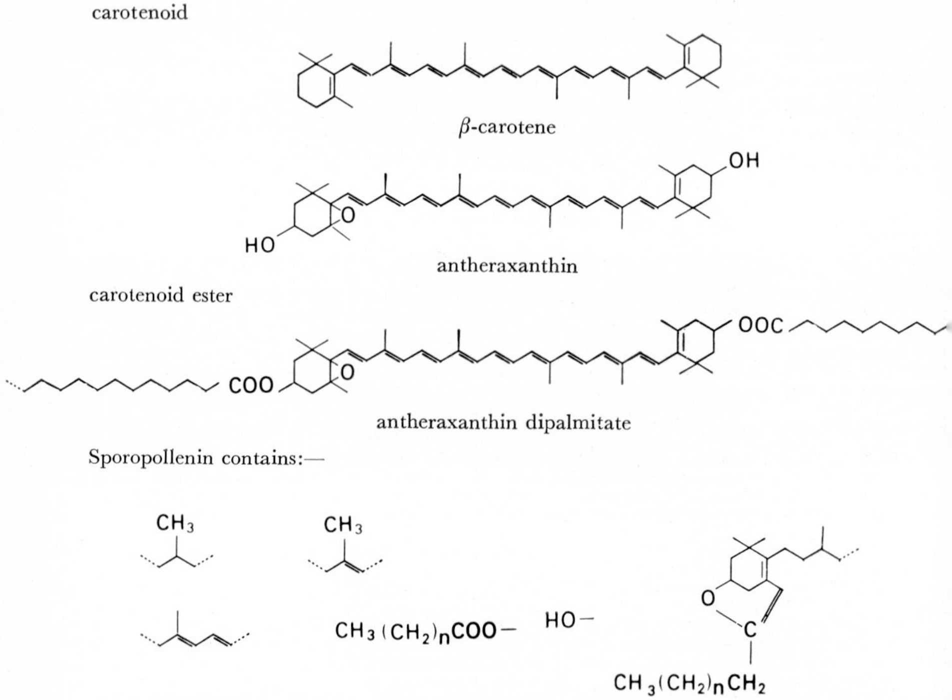

carotenoid

β-carotene

OH

HO

antheraxanthin

carotenoid ester

OOC

COO

antheraxanthin dipalmitate

Sporopollenin contains:—

CH_3 CH_3

$CH_3(CH_2)_nCOO-$ HO—

O

C

$CH_3(CH_2)_nCH_2$

Fig. 9.20. Sporopollenin—An oxidative co-polymer of carotenoid–carotenoid esters

specifically composed of pollen and spore residues, in turn shown to be sporopollenin. Fimmenites, which occur in sediments throughout Europe, are deposited in the inundation zones of basins and during sedimentation all other organic matter is degraded, except the sporopollenin which concentrates to form the insoluble organic material of the sediment. Other similar geological occurrences of enriched sporopollenin deposits have been described from the Eocene deposits of Geiseltal near Halle in East Germany, "Draved Deposits" of Denmark and in the world-wide occurrence of Tasmanites.

Geochemical analysis (Brooks and Shaw, 1968–1971) of sedimentary insoluble organic matter (especially from Precambrian rocks) and comparison of its degradation products with analogous materials from sporopollenins derived from known modern biological sources, showed a very close similarity (Table 9.06 and Fig. 9.19). These experiments suggested that much of the insoluble organic matter in Precambrian rocks could have derived from sporopollenin produced by lower plants. This led further to the suggestion that sporopollenin could be used as a "biological marker" which would provide good evidence for the presence of former living systems.

Sporopollenins (and hence the insoluble organic matter in ancient rocks) vary somewhat in chemical properties in a manner which depends both on the source of the material and on its diagenetic history. These variations are entirely to be expected and may to some extent be classified:

(i) a sporopollenin derived mainly from carotenoid esters (Fig. 9.20) will have an "aliphatic-type" structure, because the fatty acids forming the ester groups will provide a large source of polymethylene groups. These fatty acid portions of the ester group within the polymer will give rise to long chain acids (following oxidation) or *n*-alkanes (following mild thermal degradation), whilst acids derived from the polyenoid portions of the polymer tend to be dicarboxylic acids. After polymerization of the carotenoid ester–carotenoid precursors, some but probably not all of the ester groups are retained within the sporopollenin polymer structure even after severe treatment with powerful hydrolytic reagents such as caustic potash solution. The reasons for this are unclear but undoubtedly include steric protection of the ester group in the polymer matrix and modification of the ester group during polymerization.

(ii) a sporopollenin derived largely from free carotenoid precursors will be of the "aromatic type" because it will contain relatively few polymethylene precursor groups.

(iii) an "aliphatic type" sporopollenin material will tend to become increasingly aromatic when heated. Using chemical and radiochemical techniques Brooks and Shaw have examined (Fig. 9.21) the effects of heat (temperatures 180°–450°C) on sporopollenins and have correlated changes in chemical structure of the sporopollenin at different temperatures. Physico-chemical (Fig. 9.22) and radiochemical examination and X-ray studies show that when sporopollenin is heated there is a gradual reduction in the number of hydroxyl groups and a corresponding gradual increase in the proportion of carbon–carbon double bonds present in the structure. After heating at 400° and 450°C (for 100 h) there is a significant increase in the aromatic nature of the material. It has been known for many years that carotenoids are aromatized when

Table 9.06. Comparison of some sporopollenins and kerogens

Material	Source	Geological age (million years)	Molecular formulae[2]	Compounds formed by oxidation			Compounds formed by alkali treatment
Sporo-pollenin[1]	Pollen and spore walls	Present	Examples	Straight-chain, iso and ante-iso-acids	Di-carboxylic acids	Branched-acids	
			$C_{90}H_{134}O_{20}$ [3] $C_{90}H_{144}O_{27}$ [4] $C_{90}H_{142}O_{35}$ [5] $C_{90}H_{135}O_{30}$ [6]	C_7–C_{18}	C_2–C_{10} (33–38%) Small amounts of: Hexadecandioic acid, 7-hydroxy-hexadecandioic acid, 6 : 11 dioxohexadecandioic acid	C_7–C_{18}	*p*-Hydroxybenzoic acid *m*-Hydroxybenzoic acid Protocatechuic acid Vanillic acid
Fossil sporo-pollenin	Tasmanin	200	$C_{90}H_{136}O_{17}$				
	Geiseltal-pollenin	250	$C_{90}H_{129}O_{19}S_7N$				
Kerogen	Green River formation	50	$C_{90}H_{134}O_{25}$ (0·66% N) (0·90% S)	C_7C_{12}	C_2–C_8 Dodecandioic acid Hexadecandioic acid 10-Oxoundecanoic acid 13-Oxotetradecanoic acid	C_9–C_{10} C_{12}–C_{17} (C_{19}, C_{20})	

s

				C_{14}–C_{30}			Aromatic and cycloaromatic compounds
Kerogen	Gdovsk shale				Phthalic acid, C_3–C_{10} 38%		
Kerogen	Baltic shale	360					Presence of phenols and hydrocarbons (up to 4% from secondary formation; $KMnO_4$ oxidation showed aromatic ring is absent
Kerogen	Dictyonema shale	400			Succinic acid and higher homologues, 20%		
Kerogen	Kukersite shale	440	$C_{90}H_{136}O_{10}$	Ketonic and lactonic acids	C_2–C_8		
Kerogen	Onverwacht Chert	3400–3700	$C_{90}H_{72}O_6$	C_{10}–C_{26}	C_4–C_{17}	C_{10}–C_{26}	*p*-Hydroxybenzoic acid *m*-Hydroxybenzoic acid Protocatechuic acid

[1] The figures are for walls from which cellulose has been removed.
[2] Molecular formulae are arbitrarily recorded on a C_{90} basis to facilitate comparison with earlier references.
[3] From *Secale cereale.*
[4] From *Lycopodium clavatum.*
[5] From *Lilium henryi.*
[6] From *Chenopodium album.*

Methyl palmitate after treatment with ozone for 10 h gives a mixture of dicarboxylic acid (C_2–C_{14}) (unpublished results).

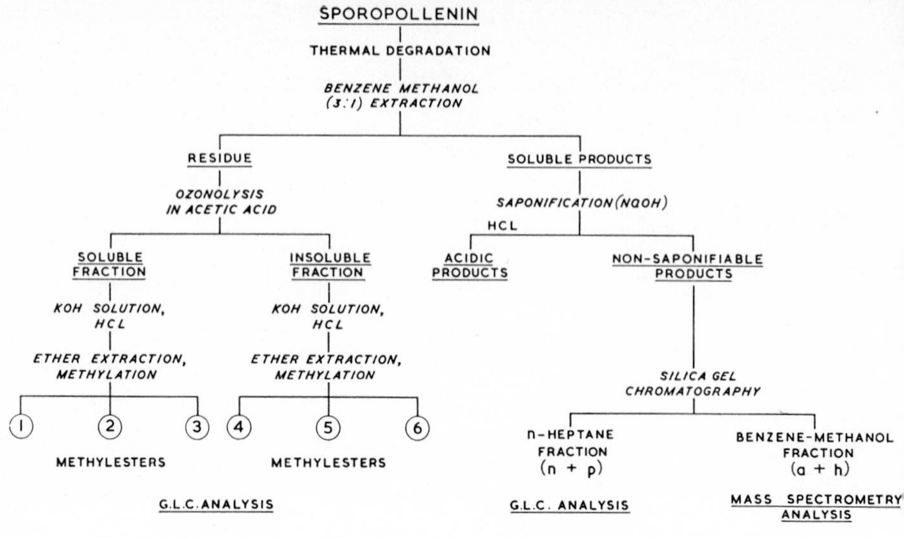

Fig. 9.21. Examination of the thermal degradation products from sporopollenin

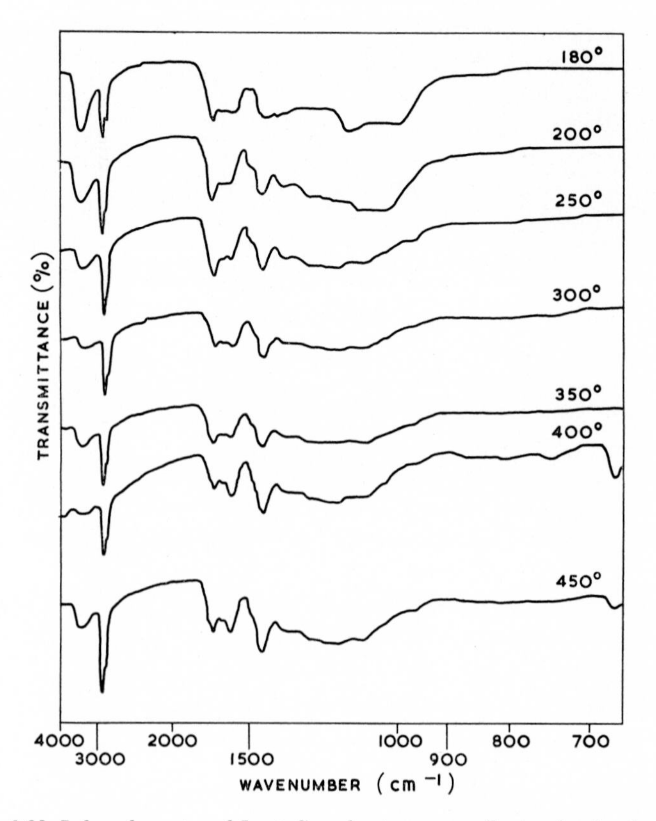

Fig. 9.22. Infrared spectra of *Lycopodium clavatum* sporopollenin after heating

heated at quite low temperatures, thus β-carotene readily cyclizes at 160° to form aromatic compounds. Calculations of the alteration of sporopollenin and its contained functional groups show that two main processes are taking place during thermal treatment: chemical dehydration of the polymer by loss of hydroxyl groups, and a gradual increase in aromatic nature. From a chemical point of view therefore polyenoid and aromatic systems are essentially equivalent.

Examination of the soluble organic products produced by thermal alteration of sporopollenins indicated an increasingly complex mixture (carbon number range C_{12} to C_{40}) of hydrocarbons and a very complex mixture (with molecular weight up to 800 for some components) of aromatic (including major amounts of phenolic compounds) and heterocyclic compounds.

The "spectra" of thermal and diagenetic products from sporopollenins are often characteristic of the soluble organic material (associated with the insoluble organic matter) extracted from Precambrian rocks and could be considered as characteristic of the former presence of sporopollenin and be used as a "biological marker". In spite of the various changes which can occur in the chemical structure of sporopollenins and in the morphology of related spore exines occasioned by heat and other diagenetic processes, sufficient of the insoluble sporopollenin basic chemical structure may often remain to enable it to be identified in Precambrian rocks and to be related to a living system. Experiments (Brooks and Shaw) with fresh spores and pollen grains show that many (e.g. poplar, ash, plane, lupin, lily, *Mucor*, *Pediastrum*) walls, after treatment to remove contents and cellulose intine, yield a material with all the typical and characteristic properties of sporopollenin yet structurally amorphous and bearing no obvious morphological relationship to the generic spore or pollen grain; so that mixtures of intact spores and pollen walls associated with amorphous material of the same general chemical composition and derivation are entirely to be expected in sediments. The most significant results, however, will naturally come from examination of minimally metamorphosed sedimentary material. Sporopollenins are far better as "biological markers" than are the simpler soluble organic chemicals, since they cannot only be considered indigenous to the sediment but can also be directly identified with known biological systems and so the value of their evidence is increased proportionally.

SUGGESTED FURTHER READING Chapter 9

Books

Abelson, P. H. (Ed.) (1967). "Researches in Geochemistry". Wiley, New York.

Breger, I. A. (Ed.) (1963). "Organic Geochemistry". Pergamon, Oxford.

Brooks, J., Grant, P., Muir, M. D., van Gijzel, P. and Shaw, G. (Eds) (1971). "Sporopollenin". Academic Press, London.

Calvin, M. (1969). "Chemical Evolution". Oxford University Press.

Craig, H. (Ed.) (1964). "Isotopic and Cosmic Chemistry". Wiley-Interscience, London and New York.

Degens, E. T. (1965). "Geochemistry of Sediments". Prentice-Hall, New York.

Eglinton, G. and Murphy, M. T. J. (Ed.) (1969). "Organic Geochemistry: Methods and Results". Springer-Verlag, Hamburg.

Faegri, K. and Iversen, J. (1964). "Textbook of Pollen Analysis". Blackwell, Oxford.

Fritsch, F. E. (1965). "The Structure and Reproduction of the Algae", vol. II. Cambridge University Press.

Gass, I. G., Smith, P. J. and Wilson, R. C. L. (1971). "Understanding the Earth". Open University Set Book. Artemis Press, Horsham, Sussex.

Holmes, A. (1965). "Principles of Physical Geology". Nelson, London.

Rankama, K. (Ed.) (1963). "The Precambrian". Wiley-Interscience, New York.

Round, F. E. (1969). "Introduction to the Lower Plants". Butterworths, London.

Swain, F. M. (1970). "Non-marine Organic Geochemistry". Cambridge University Press.

Tschudy, R. H. and Scott, R. A. (Eds) (1969). "Aspects of Palynology". Wiley, New York.

Articles

Barghoorn, E. S. (1971). The Oldest Fossils. *Sci. Am.* **224,** 30–42.

Brooks, J. and Shaw, G. (1968a). Chemical Structure of the Exine of Pollen Walls and a New Function for Carotenoids in Nature. *Nature, Lond.* **219,** 532–533.

Brooks, J. and Shaw, G. (1968b). Identity of Sporopollenin with Older Kerogen and New Evidence for the Possible Biological Source of Chemicals in Sedimentary Rocks. *Nature, Lond.*, **220,** 678.

Brooks, J. and Shaw, G. (1972a). Geochemistry of Sporopollenin. 3rd International Geochemical Congress, Moscow. July 1971. *Chem. Geol.*

Brooks, J. and Shaw, G. (1972b). The Role of Sporopollenin in Palynology. 3rd International Palynological Congress, Novosibirsk, U.S.S.R. July 1971, in press.

Cloud, P. (1968). Premetazoan Evolution and the Origins of the Metazoa. *In* "Evolution and Environment" (Ed. E. T. Drake), Yale University Press, pp. 1–72.
Cloud, P. and Licari, G. R. (1968). Microbiotas of the Banded Iron Formations. *Proc. natn. Acad. Sci. U.S.A.* **61,** 779–786.
Cloud, P. and Licari, G. R. (1972). Ultrastructure and Geologic Relations of Some Two-Aeon Old Nostocacean Algae from Northeastern Minnesota, *Am. J. Sci.* **272,** 138–149.
Glaessner, M. F. (1968). Biological Events and the Precambrian Time Scale. *Can. J. Earth Sci.* **5,** 585–590.
Gruner, J. W. (1925). Discovery of Life in the Archaean. *J. Geol.*, **33,** 151–152.
Han, J., McCarthy, E. D., Van Hoeven, W., Calvin, M. and Bradley, W. H. (1968). Organic Geochemical Studies II. The Distribution of Aliphatic Hydrocarbons in Algae, Bacteria and in a recent Lake Sediment: a Preliminary Report. *Proc. natn. Acad. Sci. U.S.A.* **59,** 29.
Hoering, T. C. (1962). The Stable Isotopes of Carbon in the Carbonate and Reduced Carbon of Precambrian Sediments. Annual Report of the Director of the Geophysical Laboratory 1961–1962. *Carnegie Instn Wash. Year Book,* 190–191.
Hoering, T. C. (1967). Conversion of Biochemicals to Kerogen and *n*-Paraffins. *In* "Researches in Geochemistry", Volume 2 (Ed. P. H. Abelson), pp. 63–86. Wiley, New York.
Hoering, T. C. (1967). The Organic Geochemistry of Precambrian Rocks. *In* "Researches in Geochemistry", Volume 2 (Ed. P. H. Abelson), pp. 87–111. Wiley, New York.
Holmes, A. (1959). A Revised Geological Time-Scale. *Trans. Edinburgh geol. Soc.* **17,** III, 183–216.
Holmes, A. (1963). Introduction. *In* "The Precambrian" (Ed. K. Rankama). XI–XXIV, Wiley-Interscience, New York.
Hunt, J. M. (1962). Geochemical Data on Organic Matter in Sediments. International Scientific Oil Conference, Budapest 1962.
Maxwell, J. R., Pillinger, C. T. and Eglington, (1971). Organic geochemistry *Quart. Rev.* **25,** 571–628.
McCarthy, E. D. and Calvin, M. (1967). Organic Geochemical Studies, I. Molecular Criteria for Hydrocarbon Genesis. *Nature, Lond.* **216,** 642–647.
Nier, A. O. and Gulbrandsen, E. A. (1939). Variation in the Relative Abundance of the Carbon Isotopes. *J. Am. chem. Soc.* **61,** 697.
Park, R., and Epstein, S. (1960). Carbon Isotopes Fractionation during Photosynthesis. *Geochim. cosmochim Acta* **21,** 110–126.
Schöpf, J. W. (1968). Microflora of the Bitter Springs Formation, Late Precambrian, Central Australia. *J. Paleontol.* **42,** 651–688.
Schöpf, J. W. (1969). Microorganisms from the Late Precambrian of South Australia. *J. Paleontol.* **43,** 111–118.
Schöpf, J. W. (1970). Electron Microscopy of Organically Preserved Precambrian Microorganisms. *J. Paleontol.* **44,** 1–6.

Schöpf, J. W. (1970). Precambrian Microorganisms and Evolutionary Events prior to the Origin of Vascular Plants. *Biol. Rev.* **45,** 319–352.

Schöpf, J. W. (1971). Biogenicity and Significance of the Oldest Known Stromatolites. *J. Paleontol.* **45,** 477–485.

Schöpf, J. W., Barghoorn, E. S., Maser, M. D. and Gordon, R. O. (1965). Electron Microscopy of Fossil Bacteria Two Billion Years Old. *Science, N.Y.* **149,** 1365–1367.

Shaw, G. (1970). Sporopollenin. *In* "Phytochemical Phylogeny" (Ed. J. B. Harborne), Academic Press, London.

Tyler, S. A. and Barghoorn, E. S. (1954). Occurrence of Structurally Preserved Plants in Precambrian Rocks of the Canadian Shield. *Science, N.Y.* **119,** 606–608.

Vassoyevick, N. B., Vysotskiy, I. V., Sokolov, B. A. and Tatarenko, Ye. I. (1970). Oil-gas Potential of Late Precambrian Deposits. Int. Geol. Rev. **13**, 407–418.

Walcott, C. D. (1899). Precambrian Fossiliferous Formations. *Bull. geol. Soc. America* **10,** 199–244.

Warman, H. R. (1971). Future Problems in Petroleum Exploration. *Petroleum Rev.* August, 1971, 96–101.

White, D. (1933). The Role of Water Conditions in the Formation of Common (Banded) Coals. *Econ. Geol.* **28,** 566–570.

10

The Precambrian Fossil Record. II

The Fig-Tree and the Onverwacht Systems

INTRODUCTION

In that part of the eastern Transvaal which lies in the north-eastern corner of South Africa near the Swaziland border (Fig. 10.01) stretches

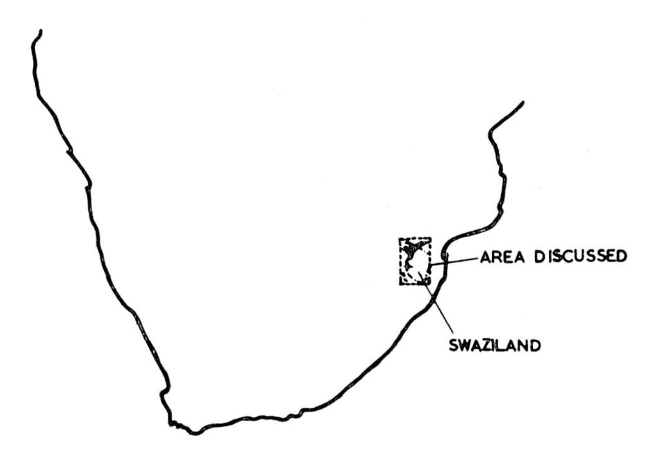

Fig. 10.01. Southern Africa

the Barberton Mountain Land. This could be the most important few hundred square miles of land on this planet, for in these hills reside the oldest known rocks on earth, the Fig-Tree and Onverwacht cherts whose ages date back some $3{\cdot}7 \times 10^9$ years to the very beginning of life on this planet. These rocks may ultimately help provide us with the answer to one of man's most ancient problems, the origin of the living system on this planet. The reader might think therefore that there would be a constant trek into the area, a veritable mine of scientific activity. Nothing could be farther from the truth. It is doubtful whether more than a handful of scientists other than geologists have ever heard of these rocks let alone considered their significance.

THE SWAZILAND PRECAMBRIAN SEDIMENTARY SYSTEM

We have seen in Chapter 9 that early Precambrian rocks contain geological and geochemical information about the existence and nature of the earliest forms of life and the sediments within the formations provide some evidence about the earth's history, about the early biological development of living systems and may even contain evi-

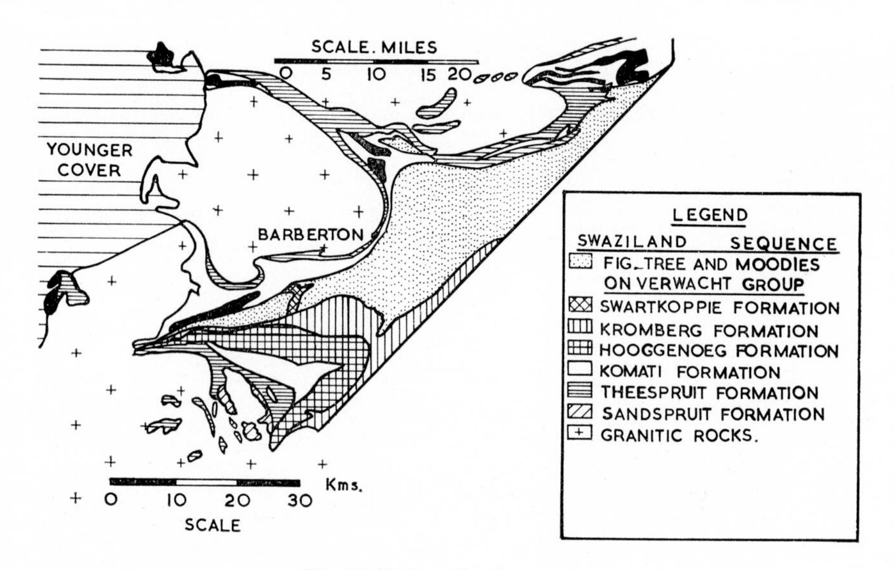

Fig. 10.02. Swaziland sequence

dence which gives some indication of the nature of the Earth's primitive crust.

Regions of Southern Africa contain Precambrian sedimentary rocks which are known to have existed for at least $3{\cdot}7 \times 10^9$ years. Probably the most important rocks of these formations are those of the Barberton Mountain Land of the Eastern Transvaal (Fig. 10.02). These formations, which have been shown to consist of a complex of folded sediments with

(a)

(b)

Fig. 10.03. The Barberton mountain land:

(a) General view of Barberton mountain land showing the low-lying Kaap Valley. The Highland Ranges surrounding the valley comprise mainly of Fig-Tree and Moodies Sediments. The town of Barberton is shown to the left of the photograph, immediately below the main range in the background (b) The Msauli Asbestos Mine. The lower parts of the valley are probably of Upper Onverwacht (Kromberg Formation), the middle slopes of the mountains comprise of Fig-Tree Shales, Cherts, etc., and the upper part comprise the Moodies System

(c)

(d)

Fig. 10.03. The Barberton mountain land:

(c) The Hooggenoeg formation of the Onverwacht series. An acid volcanic and chert zone (shown as the white layer in the middle distance) surrounded by smooth grey hills of basic volcanics. Richard Viljoen, who (with his brother Morris) is one of the geological experts of the Barberton Mountain Land Area is standing in the foreground on a ridge of banded black and white chert overlying a very large acid zone at the top of the Hooggenoeg formation (d) The termination of the Hooggenoeg formation. A typical volcanic cycle showing basic lave underlying smooth grey slopes, narrow acid lava zone and narrow somewhat carbonaceous chert to the left

ages of up to at least $3{\cdot}7 \times 10^9$ years old, have supercrustal layered rocks which are amongst the oldest known rocks on earth and lay adjacent to what is considered the basement rock formed from the primitive reworked earth's crust (dated $3{\cdot}8$–$4{\cdot}0 \times 10^9$ years old). The Barberton Mountain Land (Fig. 10.03) represents one of the many ancient greenstone belts (these are altered basic igneous rocks which owe their

(e)

Fig. 10.03. The Barberton mountain land:

(e) The Hooggenoeg formation. Chert and acid volcanic horizon constituting the top of a basic volcanic cycle. The basic volcanics form the smooth grey slopes

colour to the presence of chlorite, hornblende and epidote) within the crystalline shield of southern Africa.

In the early 1880's reef gold was discovered in the Barberton area and soon afterwards gold deposits were found in the Witwatersrand sediments. It is the continuation of these mining and prospecting activities to the present that has mainly given rise to many of the geological studies of this area and also of the rock formations which extend across the Transvaal border into Swaziland. The earliest reports on the geology of the Barberton gold fields were made by Hall (1918) on the Archaean rocks; he classified the system according to increasing age (see Fig. 10.04) and these formations were collectively referred to as the Swaziland System.

3. OLDER GRANITE	*Nelspruit Type and De Kaap Valley Type*—granite and related rocks.
2. SWAZILAND SYSTEM	*Jamestown Series*—containing talc, hornblende and related basic metamorphic rocks.
	Moodies Series—containing slates, greywacks, conglomerates, banded ironstones and quartz.
1. SWAZILAND SYSTEM	*Onverwacht Volcanic Series*—containing basic lavas, tuffs and other extrusive contemporaneous rocks.

Fig. 10.04

Further geological examination of the area by the Geological Survey of South Africa (1956) suggested various important changes in the above classification. The upper parts of the Moodies Series was redefined as the Moodies System (Fig. 10.05a) and the lower parts grouped together to form the Fig-Tree Series (Fig. 10.05b) and these, together with the underlying Onverwacht Series, composed the Swaziland System. The Jamestown Series was also re-named the Jamestown Igneous Complex and considered to be intrusive layers into the Moodies and Swaziland Systems. The exact classification of these formations in the Barberton area has not yet been fully agreed, since there is a lack of time–stratigraphic controls in the form of fossils and accurate radiometric dating. It has, however, been suggested by the International Sub-Commission on Stratigraphic Terminology that a lithostratigraphic nomenclature be used. In this new classification the Swaziland System becomes the Swaziland Sequence, and the Onverwacht and Fig-Tree Series and the Moodies System now become Groups.

Deposition of the Swaziland Sequence Sediments

Examination of the conditions of deposition of this group of sediments (mainly the Fig-Tree and Moodies Group) has shown, in addition to their chemical composition, certain physical characteristics including rhythmic banding, bedding, current-bedding, the presence of current-ripples and grading.

The Fig-Tree Group consists of a thick formation (with maximum thickness 7000 ft) of cleaved fine-grained and coarse-grained slate and greywacke but includes horizons of banded chert, banded ironstone and banded jasper in a variety of colours. The lowermost layer of the Fig-

(a)

(b)

Fig. 10.05. The Moodies and Fig-Tree systems. (a) Moodies System. Cross bedding in vertical Moodies System quartzites (b) Fig-Tree System. Banded cherts in the Fig-Tree sediments. Photographs were kindly provided by Professor John Ramsey

Tree Group (the Sheba Formation) which overlays the Swartkoppie Formation of the Upper Onverwacht Group is a fine-grained sediment which forms continuous individual layers over long distances. These features suggest that the formation was deposited in a bathyal environment in deep water. The upper portions of the Fig-Tree Group (Schoongezight Formation) is characterized by coarse-grained material and conglomeratic beds, which suggest deposition in shallow water. The banded cherts which occur through the Sequence are of great importance both structurally and economically. In the mass of soft contorted rocks these chert bands form the only reliable markers (often called "Bars"), and also act as the locus of much gold and sulphide deposition.

Carbon is of widespread occurrence in the Sequence, and some shales have been converted to graphitic schists by metamorphism. The Sequence also contains relatively large amounts of organic matter and this is of widespread occurrence. The black cherts which occur through the Moodies, Fig-Tree and Onverwacht Groups contain upwards of 0·20% organic material and some layers contain so much carbon material that they have been converted to graphitic slates.

The Sequence (mainly the Fig-Tree Group) contains banded ironstones in a variety of rocks ranging from highly ferruginous shaly material interbedded with bands of crystalline chert to rocks composed of alternating layers of chert and iron oxide with negligible amounts of argillaceous material. The formation of these banded ironstones associated with banded cherts shows that organic acids and carbonic acids must have been present in aqueous solution during the period of sedimentation of the formations, in order to dissolve the ferric oxide and silica. In addition, it is suggested that organic matter must also have been present in the water to prevent the two colloids from co-precipitating. Further, it is suggested that an electrolyte was present in the ocean or lake into which the water drained for sedimentation, in order to precipitate the iron and silica colloids.

THE ONVERWACHT GROUP

The Onverwacht Group occupies about 400 square miles in the southern part of the Mountain Land and recent investigations by M. J. and R. P. Viljoen (1970) have revealed the existence of an hitherto unrecognized assemblage of remarkably well-developed, preserved and exposed rocks. The Group has a thickness of 50,000 ft, mainly dips in a vertical direction and is described as having been folded about a number of major axes. Six formations (Figs 10.06 and 10.07) each with its

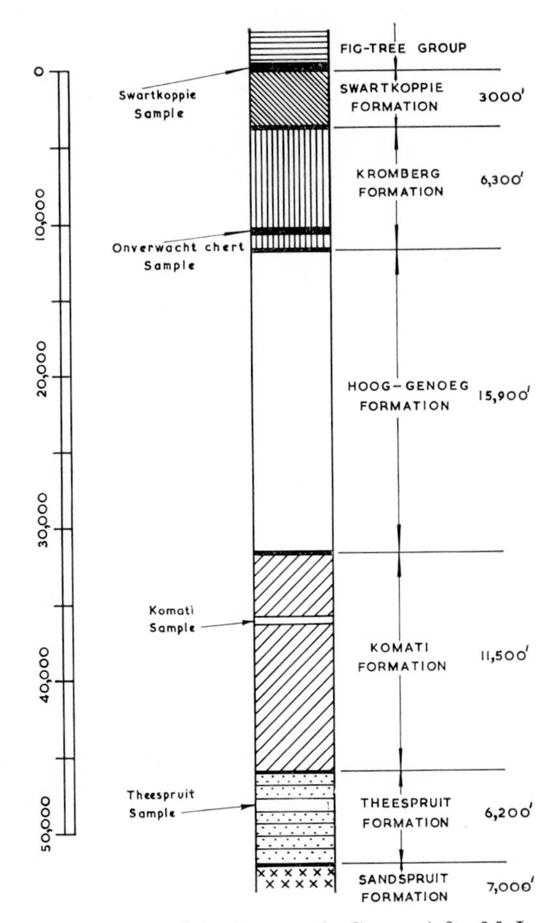

Fig. 10.06. Stratigraphic column of the Onverwacht Group, (after M. J. and R. P. Viljoen, compiled November 1968)

Group	Unit	Formation
		SWARTKOPPIE (3000 ft)
	MAFIC to FELSIC	KROMBERG (6300 ft)
		HOOGGENOEG (15,900 ft)
ONVERWACHT		. . . MIDDLE MARKER . . .
		KOMATI (11,500 ft)
	ULTRAMAFIC	THEESPRUIT (6200 ft)
		SANDSPRUIT (7000 ft)

Fig. 10.07

(a)

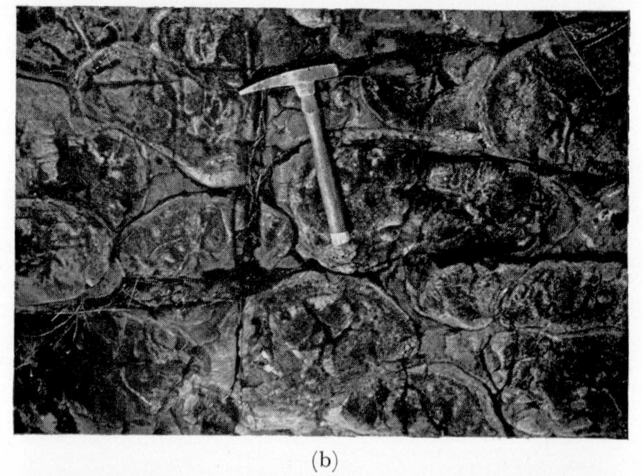

(b)

(c)

own characteristic assemblages and association of rock and mineral types, have been recognized.

The Sandspruit formation is the lowest recognizable formation and forms the basement of the ultramafic unit. It consists largely of ultramafic with some mafic material. The overlying Theespruit formation (Fig. 10.08a) is characterized by the presence of water-laid felsic tuffs (now quartz–sericite schists). These two lower formations are interlayed with mafic and ultramafic horizons as well as a variety of talc, chlorite and carbonate schists. The Komati formation (Fig. 10.08b), which forms the uppermost layer of the lower ultramafic unit, is primarily composed of mafic material, with some ultramafic material. These three foundations are the oldest assemblage of relatively little metamorphosed volcanics known on earth.

The Hooggenoeg formation (Fig. 10.08c) is the lowermost layer of the Mafic to Felsic Unit situated just above the Middle Marker (Fig. 10.09a). It consists of a layered succession of mafic to felsic units. The composition of the overlying Kromberg formation (Fig. 10.09b and c) is very similar. The Swartkoppie formation occurs at the top of the Onverwacht Group and generally consists of sheared intermediate to felsic volcanics, but also interlayered with chert horizons. The Swartkoppie formation had previously been included in the Fig-Tree Group, but recent studies by members of the Economic Geology Research Unit of the University of the Witwatersrand suggests it is better to include the Swartkoppie formation in the Upper Onverwacht Group.

The three lower formations (Sandspruit, Theespruit and Komati) of the Onverwacht Group are distinguishable from the upper formation, in that they contain abundant ultramafic material and are characterized by poorly developed sediments. Above the ultramafic unit and interlaying the two units is an important break between the formation, a continuous sedimentary horizon called the Middle Marker. From the Middle Marker upwards, the petrology and chemistry of the rock, show marked changes; the ultramafic material decreases and felsics pyroclasts and sedimentary materials become more abundant.

Metamorphism, silicification and carbonitization have had varying effects on the original chemistry of the Onverwacht Group in many

Fig. 10.08. Formations of the Onverwacht Group. (a) Theespruit formation. Typical aluminous felsic tuff, showing coarser and finer tuff bands as well as fine-grained siliceous and somewhat carbonaceous cherty bands. These latter particularly towards the shadow area (b) Komati formation. Pillowed magnesium-rich metabasalts which are situated just below the Middle Marker Horizon (c) Hooggenoeg formation. Pillow structures showing vesicular and amygdaloidal cavities and also with well-developed interstitial meta-halagonite (dark material) and carbonate (light material). The photographs were kindly provided by Morris Viljoen

(a)

(b)

(c)

areas. The entire Barberton area has suffered only low-grade regional greenschist facies of metamorphism and within the sedimentary formations many areas show virtually no signs of even the lowest grade of metamorphism. However, the majority of the volcanic rocks of the Onverwacht Group consist almost entirely of greenschist metamorphic assemblages and contain only rare occurrences of primary phases. The chemistry of the Group indicates a remarkable trend of igneous differentiation within the basaltic rocks. These basaltic rocks are all considered to be of a "primitive" nature and the chemistry of the formations of the Mafic–Felsic Unit above the Middle Marker closely resembles that of ocean tholeiites. Most of the basalts of the three lower formations are considered to be very primitive and unlike any other known widespread class of basalts. The extrusive ultramafic unit of the three lowest formations has a distinctive chemistry which is thought to be unlike that of any well-established class of ultramafic material.

The Swaziland System appears to have given rise to two separate orogenies in early Precambrian times. The earliest resulted in the metamorphism of sediments and intrusives of Pre-Swaziland age to form the Ancient Gneiss Complex (Basement) and the production of associated gneisses by granitization. The latter orogeny is Post-Moodies in age and affected the materials that now form the Swaziland System as well as that present as the Pre-Swaziland basement. The oldest granite rocks in association with the Swaziland System have been dated using U–Pb and Rb–Sr radiometric methods as $3{\cdot}2 \times 3{\cdot}4 \times 10^9$ years old. The granite terrain which forms both the ancient low-lying gneiss and the younger high-lying homogeneous plateau north of Swaziland gives way once it reaches Swaziland and forms the underlying variety of gneisses. Hunter (1968) divided these gneisses into the "Ancient Gneiss Complex", the "Tonalitic Gneiss" and the "Granodiorite Suite" and considered the former two to constitute the Pre-Swaziland Sequence floor. However, R. P. Viljoen and M. J. Viljoen consider much of these granites to represent xenoliths of Onverwacht material and report that if this interpretation is correct, then the

---

Fig. 10.09. Formations of the Onverwacht Group. (a) "Middle Marker Horizon." This is mainly black chert, underlain by sheared magnesium-rich metabasalts to the left of the picture and massive metathal basalts to the right. The former belong to the Komati formation and the latter to the Hooggenoeg formation (b) Kromberg formation. Small "puddle" of black somewhat carbonaceous chert terminating a small felsic tuff cycle. The cherty layer has been largely removed by erosion, which is well shown just below the pen. This structure is at the section between the Kromberg Formation and Kromberg Gorge (c) Kromberg formation. Banded black chert (resistant shining layers) and impure limestone (dull grey) sediment and also showing layers of carbonaceous siliceous shale. The photographs were kindly taken by Morris Viljoen

provisional age of $3{\cdot}4 \times 10^9$ years for some of the rocks would be the minimum for the Onverwacht Group and the correct age is probably older.

MICROFOSSILS AND ORGANIC CHEMICALS IN THE FIG-TREE AND ONVERWACHT ROCKS

The search for evidence of early terrestrial life in the better preserved sediments of the Fig-Tree and Onverwacht Group ($3{\cdot}0$ to over $3{\cdot}4 \times 10^9$ years old) has shown conclusively that fossils and related carbonaceous materials unequivocally existed in these carbonaceous argillites, siltstones and cherts, which are probably the oldest, little altered sediments on Earth.

The evidence for the presence of life in these sediments has rested on:

(a) micro-palaeontological studies, which show the presence of morphologically intact and partially intact microfossils of a "spore-like" nature and of preserved thread-like filamentous materials;

(b) the presence of soluble and readily extractable organic compounds which have chemical structures and occur in ratios characteristic of compounds one might expect to have arisen from a biological system;

(c) the presence of insoluble organic matter, in which no kind of visible or microscopic structure may be discerned. This material is quantitatively by far the most important organic constituent and may amount to more than 95% of the total insoluble organic matter and has properties characteristic of sporopollenin (Chapter 9).

Since 1965, when Barghoorn of Harvard University collected cherts from several localities in the Fig-Tree Group and examined them for the presence of organic material and especially fossils, a number of other investigators including geologists, biologists, chemists and geochemists have all studied these rocks for evidence of early forms of life.

Micropalaeontological Studies

Barghoorn and Schöpf used two techniques in their studies: thin sections of Fig-Tree Chert were prepared for examination by reflected and transmitted light and polished surfaces were also made. Examination of the thin sections showed that the rock matrix contained abundant laminations of dark-coloured and virtually opaque particles of organic matter. The laminations of the rock were irregular, but usually aligned parallel to the strata of the chert, suggesting that they originally formed part of an aqueous sedimentary deposit. Barghoorn and Schöpf were unable to discern morphological bodies within the layers of organic matter from examination of the thin sections under the light microscope. However, examination of the polished surfaces of Fig-Tree Group preparations using an electron microscope showed a number of rod-like structures. These structures ranged in length from slightly less than 0·5 μm to about 0·7 μm and were 0·2 μm to a little more than 0·3 μm in diameter. They were compared, and similarity with modern bacteria in both structure and dimension was suggested.

Barghoorn and Schöpf later identified larger microfossils in their thin section preparations. These microfossils were spheroidal and examination of 28 well-defined specimens showed the majority to have diameters between 17 and 20 μm. Several microfossils showed a darkened interior and this suggested that the original cytoplasm within the spheroid may have coalesced and become "coalified". The spheroidal microfossils, when examined with an electron microscope, were found to resemble certain modern blue-green algae of the coccoid group. The rod-like structures were named *Eobacterium isolatum* by Barghoorn and Schöpf and the spheroidal structures *Archaeosphaeroides barbertonensis*. Thus for the first time, evidence has been obtained for the evidence of two species, which had successfully inhabited an aquatic environment more than $3{\cdot}0 \times 10^9$ years ago.

About the same time that Barghoorn was carrying out his studies in Harvard, Hans D. Pflug (at Justus Leibig University, Gissen, Germany) was independently examining cherts and shales from the Fig-Tree sediments for structured organic remains. Samples collected in the vicinity of the Sheba Gold Mine, near Barberton and dated using radiometric data at more than $3{\cdot}2 \times 10^9$ years old, were shown to contain assemblages of the remains of organisms. Chemical and optical examination showed that the walls consisted of organic material. Pflug identified "Globular-type A" structures which resemble cysts of flagellates and "Filamentous-type C" bodies which he suggested were similar to mostocalean blue-green algae. He also assigned other structures of problematic affinity as "Globular-type B", "Filamentous -type D" and "Irregular-type F" species. Chemical analyses of several of the fossil micro-organisms suggested that the parent algae were able to precipitate metal salts, especially iron, copper and calcium ions, from water. This observation led Pflug to suggest that similar biological processes may have been important in the formation of Precambrian sedimentary ore deposits.

The paleobiological assessment of the micro-organisms from the Fig-Tree Group indicated that they were almost certainly of biological origin and probably represented the remnants of single-celled algae-like micro-organisms. These organisms, which have an organic composition, constant morphology, limited size range and relatively good preserved conditions, show some similarity to known spheroidal algae of both modern (blue-green algal groups like the *Chroococcales*) and fossil micro-organisms. The spheroids from the Fig-Tree Group, with their characteristic morphology, are also quite similar in appearance to the reticulate, spheroidal alga-like micro-organisms from the Gunflint Cherts of Lake Superior in Western Ontario ($2{\cdot}0 \times 10^9$ years old) and are also comparable to the well-preserved blue-green algae from the Bitter Springs Formation of Central Australia ($1{\cdot}0 \times 10^9$ years old). In spite of the similarity of the micro-organisms from the Fig-Tree Group with some modern blue-green algae and other younger Pre-cambrian microfossils, care must be taken not to assign them to any particular algal group. Indeed, in view of their extreme age ($3{\cdot}0 \times 10^9$ years) there may not necessarily be any modern equivalent algal group. However, paleobotanists agree that the *Archaeosphaeroides barbertonensis* micro-organisms probably represent the remnants of unicellular, non-colonial, alga-like organisms which may be evolutionary precursors of the modern coccoid blue-green algae. The identifications and classification confirm that photosynthetic micro-organisms existed in Fig-Tree times more than $3{\cdot}0 \times 10^9$ years ago.

Examination of the Onverwacht Group sediments for microstructures by Engel and his co-workers in 1969 showed the presence of "cup-shaped" and spherical microstructures. These microstructures were examined both in thin sections of rock and in powdered preparations, and although both preparations revealed the same type of microstructures, they did not seem to be uniformly distributed in the rock matrix, but appeared rather in small isolated areas or "pockets". The microstructures were chemically resistant to the maceration process (boiling 6N hydrochloric acid followed by treatment for 2 h with boiling 48% hydrofluoric acid) which provided indirect evidence of their gross organic chemical nature. The size distribution of 590 microstructures showed a large spread varying from 6 to 193 μm, with no apparent dominant sizes. This large variation in size suggested that the particles were either non-biological organized elements or maybe deformed members of several species of simple microfossils. The Onverwacht microstructures were reported to have a simple morphology, not to fall within a narrow size range and not to possess complex features resembling specialized structures characteristic of any known living lower plants. Consequently the particles were not considered to meet acceptable criteria for biological origin.

More recent results on the rocks from Swartkoppie, Kromberg and Theespruit formations of the Onverwacht Group from the Komarti River Region by Brooks and Muir have shown the presence of both structured spheroids and filamentous micro-organisms. Slices (0·5–1·0 cm thick) were cut from samples taken from the interior parts of rocks collected from the Swartkoppie, Kromberg and Theespruit formations. These slices were treated with 20% hydrofluoric acid without any prior crushing. Previous experiments indicated that prior crushing of the rock destroyed the morphology of the indigenous organic material. The acid gradually digested the rock matrix and the insoluble organic matter separated as a concentrated dark residue (0·2–0·48% by weight of the rock). These insoluble organic residues were shown (see later) to be very similar chemically to sporopollenin, an insoluble organic polymer of known biological origin (Chapter 9). The quantities and qualities of this insoluble material make it unlikely that it could have migrated from other younger sources and it is therefore generally considered indigenous to the rock (in contrast to soluble organic matter, present only in minute quantities, and vulnerable to all the expressed criticisms of contamination).

The insoluble organic residues and samples of rock remaining after partial acid digestion were mounted on aluminium specimen holders, coated with 200 Å gold and examined in the Cambridge Scientific Instrument Co. "Stereoscan" scanning electron microscope.

The insoluble organic matter occurs in two general types:

(i) Structural material: with definite forms of shapes; sometimes broken specimens can be recognized.

(ii) Amorphous material: in which no kind of visible or microscopical structure can be discerned.

In the Onverwacht rocks, only a limited number of morphologically preserved types are recognized. It is generally accepted that structured micro-organisms, up to several microns in size, and occurring in relatively large quantities, could not pass through intercrystal spaces without considerable distortion of their form, and can be considered indigenous to the rock. Published work shows that perhaps the most remarkable characteristic of Precambrian microfossils is the perfect three-dimensional preservation of even extremely delicate structures. This preservation, and the observation that the micro-organisms often extend without distortion from one mineral crystal to another, across an intercrystal space, seem to provide incontrovertible evidence that

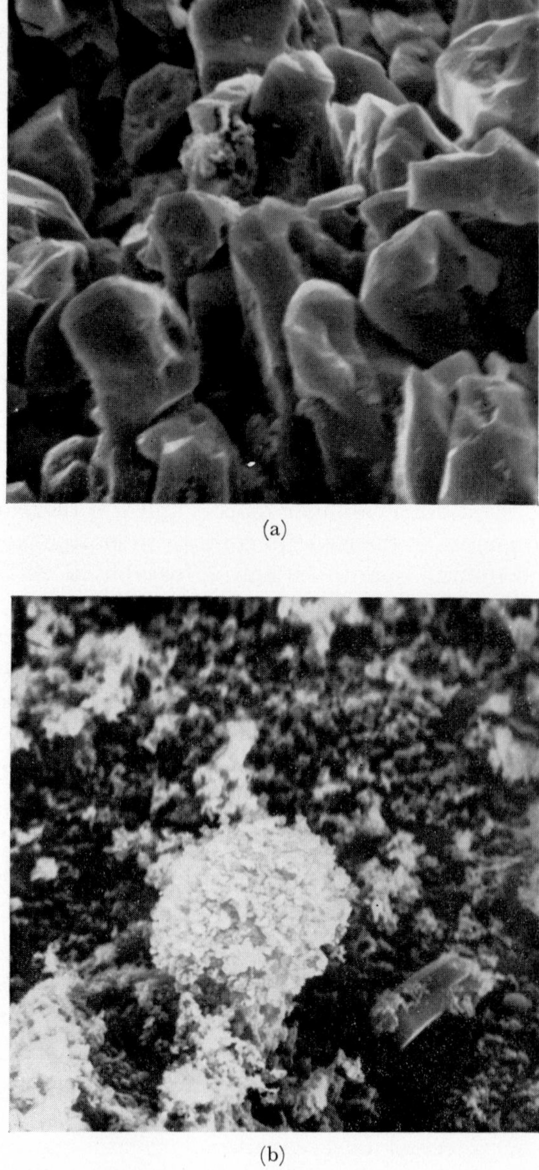

(a)

(b)

Fig. 10.10. Onverwacht Group—Threespruit formation. (a) Chert after etching with hydrofluoric acid. The organic matter shows as small irregular particles on the etched silica grains (b) Spheroid and associated organic residues released from the chert after etching. Well-defined spheroids such as this are relatively uncommon compared with similar residues from younger formations of the Onverwacht Group. Diameter of spheroid approximately 7–10 μm

(a)

(b)

Fig. 10.11. Onverwacht Group—Kromberg formations. (a) and (b) Chert after etching showing organic matter in the form of spheroids and amorphous material. The organic matter is quite abundant and can be seen on the surface of the grains of silica

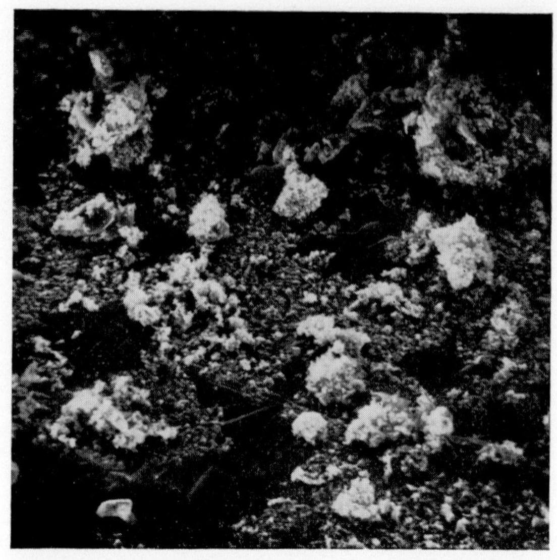

Fig. 10.12. Onverwacht Group—Swartkoppie formation. Organic material in the form of spheroids and amorphous matter present in the etched rock surface

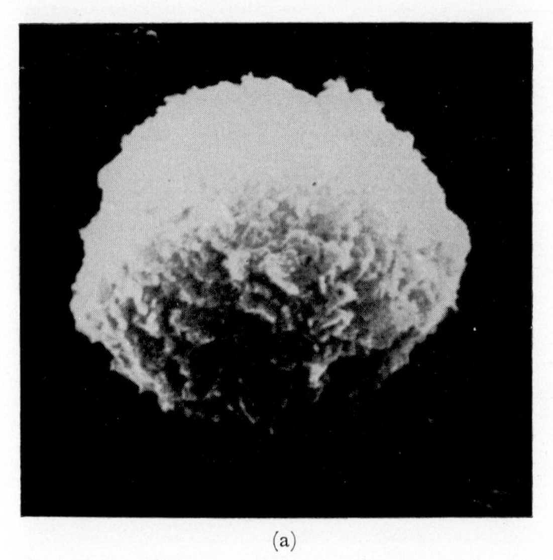

(a)

Fig. 10.13. Onverwacht Group—Swartkoppie formation

(a) Spheroidal object with a small dimple or depression at the bottom right-hand side of the specimen. Diameter 15–20 μm

(b)

(c)

Fig. 10.13. Onverwacht Group—Swartkoppie formation

(b) Spheroidal object together with a partially fragmented specimen. Both spheroidal objects are very comparable with the specimens illustrated by Schöpf (see Fig. 9.09) from the Bitter Springs formation of central Australia

(c) Low magnification view of organic spheroids, showing the relative abundance of the spheres

(a)

(b)

Fig. 10.14. Onverwacht Group—Swartkoppie formation. (a) An organic filament partly released from the siliceous matrix after etching with hydrofluoric acid. Filament approximately 20–25 μm long and 5 μm wide (b) Filamentous material which had been released from the chert after treatment with hydrofluoric and hydrochloric acids. Filament approximately 50 μm long and 10–15 μm wide

structurally preserved micro-organisms are indigenous to the rock and represent chemically little altered organisms that lived in the earliest Precambrian times. These micro-structures—both spheroids and filamentous micro-organisms—are probably the remains of microscopic organisms which could include blue-green algae, eukaryotic organisms or prokaryotic bacteria. In the Onverwacht Group, there are basically two kinds of micro-structures: spheroids and filaments. The former are the most abundant, and, contrary to the results of Engel, in a single sample, the spheroids tend to be of uniform size. Spheroids from the older Theespruit formation (Fig. 10.10) tend to rather less common and smaller (7–10 μm diameter) than those in the other Onverwacht formations. Those from the succeeding younger horizons (Kromberg (Fig. 10.11) and Swartkoppie (Fig. 10.12) formations) are larger (15–20 μm) and much more abundant and not infrequently found associated in pairs (Fig. 10.13) Filamentous bodies are rarer than spheroidal micro-organisms in the Onverwacht formations, but they have been found *in situ* in etched chert (Fig. 10.14) and also in the organic residue remaining after treatment of the Fig-Tree rock with hydrofluoric acid (Fig. 10.15).

These micro-structures are undoubtedly the remains of living organisms, and their morphology suggests that they probably represent the remains of cell walls. Since they are so very old, they are frequently assumed to be simple prokaryotic organisms which reproduced asexually probably by simple division, but there is no real evidence for this and indeed many of them are more reminiscent of eukaryotic organisms.

The Onverwacht Group rocks in which the micro-structures occur are cherts, some of which may be secondary after carbonates, which lie on top of extrusive sub-aqueous volcanic rocks. The cherts are sometimes laterally extensive, but more often occupy pockets in the surface of the lavas. The juvenile water which escapes from present-day volcanic activity, whether at the time of eruption, or at late stages such as fumarole formation, is extremely rich in nutrients for all forms of plant life. Geologically, more recent examples are the hot springs associated with the Late Tertiary volcanic activity in the Rocky Mountains (Yellowstone National Park) of the U.S.A. which contain vast amounts of varied and abundant microflora. Prokaryotic blue-green algae and bacteria are particularly common and occur in various micro-environments, depending upon pH, temperature and nutritional factors. Filamentous, rod-like and spheroidal micro-organisms have even been found to exist in waters almost at boiling point. Similar conditions may well have existed during the formation of the Onverwacht Group sediments in which organisms that could tolerate relatively high tem-

(a)

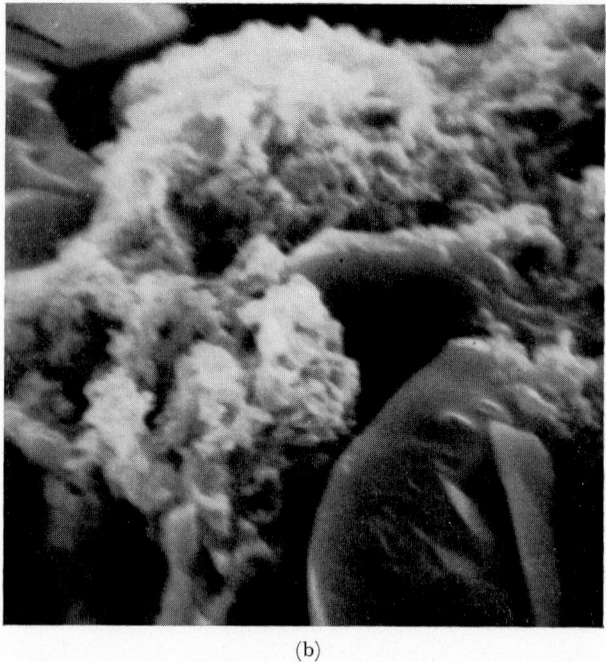

(b)

Fig. 10.15. Fig-Tree Group. (a) Organic spheroids with diameters 15–20 μm (b) Chert from the Fig-Tree group showing the abundance of organic matter protruding from the siliceous matrix after etching with hydrofluoric acid

peratures lived in shallow water lying on the surface of lavas and which, by analogy with present-day extrusives, probably maintained relatively high temperatures for considerable periods of time.

However, in addition to high temperature micro-organisms, microfloras of many volcanic pools are so abundant that they form crusts which eventually cover the entire surface of the pools. In larger volcanic lakes (e.g. Waimungu, New Zealand) extensive crusts form at the edges and extend for a considerable distance towards the centre of the lake.

In the more recent examples, although the temperature of the water is not exceptionally high, the lake waters are charged with juvenile carbon dioxide and other essential minerals such as phosphorus, sulphur and nitrogen. The recent volcanic lakes may mirror the environmental situation in Onverwacht times, and if primitive micro-organisms found themselves in such environments they would be expected to have proliferated and evolved in what could be exceptionally favourable circumstances. Such environments would also help to explain why the structured insoluble organic material occurs in small pockets with a rather random distribution, because if the microbiotas were confined to small pools, then an irregular, sporadic distribution would be expected.

The Onverwacht Group micro-organisms also show morphological similarities with micro-organisms from the overlying Fig-Tree Group which further augments their biological nature.

Soluble Organic Compounds from the Precambrian

Hydrocarbons of high molecular weight have been identified in the Nonsuch shale (1×10^9 years old), the Gunflint Iron formation (2×10^9 years old) and the Soudan Iron formation ($2{\cdot}7 \times 10^9$ years old), and shown to include the saturated isoprenoid components, pristane and phytane and these are considered to have chemical structures and to occur in ratios characteristic of compounds expected to have arisen from a biological system. The presence of these "Biological Markers" in Late and Middle Precambrian rocks has stimulated a search for their presence in the Fig-Tree and Onverwacht sediments.

Alkanes, including pristane and phytane, have been identified in organic extracts from the Fig-Tree chert. Oro and Nooner found extremely low levels (0·003–0·15 ppm) of aliphatic hydrocarbons (C_{15}–C_{25}), including relatively large amounts of pristane to be present in a sample of Fig-Tree rock collected in the Sheba gold mine near

Barberton. Although only small total amounts of alkanes were present in the rock, their carbon number distribution and the configuration of the isoprenoid components indicated that they probably represented compounds with a biogenic origin.

Various attempts have been made to identify amino acids in the Fig-Tree chert, but these compounds occur in very minor quantities (nanomoles per gram of chert) and any assigned significance must be treated with reserve, especially the reports of Schöpf, Kvenvolden and Barghoorn (1968), who extracted a powdered rock sample with 400 ml 0·5N ammonium acetate for 30 min at 80°, washed with 100 ml of hot 0·5N ammonium acetate solution, filtered and evaporated, and removed the remaining ammonium acetate by sublimation. The claim of 5·9 nanomoles of amino acids per gram of Fig-Tree chert must be judged against the method of extraction and although the authors discuss the problems of contamination and indigenousness of the amino acids in the rock, the treatment with ammonium acetate at 80° for 30 min and its removal by sublimation casts doubts on these observations since it would be quite possible that such minute amounts of the amino acids identified (almost entirely glycine) were abiogenically synthesized during the extraction process. Kvenvolden, Petterson and Pollock of the Exobiology Division at the Ames Research Centre, NASA, California, have reported that glycine and a trace of α-alanine were present in the Fig-Tree chert as free amino acids. Bound amino acids (concentration less than about 2 nanomoles per gram of chert liberated by acid hydrolysis were identified as glycine, serine, threonine, leucine, α-alanine, valine, proline, aspartic acid, glutamic acid, β-alanine, phenylalanine and isoleucine. The optical configuration of the bound amino acids was examined by capillary column gas chromatography which showed that the predominant amino acids were of the L(+)-configuration with the D(+)-forms present in very low concentrations. The finding of L(+)-amino acids in these rocks strongly suggested that biological processes had been associated with the rocks. Alternatively, it has repeatedly been shown that complete racemization of amino acids normally takes place in sediments during the earliest processes of sedimentation, so one might expect that amino acids extracted from the $3{\cdot}0 \times 10^9$ years-old Fig-Tree chert would be racemic (see also Chapter 9). These observations therefore suggest that the chert may either have provided an unusually stable environment in which racemization reactions were reduced to an absolute minimum or, more likely, that the optically active amino acids were not indigenous to the chert and were consequently much younger than the rock itself.

Han and Calvin (1969) analysed the aliphatic hydrocarbons and fatty acids extracted from a sample of only slightly metamorphosed Onverwacht chert. They found free aliphatic hydrocarbons (0·05 ppm) and fatty acids (0·04 ppm comprising 0·03 ppm "bound" to the rock matrix but released by acid treatment and 0·01 ppm in the free state). A series of *n*-paraffins with carbon numbers from C_{12} to C_{24} with an equal distribution of odd to even-numbered homologues were also identified. The extracted hydrocarbons were considered to be more characteristic products of photosynthetic micro-organisms than of higher plants, since higher plants would have been expected to contribute significant amounts of hydrocarbons in the range C_{23}–C_{35}. In addition, a series of isoprenoid hydrocarbons were identified with pristane (C_{19}) and phytane (C_{20}), less abundant than their C_{15} and C_{16} homologues. This relationship, which is unusual, led Calvin to suggest that the microbiological activity taking place in the early Precambrian sediments differed from that in sediments of younger geological age. The fatty acid fraction also contained palmitic acid (*n*-C_{16}) as the major component, and since micro-organisms are known to contain large quantities of this acid, this was taken as good evidence that the organic compounds extracted from the Onverwacht chert had a biological origin.

Brooks and Shaw examined a hydrocarbon fraction (0·10 ppm of the rock) extracted from a sample of Onverwacht Group rock of the Swartkoppie formation. Capillary column gas chromatographic analysis of the fraction showed the presence of a complex mixture of straight- and branched-chain alkanes ranging in carbon number from C_{16} to C_{32}, of apparent bi-modal distribution with maxima about *n*-C_{32} and *n*-C_{28}, and an even distribution of odd- and even-numbered carbon homologues. This distribution of carbon compounds in the hydrocarbon fraction (with major components at *n*-C_{20} and *n*-C_{28}) is significantly different from those extracted from other samples of Onverwacht chert. Stable carbon isotope ratios were measured on the soluble organic extract ($\delta^{13}C$—23·2) and this value was almost identical with the value for the carbonate ($\delta^{13}C$—3·2) present in the chert. It was reasoned that if the carbonate deposit was representative of the carbon dioxide present in the atmosphere prior to and during the Onverwacht sediment formation, then photosynthesizing micro-organisms would have fractionated the residual stable carbon isotopes in the organic matter. It appears that the $\delta^{13}C$ stable isotope ratio of the carbonate present in the Onverwacht chert may not reflect the Early Precambrian carbon dioxide in the atmosphere, but could be the result of either carbonitization of the sediment or a "biological" carbonate deposition caused by reactions taking place within the sediment (see later). These stable carbon isotope ratios must of course be interpreted with great care.

The soluble organic chemicals present in the early Precambrian sediments occur in very minor quantities and in any case are likely to have undergone movement in total or in part from their point of origin since they have a potential mobility in non-aqueous (hydrocarbon) or aqueous (emulsion) phases, so that it may not be always possible to align precisely such chemicals with a particular era of time. Following comments by Brooks and Shaw (1968), Nagy (1970) has shown that ready permeation of a sample of Onverwacht chert by a suspension of hydrocarbons in water occurred to give a rock which contained soluble organic matter (at least 0·35 g of dissolved C_9–C_{30} *n*-paraffins in water can flow through one cubic metre of the chert) of an amount similar to or greater than that normally associated with the chert. Recent studies by Sanyal, Kvenvolden and Marsden showed that three (Theespruit, Kromberg and Swartkoppie Formations) out of four (the other being the Hooggenoeg Formation) samples of the Onverwacht chert had measured permeabilities much higher than the single value reported by Nagy. These workers commented that the relationship between measured permeabilities and true permeabilities during geological time is not known, but cementation, compaction and metamorphism may have caused significant permeability reduction during the geological past. Alternatively, it is suggested that earth movements may have increased the effective *in situ* permeabilities of these rocks by inducing fracturing which may not be evident in small laboratory samples.

The permeability and porosity experiments show that it is quite likely that fluids have been able to flow through these rocks by means of both intergranular and fracture channels. The observations of Kvenvolden suggest that the flow of fluids in these rocks may have been even greater than estimated by Nagy, and consequently many of the Onverwacht cherts may have been host rocks rather than source rocks for some of the contained soluble organic compounds. The laboratory results confirm the long-held view that soluble organic matter (or at least a variable proportion) is probably brought into the sediment by solution which is known to percolate through the rocks at dates following sediment formation. The studies by Visser (1956) of Precambrian rocks in the Barberton area commented on the abundance of spring water seeping into the sediment. It is difficult and often misleading therefore to interpret the origin of soluble organic chemical "Biological Markers" in early Precambrian rocks.

Chemical Studies on the Insoluble Organic Material of Precambrian Sediments of the Fig-Tree and Onverwacht Series

Quantitatively, by far the most important organic constituent of the Fig-Tree and Onverwacht groups of sediments is the insoluble organic polymeric material, or kerogen, they contain. This insoluble organic matter, which in many rock samples can be as high as 99% of the total organic matter present in the rock, seems to be basically indigenous and is considered to have deposited with the rocks during Fig-Tree and Onverwacht times.

Scott, Modzeleski and Nagy (1970) examined various samples of Fig-Tree and Onverwacht cherts for insoluble organic matter using pyrolysis gas chromatography and mass spectroscopy. These experiments involved direct pyrolysis of ground-up rock samples (Fig-Tree, Kromberg and Theespruit formations) containing organic matter (with up to 1% total carbon). Results suggested that the Onverwacht samples contained mainly aromatic degradation products of kerogen, while the Fig-Tree sample showed an abundance of *n*-alkanes, which the workers suggest are usually associated with biological origin. Nagy commented that it is difficult to account for an aromatic-type kerogen from biological sources in early Precambrian times, long before the appearance of the abundant aromatic biochemical lignin, and concluded that the origin of the aromatic kerogen is unknown, although it has been suggested (Brooks and Shaw, 1969) that such a composition might be derived from living systems.

Chemical and geochemical studies led Brooks and Shaw to postulate that there may be a relationship between the chemical nature of the spore-like micro-organisms and the more abundant amorphous insoluble organic matter present in the sediment.

Sporopollenin is the tough highly resistant chemical substance which constitutes the major part of modern spore walls and has been shown in many cases to be chemically identical with the most resistant insoluble organic matter or "kerogen" present in Precambrian sediments (Chapter 9), and it was suggested that the presence of sporopollenin could be used as a precise marker for biological materials in early Precambrian rocks. These suggestions were based on a comparison of both the physical and chemical properties of natural fresh and fossil spores and synthetic sporopollenins which have been examined with those of a variety of sedimentary insoluble organic residues. In particular they each possess characteristic resistance to anaerobic biological, and non-oxidative chemical decay, they survive acetolysis, they are unsaturated and produce very similar products (mono- and di-carboxylic acids) on oxidation and have similar chemical analyses. Also the frequent presence of nitrogen and sulphur in fossil sporopollenins can be readily duplicated in the laboratory by mild treatment of fresh

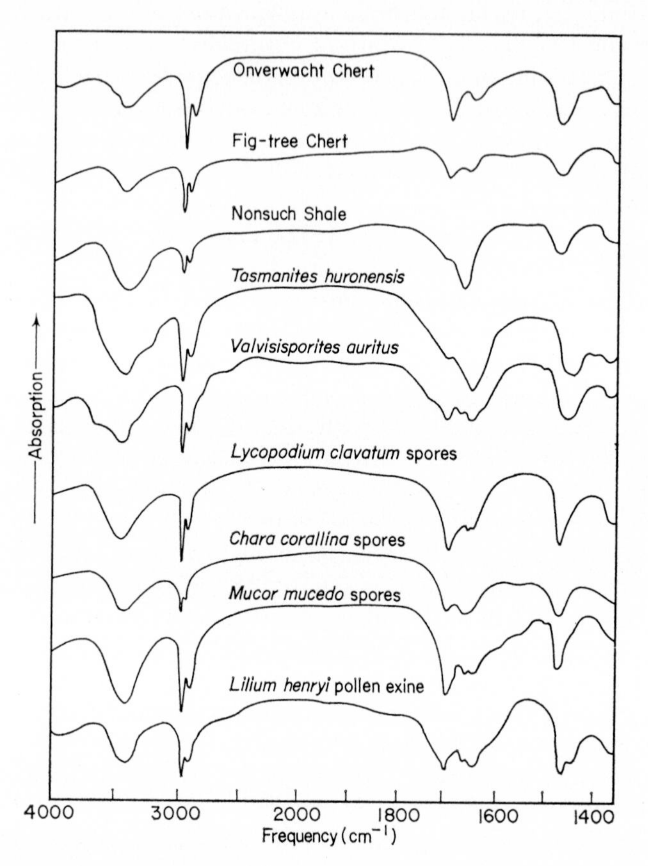

Fig. 10.16. The infrared spectra of some insoluble organic matter from various cherts, shales, fossil spores and modern pollen and spores

sporopollenin with various nitrogen and sulphur-containing compounds and with elemental sulphur.

Sporopollenin derived from modern *Pediastrum duplex* (a colonial algae), modern *Chara corallina* (a green algae), the sexual (±) zygospore of the modern fungus *Mucor mucedo*, those from many modern pollen grain exines, and synthetic material prepared by oxidative polymerization of carotenoids, were compared with the insoluble organic matter present in the Fig-Tree and Onverwacht Series cherts and with morphologically intact planktonic algal microfossils *Tasmanites punctatus* ($2{\cdot}5 \times 10^8$) and *Tasmanites huronensis* ($3{\cdot}5 \times 10^8$) and a Carboniferous megaspore, *Valvisisporites auritus* ($2{\cdot}5 \times 10^8$ years old).

The analytical techniques used for detailed comparison included:

(i) Infra-red spectroscopy (Fig. 10.10) which reveals in all materials a readily repeatable and quite characteristic spectral pattern. In particular the broad bands at 3500–3000 cm^{-1} may be assigned to hydrogen bonding of hydroxyl groups and are characteristic of hydroxyl-

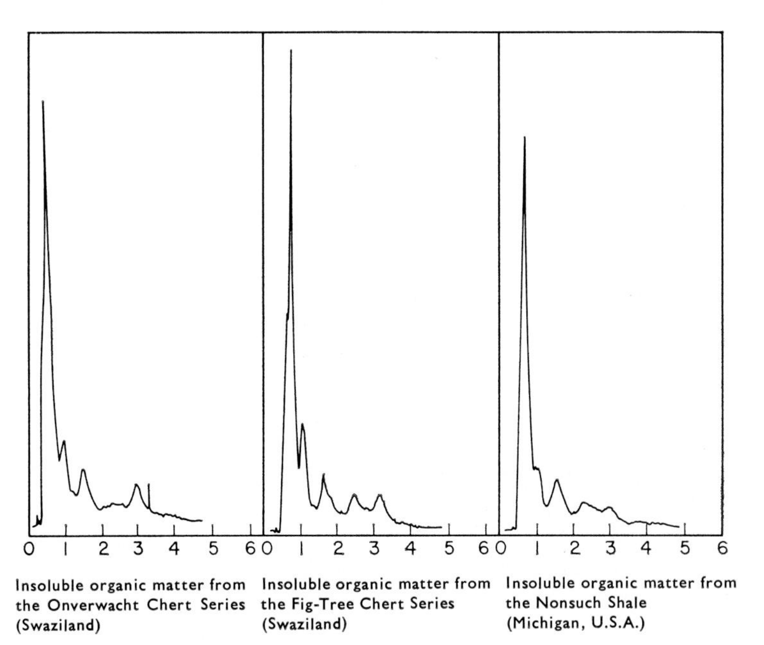

Fig. 10.17. Pyrolysis–gas chromatograms of some insoluble organic matter from Precambrian sediments. (Pyrolysis temperature 610°C)

containing polymers. The group of bands between 3000 and 2800 cm^{-1} represent carbon–hydrogen bond stretching vibrations of $—CH_3$, $—CH_2$ and —CH groups. The broad band at ~1700 cm^{-1} and part of the band at 1650 cm^{-1} is attributed to carbonyl (C : 0) absorption and the latter may include conjugated C=C absorption. The absorption at 1450 cm^{-1} is due to either C-methyl or $—CH_2$ groups and the band at 1380 cm^{-1} to C-methyl groups.

(ii) Pyrolysis gas chromatography (Fig. 10.17); here similar chromatograms are obtained, and variations of pyrolysis temperature results in similar changes in the chromatogram patterns.

(iii) Fusion with potassium hydroxide and thin-layer chromatography of the products (Fig. 10.18), which in all cases reveals the presence of a mixture of *m*- and *p*-hydroxybenzoic and protocatechuic acids in varying amounts with smaller amounts of other phenolic acids. Contrary to some published work, it is interesting to note that phenolic acids of this type have origins other than lignin.

(iv) Resistance to acetolysis.

(v) Chemical unsaturation and hence susceptibility to oxidation. The

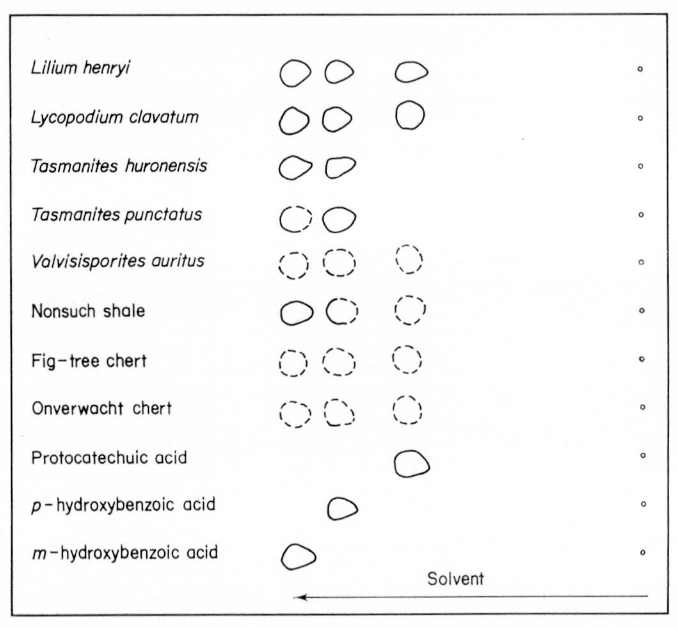

Fig. 10.18. Thin-layer chromatography of the potash fusion products of some Precambrian insoluble organic matter, fossil and modern sporopollenins

most useful degradative technique involved oxidation with ozone. Analysis of the major degradation products by capillary column gas chromatography (Figs. 10.19 and 10.20) indicated the presence of mixtures of mono- and of di-carboxylic acids. The mono-carboxylic acids contained major components corresponding to palmitic (n-C_{16}) and stearic (n-C_{18}) acids, and the di-carboxylic acids contained C_6–C_{17} components with major identified components C_6–C_{11} (including probable C-methyl and gem-dimethyl substituted derivatives C_6 to C_8). These products are very characteristic of the sort of compounds produced by ozone degradation of sporopollenins. Small differences occurring between the various types of sporopollenin are not only understandable, but entirely to be expected in terms of its known chemistry and geochemistry. The somewhat enhanced production of di-carboxylic acids obtained from the insoluble organic matter derived from the Onverwacht chert compared with most typical fresh sporopollenins may be ascribed to some aromatization induced in the Precambrian material by subjection to mild thermal processes (estimated not much greater than 200°) Brooks and Shaw have shown that sporopollenin when heated in ground rock samples did not dehydrate at

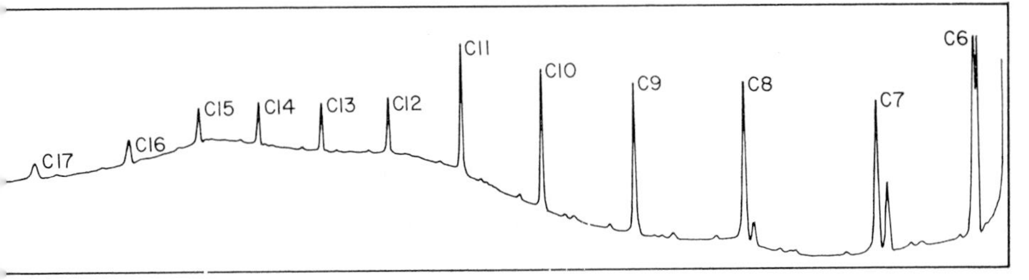

Fig. 10.19. Chromatogram of di-carboxylic acids (methyl esters) from ozonolysis of the insoluble organic matter extracted from the Onverwacht chert

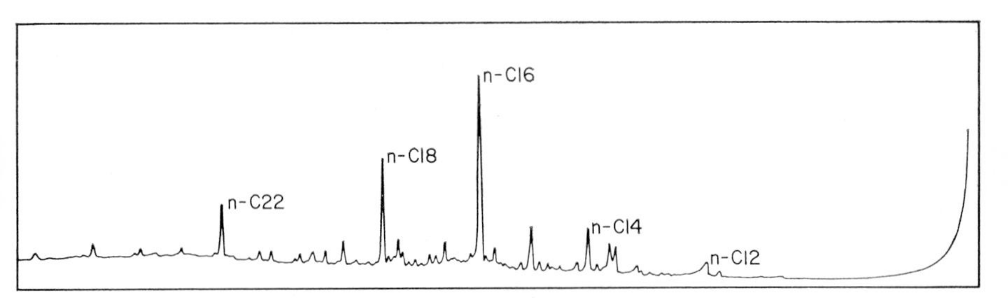

Fig. 10.20. Chromatogram of mono-carboxylic fatty acids (methyl esters) from ozonolysis of the insoluble organic matter extracted from the Onverwacht chert

temperatures less than 200°, but at higher temperatures (200° up to 400°C) rapid dehydration accompanied by enhanced aromatization occurs as indicated by elemental analysis, infrared spectra and radiochemical analysis (see also Chapter 9).

The stable carbon $\delta^{13}C$ isotope ratio for the Onverwacht Group (Kromberg Formation) chert carbonate, insoluble organic and soluble organic matter (Table 10.01) shows significant differences from the corresponding values of the Fig-Tree chert. Since we may regard the soluble material as probable contamination, examination of the insoluble organic material and carbonate, considered indigenous to the chert, is discussed.

Table 10.01. Analyses of some Precambrian sedimentary rocks

	Age ($\times 10^9$ years)	% carbonate		% soluble organic matter		% insoluble organic matter		Elemental analysis of organic matter				
			$\delta^{13}C$		$\delta^{13}C$		$\delta^{13}C$	%C	%H	%O	%N	%S
Nonsuch shale	1·0	13·1	−8·3	0·06	−28·1	4·00	−28·1	—				
Fig-Tree chert	3·1	1·6	−2·2	0·4 ppm	−27·5	0·22	−26·9	74·02	6·98	14·56	1·94	2·18
								$C_{90}\ H_{103}\ O_{13}\ N_2\ S$				
Onverwacht chert	3·4–3·7	0·05	−3·2	0·3 ppm	−24·2	0·24	−15·8	85·88	5·76	7·70	0·60	—
								$C_{90}\ H_{72}\ O_7\ N_{0.5}$				

If one takes the Fig-Tree sediments as an example of an unmetamorphosed carbonaceous sediment, the $\delta^{13}C$ value of the insoluble organic material is in agreement with reported values ($\delta^{13}C$–26) for other Precambrian (marine) deposits and the $\delta^{13}C$ value for the car-

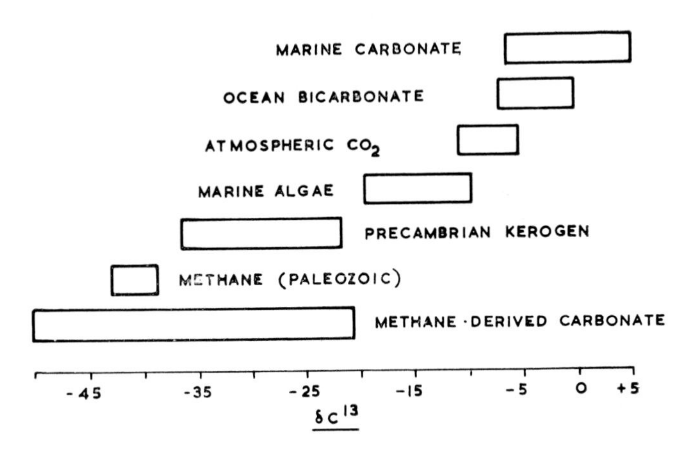

Fig. 10.21. $\delta^{13}C$ in various types of organic matter and carbonates

bonate is representative of a marine or oceanic carbonate ($\delta^{13}C$ 0 to -5) (Fig. 10.21). The corresponding values for the Onverwacht insoluble organic matter ($\delta^{13}C$–15·8) and carbonate ($\delta^{13}C$–3·2) are very unlike the value for normal Precambrian sediments.

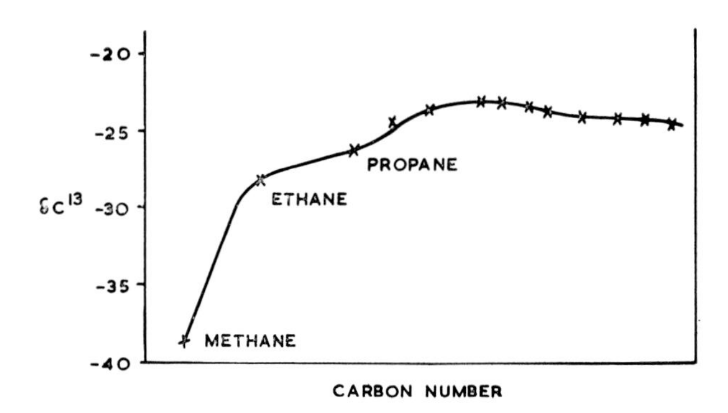

Fig. 10.22. δ^{13} values of low molecular weight hydrocarbon gases evolved from organic matter

Silverman has shown that when organic matter is heated, evolution of low molecular weight hydrocarbons C_1–C_4 (Fig. 10.22) occurs and these hydrocarbons show an enrichment of the ^{12}C isotope (methane

CH_4 $\delta^{13}C$–38) value (towards $\delta^{13}C \rightarrow 0$) of the residual organic matter. Since the Onverwacht insoluble organic material has been heated to at least 200° at some time during its geological history, one might expect that both low molecular weight hydrocarbons and probably carbon dioxide were evolved. This would result in slow alteration of the insoluble organic residue (typical Precambrian value $\delta^{13}C \sim -26$) and the recorded value ($\delta^{13}C$–15·8) would result from the prolonged evolution of low molecular weight hydrocarbon.

Hathaway and Degens in studies with carbonate-bearing rocks have found $\delta^{13}C$ values for carbonates in a range up to $\delta^{13}C$–60. The changes in $\delta^{13}C$ value for marine or oceanic carbonate (normal values $\delta^{13}C$ 0 to −5) to values of $\delta^{13}C$–60 are caused by chemical oxidation of methane which has been evolved from biological material,

$$CH_4 + 2O_2 \rightarrow CO_2 + 2H_2O$$

and this reaction is favoured in lower temperature environments. The carbon dioxide formed by this reaction would dissolve in water and be precipitated as calcium carbonate. Carbonates deposited in this way are known to have $\delta^{13}C$ values within the range which include the Onverwacht Group chert carbonate.

The processes taking place in the Onverwacht Group sediments may be summarized as follows: insoluble organic matter of biological origin similar in nature to material present in the Fig-Tree sediment underwent mild thermal alteration resulting in evolution of low molecular weight hydrocarbons and some aromatization of the residual insoluble organic matter.

Together with stereoscan micrographs of the structured spheroidal and filamentous micro-organisms present in the Onverwacht chert, chemical studies strongly augment and, we believe, provide powerful evidence for the existence of life not only throughout the Middle and Late Precambrian (Chapter 9), but also in the oldest known rocks on earth—the early Precambrian Onverwacht Sediments.

SUGGESTED FURTHER READING Chapter 10

Books

Brooks, J., Grant, P. R., Muir, M. D., van Gizjel, P. and Shaw, G. (Eds) (1971). "Sporopollenin". Academic Press, London.

Calvin, M. (1969). "Chemical Evolution". Oxford University Press.

Craig, H., Miller, S. L. and Wasserburg, G. T. (Eds) (1964). "Isotopic and Cosmic Chemistry". North-Holland, Amsterdam.

Harbone, J. B. (Ed.) (1970). "Phytochemical Phylogeny". Academic Press, London.

Gass, I. G. and Clifford, T. N. (Eds) (1970). "African Magmatism and Techtonics". Oliver and Boyd, Edinburgh.

Haughton, S. H. (1969). "Geological History of Southern Africa". Geological Society of South Africa.

Margulis, L. (1970). "Origin of the Eukaryotic Cell". Yale University Press.

Rutten, M. G. (1971). "The Origin of Life by Natural Causes". Elsevier, Amsterdam.

Articles

Barghoorn, E. S. (1971). The Oldest Known Fossils. *Sci. Am.* **224,** 30–42.

Barghoorn, E. S. and Schöpf, J. W. (1966). Microorganisms Three Billion Years Old from the Precambrian of South Africa. *Science, N.Y.* **152,** 758–763.

Brock, T. D. and Darland, G. R. (1970). Limits of Microbial Existence: Temperature and pH. *Science, N.Y.* **169,** 1316–1318.

Brooks, J. (1971). Some Chemical and Geochemical Studies on Sporopollenin. *In* "Sporopollenin" (Eds J. Brooks *et al.*). 351–407. Academic Press, London.

Brooks, J. and Muir, M. D. (1971). Morphology and Chemistry of Organic Insoluble Matter from the Onverwacht Series Precambrian Chert and the Orgueil and Murray Carbonaceous Meteorites. *Grana* **11,** 9–14.

Brooks, J. and Muir, M. D. (1971). "Early Precambrian Microorganisms in the Onverwacht Group ($3{\cdot}4$–$3{\cdot}7 \times 10^9$ Years) from the Swaziland Sequence of Southern Africa". Third International Palynological Conference, Novosibirsk, U.S.S.R. (in press).

Brooks, J. and Muir, M. D. (1972). "Chemistry and Morphology of the Micro-organisms in the Early Precambrian Rocks of the Onverwacht Group". IUPAC International Symposium on Chemistry in Evolution and Systematics held in Strasbourg, France, July 3–8.

Brooks, J. and Shaw, G. (1968). Identification of Sporopollenin with Older Kerogen and New Evidence for the possible Biological Source of Chemicals in Sedimentary Rocks. *Nature, Lond.* **220,** 678–679.

Brooks, J. and Shaw, G. (1971). Evidence for Life in the Oldest Known Sedimentary Rocks—the Onverwacht Series Chert, Swaziland System of Southern Africa. *Grana* **11,** 1–8.

Brooks, J. and Shaw, G. (1972). Geochemistry of Sporopollenin. International Geochemical Congress, Moscow, U.S.S.R. *Chem. Geol.*

Button, A. (1972). Early Proterozoic Algal Stromatolites of the Pretoria Group Transvaal Sequence. *Trans. Geol. Soc. South Africa,* 201–210.

Engel, A. E. J. *et al.* (1968). Algal-like Forms in Onverwacht Series, South Africa: Oldest Recognized Lifelike Forms on Earth. *Science, N.Y.* **161,** 1005–1008.

Han, J. and Calvin, M. (1969). Occurrence of Fatty Acids and Aliphatic

Hydrocarbons in a 3·4 Billion-year-old Sediment. *Nature, Lond.* **224,** 576–577

Hall, A. L. (1918). The Geology of the Barberton Gold Mining District. *Mem.* **9,** *Geol. Surv. S. Afr.*

Hathaway, J. C. and Degens, E. T. (1969). Marine-derived Methane Carbonates of Pleistocene Age. *Science, N.Y.* **165,** 690–692.

Hunter, D. R. (1968). The Precambrian Terrain in Swaziland with Particular Reference to the Granite Rocks. PhD. Thesis, University of Witwatersrand, Johannesburg, South Africa.

Hurley, P. M., Pinson, W. H., Nagy, B. and Teska, T. M. (1972). Ancient Age of the Middle Marker Horizon, Onverwacht Group, Swaziland Sequence, South Africa. *Earth Planet. Letters* **14,** 360–366.

Nagy, B. and Nagy, L. A. (1969). Early Precambrian Onverwacht Micro-Structures, possibly the Oldest Fossils on Earth. *Nature, Lond.* **223,** 1226–1229.

Nagy, B. (1970). Porosity and Permeability of the Early Precambrian Onverwacht Chert: Origin of the Hydrocarbon Content. *Geochim. cosmochim. Acta* **34,** 525–527.

Oehler, D. Z., Schöpf, J. W. and Kvenvolden, K. A. (1972). Carbon Isotopic Studies of Organic Matter in Precambrian Rocks. *Science, N.Y.* **175,** 1246–1248.

Oró, J. and Nooner, D. W. (1967). Aliphatic Hydrocarbons in Precambrian Rocks. *Nature, Lond.* **213,** 1082–1085.

Pflug, Hans. D. (1967). Structured Organic Remains from the Fig-Tree Series (Precambrian) of the Barberton Mountain Land (South Africa) *Rev. Palaebot. Palynol.* **5,** 9–29.

Sanyal, S. K., Kvenvolden, K. A. and Margden, S. S. (1971). Permeabilities of Precambrian Onverwacht Cherts and Other Low Permeability Rocks. *Nature, Lond.* **231,** 325–327.

Schöpf, J. W. and Barghoorn, E. S. (1967). Alga-like Fossils from the Early Precambrian of South Africa. *Science, Lond.* **156,** 508–512.

Schöpf, J. W., Kvenvolden, K. A. and Barghoorn, E. S. (1968). Amino Acids in Precambrian Sediments: an Assay. *Proc. natn. Acad. Sci. U.S.A.* **59,** 639–646.

Scott, W. M. *et al.* (1970). Pyrolysis of Early Precambrian Onverwacht Organic Matter. *Nature, Lond.* **225,** 1129–1130.

Silverman, S. R. (1964). Investigations of Petroleum Origins and Evolution Mechanisms by Carbon Isotope Studies. *In* "Isotopic and Cosmic Chemistry" (Ed. H. Craig), 92–102, North-Holland, Amsterdam.

Smith, J. W., Schöpf, J. W. and Kaplan, I. R. (1970). Extractable Organic Matter in Precambrian Rocks. *Geochim. cosmochim. Acta* **34,** 659–675.

Sylvester-Bradley, P. C. (1971). Environmental Parameters for the Origin of Life. *Proc. geol. Ass.* **82,** 87–136.

Sutton, J. (1967). The Extension of the Geological Record into the Precambrian. Presidential Address to the Geological Society. *Proc. Geol. Ass.* **78,** 493–534.

Transactions of the Geological Society of South Africa (1972). South African Code of Stratigraphic Terminology and Nomenclature, Introduction, pp. 111–129.
Viljoen, M. J. and Viljoen, R. P. (1969). Upper Mantle Project, 11 parts. Geological Society of South Africa, Special Publication No. 2.
Visser, D. J. L. (1956). The Geology of the Barberton Area. Geological Survey of South Africa, Special Publication No. 15.

11
Meteorites

INTRODUCTION

The solar system contains many solid bodies with sizes varying from those of the planets through the smaller but still large satellite moons and asteroids to microscopically small dust particles. This solid matter follows various orbits round the sun and the smaller particles especially are constantly dragged into the gravitational field of their fellow bodies, including the earth. Much of this solid matter after entering the earth's

Fig. 11.01. Meteor shower. The Leonid Meteor Shower which occurred on 17 November 1966. This photograph was taken by D. McLean at Kitt Peak. Photograph (RAS 697) was kindly provided by the Royal Astronomical Society, London

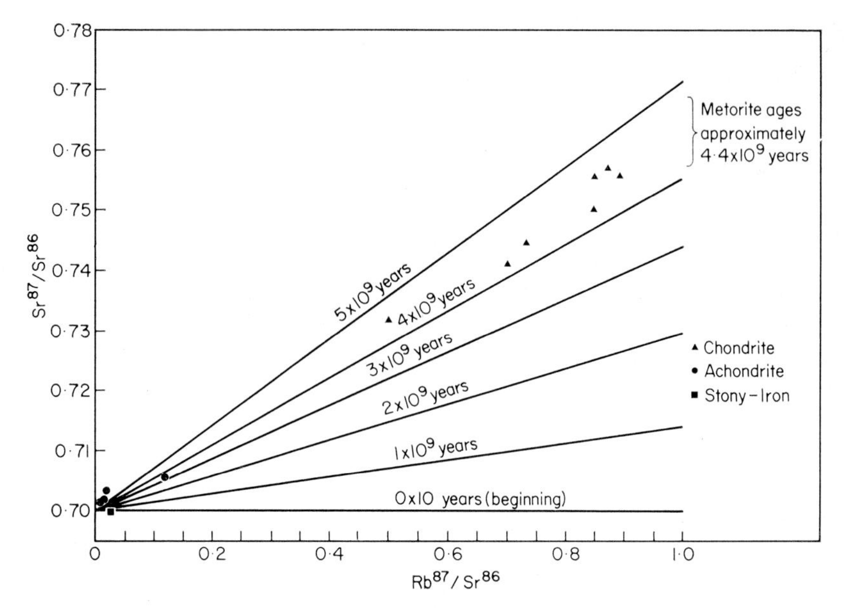

Fig. 11.02. Strontium[87] and Rubidium[87] abundances in meteorites

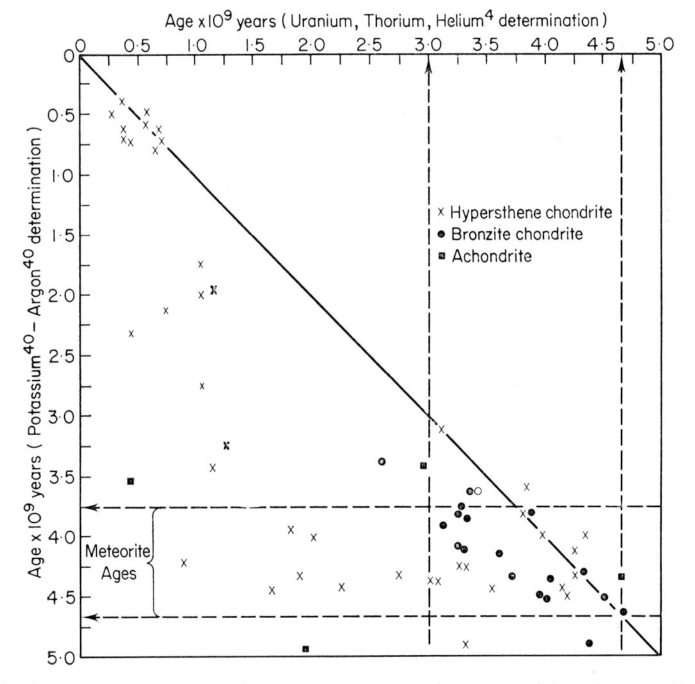

Fig. 11.03. Age determination of stony meteorites (cf. U, Th–He^4 and K^{40}–A^{40} results)

upper atmosphere is rapidly destroyed by frictional heat and combustion. Such material constitutes *meteors* (Fig 11.01). On the other hand, larger particles may survive the heat and combustion and eventually land on earth. These are the *meteorites*, and although their outermost parts are heated to high temperatures during descent, nevertheless the low conductivity of their substance to heat frequently means that, especially in reasonably large specimens, their internal parts have remained unaltered.

They are currently the only extra-terrestrial material containing organic matter available to us, and they are also the oldest known matter. Radiochemical dating (Chapter 2) has repeatedly shown meteorites to be $4{\cdot}5 \times 4{\cdot}7 \times 10^9$ years old (Figs 11.02 and 11.03), which makes them at least $0{\cdot}5 \times 10^9$ years older than the oldest rocks so far found on earth. Meteorites thus contain an ancient record whose counterpart may have been destroyed on the earth's surface by geological processes.

CLASSIFICATION OF METEORITES

It has proved difficult to achieve an agreed, satisfactory classification of meteorites. The main bases used include mineralogical composition, chemical composition, metal content, elemental composition, internal structure and colour. The various combinations of these parameters has resulted in both simple and complex classifications.

Currently meteorites are divided into three broad classes, which differ in their metal silicate content and are classified as: *Irons*, *Stones* and *Stony-Irons*.

In 1965, 1759 meteorites were known and by 1971 the total number of meteorites had reached 2163 (Table 11.01). Less than half of these are

Table 11.01. Total known meteorites (March 1971) 2163

Irons	688	31·8%
Stony-Irons	78	3·6%
Stones (Chondrites)	1305	60·3%
Unclassified	92	4·2%

meteorites of known falls (987 meteorites), whilst the others were found (1176). Stony meteorites (Tables 11.02 and 11.03) comprise by far the largest proportion of the known falls (77% of the 987 meteorites), whilst iron meteorites (Table 11.04) comprise 7·8% and stony-irons 1·3% of

Table 11.02. Stony meteorites

Type	Falls	Finds	Paired	Total	
CHONDRITES					
Enstatites	11	6	—	17	1·5%
Olivine–bronzites	232(1)	216(6)	3	458	38·4%
Olivine–hypersthene	285(2)	190(3)	8	488	41·0%
Amphoterites	40	10	—	50	4·2%
Carbonaceous					
Type C3 and C4 olivine–pigeonite	10(1)	2	—	13	1·1%
Type C1 and C2	22	—	—	22	1·9%
Unclassified	2	—	—	2	0·1%
Unclassified	44(1)	22(3)	2	72	6·1%
ACHONDRITES					
Enstatites	8	1	—	9	0·7%
Hypersthenes (diogenites)	8	—	—	8	0·6%
Olivine (Chassignites)	1	—	—	1	<0·1%
Olivine–pigeonites (ureilites)	2	3	—	5	0·5%
Diopside–olivines (nakhlites)	1	1	—	2	0·1%
Augite (angrites)	1	—	—	1	<0·1%
Plagioclase–pigeonites (eucrites)	25(1)	3	—	29	2·5%
Plagioclase–hypersthene (howardites)	13	1	—	14	1·2%
Unclassified	1	1	—	2	0·1%
	712	468	13	1193	

Table 11.03. Other stony meteorites

Type	Falls	Finds	Paired	Total	
Whitleyite	1	—	—	1	0·9%
Calcarite	(1)	—	—	1	0·9%
Amathosites	—	(2)	—	2	1·8%
Unclassified	42(36)	17(8)	1	104	96·3%
	80	27	1	108	

Table 11.04. Iron meteorites

Type	Falls	Finds	Paired	Total	
Nickel-rich ataxites	2	43	—	45	6·5%
Finest octahedrites	—	21	1	22	3·2%
Fine octahedrites	4	65	1	70	10·2%
Medium octahedrites	16	197(1)	13	227	33·0%
Coarse octahedrites	4	84(2)	5	95	13·8%
Coarsest octahedrites	2	22	—	24	3·5%
Octahedrites (not further classified)	8(2)	39	2	51	7·4%
Hexahedrites	6	32(1)	11	50	7·3%
Nickel-poor ataxites	1(1)	23	3	28	4·0%
Brecciated octahedrites	2	5	—	7	1·0%
Metabolites	—	3	—	3	0·4%
Unclassified irons	4(15)	33(12)	2	66	9·6%
Totals	49(18)	567(16)	38	688	

( ) doubtful classification.

Table 11.05. Stony-iron meteorites

Type	Falls	Finds	Paired	Total	
Pallasites	4(1)	39(3)	2	49	62·8%
Siderophyre	—	1	—	1	1·3%
Lodranite	1	—	—	1	1·3%
Mesosiderites	7	19	—	26	33·3%
Unclassified		(1)		1	1·3%
	12	64	2	78	

the known falls (Table 11.05). The stony meteorites are usually divided into two sub-classes: *Chondrites* and *Achondrites.*

Chondrites are the main type of meteorite (84% of known falls) and contain rounded mineral formations (Fig. 11.04a and b) (globules) of silicate (called chondrules from the Greek word *chondrós* meaning grain), with diameters from 0·5 to 1–2 mm but not exceeding 3–4 mm.

Achondrites are chondrule-free stones, and also differ from chondrites in both texture and composition.

Chondrites are the most abundant meteorites and generally considered

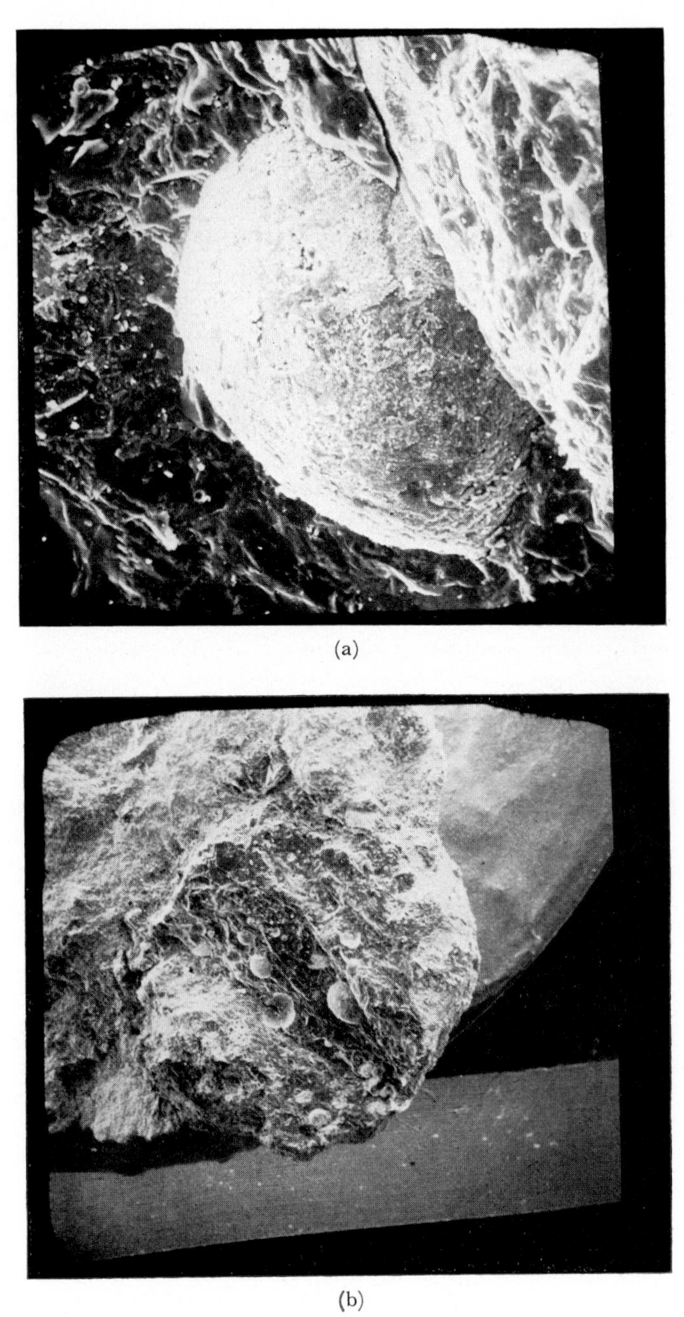

(a)

(b)

Fig. 11.04. Meteorite structure showing chondrules. (a) and (b) The surface of the Allende meteorite showing the rounded mineral formation called chondrules

to have the most primitive compositions. They are divided into five sub-classes on the basis of the degree of oxidation of their contained iron. All chondrites contain between 20 and 30% iron (Fe) which is partitioned between metal (Ni–Fe), troilite (FeS), and the silicates olivine $(Fe, Mg)_2SiO_4$ and pyroxene $(Fe, Mg)SiO_3$.

The enstatite chondrites are highly chemically reduced, and contain iron mainly in the form of the free metal or as troilite. They are rare among the chondrites, and their name is derived from the principle silicate constituent, enstatite ($MgSiO_3$).

The *Olivine–Bronzite* chondrites are an abundant group characterized by having a high free iron content. They contain mainly olivine (25–40%) an ortho-rhombic pyroxene (20–35%), together with varying amounts of nickel–iron (16–21%), plagioclase, maskelynite, troilite, chromite, diopside and occasionally small amounts of phosphate.

The *Olivine–Hypersthene* chondrites are the commonest of the classes and characterized by having their iron present as ferromagnesium silicates. They are otherwise similar to the olivine–bronzite chondrites, the major constitutents being olivine (35–60%) and hypersthene (25–35%) with minor amounts of plagioclase, maskelynite, nickel–iron, troilite and varying amounts of chromite and phosphate.

The *Olivine–Pigeonite* chondrites, originally classified with the olivine–hypersthene chondrites, are not classified in a small separate group since it was found that they contained pigeonite and not, as previously thought, hypersthene. They also have a higher total iron content. They are characterized by having iron present, as in ferromagnesium silicates, and by the presence of large amounts of olivine (approximately 70%) associated with pigeonite, plagioclase, troilite and minor amounts of nickel–iron. Many olivine–pigeonite chondrites (Felix, Grosnaja, Kaba, Lance, Mokoia and Vigarano chondrites) are black from the presence of carbonaceous material, although the carbon content does not generally exceed 0·5%.

The most interesting group of chondrites are the *carbonaceous chondrites*, which are of special importance because of their contained organic matter and associated hydrated minerals (Tables 11.07 and 11.08).

An alternative classification of meteorites proposed by Van Schmus and Wood retains the well-established chemical groupings, but in addition considers the mineralogical and textural properties of the materials which in turn reflect the degree of metamorphism experienced by the meteorite during its history. The chemical groups are identified as follows:

Type E—The Enstatite chondrites.
Type C—The Carbonaceous chondrites.

Types H.L.—The high-iron and low-iron chondrites which include, in addition to the olivine–bronzite (H) and olivine–hypersthene (L) chondrites, a number of the unequilibrated chondrites.
Type LL—A group established to contain the amphoteric chondrites as well as various unequilibrated chondrites.

ORIGIN OF METEORITES

Chance observations by random observers of meteorite orbits over the years have led to many and varied conclusions. The main hypotheses concerning the origin of meteorites are based on the following available data:

(1) The computed orbits of the Sikhote–Alin meteorite which fell in the Soviet Union on 12th February 1947 and of other meteorites, especially the calculations made by Ceplecha on the Pribam meteorite. The Pribam chondrite which fell in Czechoslovakia during the night of 7th April 1959 is the only meteorite to be adequately observed from two different points and simultaneously photographed. These observations indicated that the meteorite travelled in an elliptical orbit which commenced in the asteroidal belt within the solar system.
(2) Chemical and mineralogical examination of the various types of meteorites suggests that they all originated from a common parent substance.
(3) The distribution of elements in the stony meteorites is very similar to that in the sun, and this has led some workers to suggest a generic relationship between the two.
(4) The majority of meteorites (about 86%) belong to one group called chondrites.
(5) The density of the chondrites is often low, which suggests that they may have been formed in a weak gravitational field.
(6) If carbon, hydrogen, sulphur, nitrogen and oxygen are excluded from analyses the elemental composition of the chondrites are generally quite uniform.
(7) The specific mineral composition of chondrites is also uniform.
(8) Material present in meteorites is generally in a relatively highly reduced form.
(9) The mineral composition of the meteorites suggests that they have been formed at high temperatures. Carbonaceous chondrites, however, show evidence for the co-existence of both high-temperature and low-temperature minerals.

(10) The isotopic compositions of carbon, sulphur and oxygen are similar in the different types of meteorites, with the exception of the carbonaceous chondrites and the ureilites.
(11) All meteorites appear to have an age of $4{\cdot}5$–$4{\cdot}7 \times 10^9$ years.

In 1794 Chladni suggested that meteorites are the result of the disruption of a planetary body, and this has formed the basis for most theories of meteorite origin. In 1803 Olbers suggested that asteroids are fragments of a disrupted planet which had previously existed between Mars and Jupiter. Meteorites have subsequently been reasonably considered as asteroids or fragments of asteroids that have been captured by the gravitational field of the earth. The asteroid/meteorites may have originated in a catastrophic collision between two small planets, or may represent primoidial matter which failed to aggregate into a planet, or both. Most astronomers appear to agree that meteorites originate within our solar system and that none come from outside, but this must be pure speculation. The majority of investigators also suggest that meteorites come from the asteroidal belt between Mars and Jupiter.

Table 11.06. Other meteorites

Type	Falls	Finds	Paired	Total	
Copper	(1)	—	—	1	1·1%
Unique	—	1	—	1	1·1%
Unclassified	10(67)	3(12)	—	92	97·8%
	78	16	—	94	

Table 11.07. Carbonaceous chondrites

Type	Falls	Finds	Total	Type
C1	4	—	4	10·8%
C2	18	—	18	48·6%
C3	8	2	10	27·0%
C4	1	1	2	5·4%
Doubtful			1	2·7%
Unclassified			2	5·4%
Total carbonaceous chondrites			37	

Carbonaceous chondrites 1·71% of total meteorites, 2·83% of total chondrites (stones)

METEORITE DISCOVERIES

When meteorites travel through the earth's atmosphere they are frequently ablated and broken into many fragments. The meteorite "finds" (meteorites found but not observed to fall) outnumber the meteorite "falls" (meteorites recovered after being observed to fall and subsequently identified) (Tables 11.06, 11.07).

The largest meteorite discovered still lies where it was found near Grootfontein, Southwest Africa. Originally this meteorite—the Hoba West meteorite—lay buried in the ground with no trace of a crater and surrounded by a superficial calcareous *tufa* apparently deposited about it since it fell. The earth around the meteorite has been excavated sufficiently to allow examination and measurement of its size. It measures 2·95 by 2·84 m on the flat upper surface, and ranges in thickness from 0·5 to 1·2 m. Although its present weight has been calculated to be about 60 tons, it is estimated that at the time of fall it would have been about 100 tons. The Hoba West meteorite is thought to have fallen in pre-Pleistocene time, but evidence of the impact will have been removed by erosion.

The second largest known meteorite is the Ahnighito (Eskimo for "the tent"), which was taken to New York in 1897 by Admiral Peary (who later discovered the North Pole) and is now one of the three Cape York Irons of West Greenland (Ahnighito 31 tons; Woman 3 tons and Dog 408 kilograms) in the meteorite collection of the American Museum of Natural History. Several other Iron meteorites have been found with weights between 10 and 30 tons.

Mexico has a number of large meteorites, the largest being the Bacubirito meteorite which still lies where it was found in the State of Sinaloa and estimated to weigh about 27 tons. Other large meteorites have been discovered in Tanzania (the Mbosi weighing between 25 and 27 tons); in Armanty, Outer Mongolia (weighing about 20 tons); in an Oregon forest (the Willamette Iron meteorite weighing 14 tons). The Chupaderos Iron meteorite found in Chihuahua, Mexico, also weighs about 14 tons and was found together with another Iron meteorite weighing 6·5 tons.

In many cases larger meteorites have exploded as they hit the ground or just before contact, and have subsequently been gathered as separate pieces. Sometimes these fragments fit together, and the original shape and form may in part be reconstructed. Meteorite fragmentation is well illustrated by the Barwell meteorite (Fig. 11.05) which fell on Christmas Eve, 24th December 1965 on the Common leading to the centre of Barwell, Leicestershire, England. The total weight of all the fragments

Table 11.08. Carbonaceous chondrites

Meteorite	Place of fall	Date of fall		Weight recovered
Type C1				
Alais	France	15 March	1806	6 kg
Ivuna	Tanzania	16 December	1938	704·5 g
Orgueil	France	14 May	1864	at least 10 kg
Tonk	India	21 January	1911	7·7 g
Type C2				
Al Rais	Arabia	10 December	1957	160 g
Bells	U.S.A.	9 September	1961	283·5 g
Bonskino (Staroye Boriskino)	U.S.S.R.	20 April	1930	1165·6 g
Cold Bokkeveld	Southern Africa	13 October	1838	at least 3 kg
Crescent	U.S.A.	17 August	1936	78·4 g
Erakot	India	22 June	1940	113 g
Haripura	India	17 January	1921	at least 315 g
Kaba	Hungary	15 April	1857	about 3 kg
Mighei (Migei)	U.S.S.R.	18 June	1889	at least 8 kg
Mokoia	New Zealand	26 November	1908	about 45 kg
Murchison	Australia	28 September	1969	about 225 kg
Murray	U.S.A.	20 September	1950	7 kg
Nawapali	India	6 June	1890	about 60 g
Nogoya	Argentina	30 June	1879	about 4 kg
Pollen	Norway	6 April	1942	253·6 g
Renazzo	Rumania	31 March	1824	10 kg
Revelstoke	Canada	31 March	1965	about 1 g
Santa Cruz	New Zealand	3 September	1939	at least 30 g
Type C3				
Allende	Mexico	7 February	1969	at least 1000 kg
Efremovka	U.S.S.R.	*found*	1962	21 kg
Felix	U.S.A.	15 May	1900	about 3·3 kg
Groznaya	U.S.S.R.	28 June	1861	about 3·5 kg
Kainsaz	U.S.S.R.	13 November	1937	at least 200 kg
Lance	France	23 July	1872	about 51·8 kg
Leoville	U.S.A.	*found*	1961–62	8·1 kg
Ornans	France	11 July	1868	about 6 kg
Vigarano	Rumania	22 January	1910	15 kg
Warrenton	U.S.A.	3 January	1877	about 45 kg
Type C4				
Coolidge	U.S.A.	*found*	1937	4·5 kg
Karoonda	Australia	25 November	1930	41·4 kg
Unclassified				
Essibi (Cl–C2 ?)	Congo	28 July	1957	Not known
Bali	Cameroons	22 December	1907	9 g in collections
Doubtful				
Simonod	France	13 November	1835	None preserved

of the Barwell meteorite (classified as a Crystalline Globular Chondrite, Type L) amounts to at least 44 kg, and this compares with the heaviest known fall in the British Isles of Limerick County in 1813, which weighed at least 47 kilograms. Twenty-six different fragments were collected over about $\frac{3}{4}$ square mile, and several of these have been fitted together perfectly along cleavage surfaces (Fig. 11.05).

Fig. 11.05. The Barwell meteorite. Various fragments of the Barwell meteorite have been joined together along cleavage surfaces. Photograph by kind permission of the British Museum (Natural History), London

In sharp contrast to the Iron meteorites, no very large Stone meteorites have been discovered. The largest individual Stone meteorite, the Norton County Enstatite Achondrite which fell in Nebraska on 18th February 1948 weighs just over one ton. Other reasonably large Stone meteorites have been found at Long Island, U.S.A. (564 kg); Paragould, U.S.A. (372 kg); Finland (330 kg); Hugoton, U.S.A. (325 kg); Clovis, U.S.A. (283 kg) and Knyyahirya, U.S.S.R. (239 kg). Similarly no very large Stony–Iron meteorites are known.

At the opposite end of the size dimension scale there are the smallest fragments that can be recognized and identified as individual meteorites. Fragments less than 1 mm in diameter and weighing as little as 0·3 mg have been identified as meteoric fragments from the Sikhote-Alin meteorite fall in the U.S.S.R. Numerous small fragmented stones (with diameters about 2 mm and weights about 18 mg) have been collected from the Holbrook meteorite which fell in the U.S.A. Similar meteorite fragments are probably present in most meteorite falls.

THE CARBONACEOUS CHONDRITES

The average carbon content in chondrites is roughly 0·04%. Carbon occurs in various forms in the different types of meteorite and its non-uniform distribution makes accurate analysis difficult.

The ordinary chondrites (Olivine–Bronzite, Olivine–Hypersthene and Olivine–Pigeonite chondrites) contain least carbon (0·01%, Rich Mountain Meteorite, U.S.A., to 0·37% Breitscheid Meteorite, West Germany), present mostly as graphite. Enstatitic chondrites contain larger amounts of carbon (0·10%, Sain Sauveur Meteorite, France, to 0·43%, Indarkh Meteorite, U.S.S.R.) also present as graphite, although in some cases (e.g. Abee Meteorite, Canada) the carbon is reported to occur as cohenite.

The carbonaceous chondrites have the highest carbon content, up to 4·8% (Orgueil Meteorite, France), and present primarily as organic matter (up to 7% in the Orgueil meteorite).

The achondrites also contain varying amounts of carbon, present, however primarily as diamond and graphite, often concentrated in the metallic phase.

The carbonaceous chondrites represent about 3% of the total observed falls and 5% of all falls of chondrites. It has been pointed out that perhaps only 150 kg of carbonaceous chondrites have been collected and these species are irreplaceable as a source of extra-terrestrial organic material.

The largest carbonaceous chondrite (Allende Meteorite, Mexico) weighs at least 1000 kg, and the smallest (Revelstoke Meteorite, Canada) about 1g. Although carbonaceous chondrites are few in number they are readily distinguished from other meteorites both in their high organic content and by the presence of hydrated minerals of low-temperature origin, and the high degree of iron oxidation and a low content or absence of ferronickel alloys.

Some workers regard these meteorites as primary matter from which all other meteorites and the planets were formed.

All classified carbonaceous chondrites have been seen to fall and were recovered shortly after landing.

Carbonaceous chondrites, like other chondrites, possess a thin, dull black, fused crust about 1 mm thick, the result of entry through the atmosphere. They are frequently brittle and easily broken with the fingers. Their specific gravity is somewhat lower (2.2–3·6) than that of ordinary chondrites (3·7–3·8) and their chondrite structure ("crumb-like" chondrules—globules with diameter usually 0·1–0·5 mm) accounts for 10–90% of their volume. Some carbonaceous chondrites, however, are relatively rich in carbon material (e.g. the Orgueil Meteorite) and do not contain chondrules.

Carbonaceous chondrites are generally classified according to a system proposed by Wiik (Table 11.09). Wiik made complete analyses

Table 11.09. Wiik's classification of carbonaceous chondrites

	% weight				
	SiO_2	MgO	C	H_2O	S
Type I	22·56	15·21	3·54	20·08	10·32
Type II	27·57	19·18	2·46	13·35	5·41
Type III	33·58	23·74	0·46	0·99	3·78

of 13 meteorites and included the analyses of three others from the literature.

Analyses of the chondrites were shown to fall into three groups, namely:

Type C1 chondrites—contain the greatest amount of water and organic matter and only traces of high temperature minerals. They have a low density (approximately 2·2), a high amorphous hydrated silicate content, much sulphur present as water-soluble sulphate, no chondrules and are magnetic (mostly as spinel).

Type C2 chondrites—contain less water and organic matter than type C1 chondrites. Density 2·5 to 2·9, high serpentine content, weakly or non-magnetic, much of the sulphur in the free state, chondrules of olivine and enstatite. Traces of nickel–iron may also be present.

Type C3 chondrites—contain the least water and organic matter and have an abundance of high temperature minerals and also contain some metallic constituents. This class is also named the olivine–pigeonite chondrites and contain mainly olivine, some pigeonite, oligoclase, troilite and some nickel–iron. Some contain sufficient carbonaceous material to give them a black colour.

Ages of Carbonaceous Chondrites

The radio-chemical age determined by Potassium–Argon (K–Ar) dating methods on carbonaceous chondrites varies from 1·2 to $4{\cdot}6 \times 10^9$ years. The low radiogenic ages of some carbonaceous chondrites is associated generally with chondrites with a high carbon content in which loss of argon can occur more readily than in those with low carbon content. Type C3 chondrites are denser than the C1 and C2 types and hence loss of argon will be reduced. Their radiogenic age is consequently close to the radiogenic age of many ordinary chondrites.

Age determination using the Rubidium–Strontium (Rb–Sr) method on the Orgueil (type C1), Murray (type C2), Mokoia (type C3) and Lance (type C3) carbonaceous chondrites has shown their ages to be $4{\cdot}5\text{–}4{\cdot}7 \times 10^9$ years. These results agree with the data obtained using the Uranium–Lead (U–Pb) method for the Orgueil ($4{\cdot}6 \times 10^9$ years), Murray ($4{\cdot}7 \times 10^9$ years) and Mokoia ($4{\cdot}6 \times 10^9$ years) chondrites.

History of the Carbonaceous Chondrites

All carbonaceous chondrites contain organic matter, but a severe high temperature history would be incompatible with the preservation of any indigenous biological material of earlier origin. The thermal history of meteorites is therefore of special importance.

Meteorites may have three periods of thermal history, namely:

(1) metamorphism and other "parent-body" formation effects,
(2) the temperature of the meteorite in space, and
(3) the temperature of the meteorite during its fall to earth.

The presence of water of non-terrestrial isotopic composition, which would have been lost at temperatures above 180°C, shows that very little metamorphism has taken place in carbonaceous chondrites. The devitrification of strained glass fragments found in type C2 chondrites has also provided evidence that these meteorites could not have been subjected to a temperature of 180°C for longer than two weeks, or 250° for more than a few minutes and never to a temperature of 300°. The distribution of nickel in the metallic grains of carbonaceous chondrites has shown that after accretion of the chondrite material to its present form it was not subjected to any high temperature history at a later time.

The temperature achieved by meteorites during their passage through space has been estimated from a study of their rare gas contents. The quantities of stable decay products such as ^{4}He and ^{40}Ar derived from radiochemical disintegrations relative to their parent elements gives a measure of the time elapsed since a meteorite first formed, provided that the rare gas content had not been removed by heating. The experiments showed that the helium losses were quite small and led to the conclusion that "chondrites and achondrites spent the major part of their history at temperature below about 40°C".

The variation of meteorite velocity and temperature during fall depends primarily on size. Meteorites with pre-atmosphere masses of less than one ton completely lose all their cosmic velocity during passage through the earth's atmosphere and reach the earth's surface in free fall. The time required for the meteorite to decelerate is very short and the accompanying heating small. Thermoluminescent studies of the effects of atmospheric heating on the Holbrook chondrite indicated that the heating affected the surface only to a depth of 10–15 mm.

Meteorites less than one ton in weight have very little cosmic velocity through the earth's atmosphere, and the thermal effects caused by the energy of impact are very small.

It is also possible to obtain an approximate estimate of the thermal history of carbonaceous chondrites from an examination of contained organic compounds. There is little evidence for pyrolysis, suggesting that thermal effects have not been severe.

Organic Matter in Carbonaceous Chondrites

There is a close parallel between the pattern and type of organic matter present in terrestrial sediments and in carbonaceous chondrites. In each case the organic constituents are:

(1) organic compounds which are readily soluble in organic and aqueous solvents, and
(2) insoluble organic matter which makes up the major proportion

and consists mainly of an amorphous insoluble substance (usually called "kerogen" in terrestrial sediments), but may also contain structured material with characteristic shapes.

In carbonaceous chondrites most of the carbon is present as organic material. In type C1 and type C2 chondrites about 30% of this organic matter is readily extracted with aqueous and organic solvents. The remaining amount up to 70% is present as insoluble organic matter which cannot be solubilized by organic solvents and is inert to all non-oxidizing chemical reagents (Table 11.10).

History of the Study of Organic Compounds in Meteorites

The first observed carbonaceous chondrite fell on 15th March 1806 in Alais, France. This type C1 chondrite, called the Alais Meteorite, was studied chemically in 1834 by the Swedish chemist Berzelius, who later, describing an aqueous extract and pyrolysis of the contained organic material, commented:

> ". . . there can thus be no doubt that the stone under investigation, despite all its external differences, is a meteorite, which in all probability hailed from the usual home of meteorites".

The results which were obtained by Berzelius led him to query:

> "Does this carbonaceous stone contain humus or other organic substances?", and, "Does this possibly give a hint concerning the presence of organic structures in other planetary bodies?"

It is only during very recent years that analytical techniques have been developed to allow these problems to be carefully and systematically studied to the point where a definite answer may be attempted.

In 1858–1859, Wohler studied the carbonaceous material in the Cold Bokkeveld (Type C2) and Kaba (Type C3) carbonaceous chondrites. Evaporation of alcoholic extracts of these meteorites gave yellowish or colourless resin—wax-like materials (0·25% by weight of the meteorite) with an aromatic odour. Further treatment produced a volatile crystalline substance whose condensate was "petroleum-like" and the remaining residue was described as a "coal-like" substance. It was insoluble in alkali and described by Wohler as similar to ozokerite and scheererite, whilst other researchers suggested new names like kabaite and "meteoritic petroleum" for this carbonaceous matter.

The carbonaceous matter in the Alais meteorite was examined in 1863 by Roscoe, who from an ether extract obtained 0·64% organic

Table 11.10. Types of carbon present in various meteorites

Meteorite	Total carbon % wt	Total carbon $\delta^{13}C$	% Soluble organic compounds	$\delta^{13}C$	% carbonate	$\delta^{13}C$	Insoluble % wt residue after H_3PO_4	Insoluble % wt residue after HCl+HF treatment	$\delta^{13}C$	% insoluble organic of total carbon content
Ivuna type C1	4·03	−7·5	0·07	−24·1	0·20	+65·8	2·54	1·57	−17·1	38·95
Orgueil type C1	3·75	−11·6	0·11	−18·0	0·13	+70·2	—	2·15	−16·9	57·33
Mighei type C2	2·85	−10·3	0·09	−17·8	0·21	+41·6	—	0·72	−16·8	25·26
Cold Bokkeveld type C2	2·35	−7·2	0·11	−17·8	0·07	+50·7	1·71	1·27	−16·4	54·04
Erakot type C2	2·30	−7·6	0·13	−19·1	0·05	+44·4	1·54	0·89	−15·1	38·69
Murray type C2	2·24	−5·6	0·11	−5·3	0·13	+42·3	1·72	1·06	−14·8	47·32
Mokoia type C3	0·74	−18·3	0·96	−27·2	zero	—	0·61	0·47	−15·8	63·51

Smith, J. W. and Kaplan, I. R. *Science*, **167,** p. 1368.)

matter with elemental composition, %C = 28·57; %H = 5·29; %S = 66·14. After excluding the free sulphur the organic matter corresponded to a hydrocarbon mixture %C = 84·37 and %H = 15·63.

In 1876 and 1879, Smith obtained crystalline colourless organic products from both the Alais and the Orgueil meteorites. The crystals, which had an aromatic odour, contained a high percentage of sulphur and produced a coal-like substance when heated and were called celestialite.

An ozokerite-like substance was extracted from the Migei meteorite in 1889 by Meunier and at the same time Simashko also extracted 4% of a viscous yellow gum (erdelite) from the same meteorite. Simashko even postulated that this carbonaceous matter was formed chemically in the meteorite under relatively low temperature conditions.

Further records of solvent extracts of meteorites were made by Friedheim (1888) on the Nogoya meteorite (type C2 which fell in Argentina on 30th June 1879); by Plon and Tschermak (1878) on the Grozmaya meteorite (type C3 which fell on 28th June 1861 in the Soviet Union) and each researcher reported the isolation of paraffin-like hydrocarbons.

Some workers attempted to isolate organic compounds from carbonaceous chondrites by methods other than solvent extraction. Cloez, in 1864, after treating the Orgueil meteorite successively with concentrated hydrochloric acid and a weak solution of potassium hydroxide obtained a carbonaceous residue (6·41% by weight of the meteorite). He described this material as being of an amorphous nature similar to terrestrial humic substances with elemental analysis: %C = 63·45; %H = 5·98 and %0 = 30·57.

Thus, even during the last century organic matter had been isolated from several carbonaceous chondrites. The extracts from these chondrites generally were viscous, colourless to yellow to brown in colour, occasionally fractionated into acicular crystals and had a "bituminous-like" aromatic smell. Elemental analysis of the extracts suggested that they had a hydrocarbon-like composition. These preliminary studies led many researchers to assume that the extractable organic matter represented an individual compound and empirical formulae were proposed for the material.

Re-examination of the early studies by Cohen in 1894 led him to postulate that the solvent extractable organic matter in meteorites was present in at least three different forms:

(1) organic compounds consisting of carbon and hydrogen,
(2) compounds containing carbon, hydrogen and sulphur,
(3) compounds containing carbon, hydrogen and oxygen.

Only recently, since the application of sophisticated analytical techniques, has it been realized that the soluble organic material in meteorites is a very complex mixture of many different classes of compound.

Although almost all the investigations were carried out on the readily extractable organic matter, various workers did nevertheless comment on the organic matter which could not be extracted.

A few weeks after the famous Orgueil meteorite fell in France, Cloez (1864) found 6·41% insoluble organic matter in the chondrite. In 1956 Wiik found 6·96% of insoluble material in the Orgueil meteorite. Accepting that care must be taken in comparing these results too closely because Cloez's experiments were of a limited nature, the values are meaningful enough to show that the Orgueil meteorite did not become contaminated with insoluble organic matter to any significant extent during 92 years of storage. For all practical purposes, there was an almost complete cessation of the studies of organic compounds in meteorites between 1899 and 1953, when not one single paper appeared on the subject. This can probably be attributed to the lack of suitable microanalytical methods for studying the small amounts of organic compounds available.

In 1953 Mueller reported elemental analyses of organic matter extracted from two samples of Cold Bokkeveld by organic solvents. After removal of the majority of the elemental sulphur the products showed %C = 19·84; %H = 6·64; %N = 3·18; %S = 7·18; %Halogen (assumed Cl) = 4·81 and % Remainder O = 40·0. The relatively high halogen content, assumed to be chlorine, may have been introduced from the chlorinated solvents (chloroform and carbon tetrachloride) used in the extractions. Subsequent examination of the organic matter present in the Murray meteorite by direct volatilization of the compounds in a mass spectrometer and analysis of their high resolution mass spectra for the presence of hetero-elements showed that nitrogen, sulphur and oxygen were present in significant quantities whilst chlorine was barely detectable and silicon, phosphorus, bromine and iodine were undetected.

In 1961 Nagy and his co-workers analysed mixtures of saturated hydrocarbons isolated from the Orgueil meteorite using infrared, ultraviolet and mass spectrometric techniques. A series of *n*-alkanes were detected apparently similar to those obtained by Berthelot in 1868. In addition, cyclo-alkanes were identified, but branched-chain alkanes were either absent or present in very low concentrations. It was suggested that some of the cyclo-alkanes resembled fragmentation products of tetracycloalkane analogues of sterols (e.g. steranes, 18, 19-dimethyl cyclopentanoperhydrophenanthrenes). The distribution of a number of the hydrocarbons detected in the Orgueil meteorite resembled to some extent the distribution of naturally occurring terrestrial hydrocarbons. They are similar to those found in recent sediments and in certain biological substances, and this led Nagy, Meinschein and Hennessy to suggest that the Orgueil hydrocarbons may be of biological origin.

Biological material must presumably have evolved in an aqueous environment, hence inorganic constituents should have been subjected to the sorting effects resulting from the presence of water, as is the case in terrestrial sediments. However, meteorite workers were convinced that this was not so since the chemical composition of meteorite minerals strongly suggested that no such sorting of mineral has occurred.

More recent studies of carbonaceous chondrites using modern analytical methods have revealed the presence of other particularly interesting organic chemicals by more detailed analyses of hydrocarbon fractions.

A benzene extract of the Orgueil meteorite was very similar to that from a typical brown alga and contained material analogous to naphthenic acid. Similar material was not present in the Bruderheim ordinary chondrite when extracted under the same conditions and with the same solvent. The question immediately asked about these results was whether contamination of the Orgueil meteorite could have been produced by these compounds with biological characteristics. It was conceded by the investigators that the meteorite would have been biologically contaminated when it fell in France, but the museum shelf did not appear to be a particularly likely place for the growth of algae, particularly when the material contains magnesium sulphate which in the presence of small amounts of adsorbed moisture would give a strong solution of magnesium sulphate, which would tend to inhibit microbiological activity. However, the possibility of contamination cannot be excluded.

Independent gas–liquid chromatographic analyses of an *n*-heptane fraction of the Orgueil meteorite by Meinschein and Oro, both showed an *n*-alkane distribution with a maximum component at *n*-C_{19}. However, similar independent studies of the Murray meteorite by the same investigators failed to produce agreement. Meinschein found that *n*-alkane distribution had its maximum component at C_{17}, whilst Oro found the maximum component at C_{21}. Both reports are in disagreement with earlier mass spectrometric results which show a maximum value at C_{23} and the results of direct volatilization–gas chromatographic studies which show a lower distribution of hydrocarbons with a maximum value at C_{15}. There has been no shortage of discussion and explanation for these discrepancies in the *n*-alkane distributions in the Murray meteorite, but apart from a suggestion that the C_{23} mode distribution may be due to contamination, there has been no satisfactory explanation for the heterogeneity of the results.

Aliphatic hydrocarbons have been identified in varying amounts (from below 0·1 to 415 ppm) in many meteorites. Their origin is unknown, but it is known that the Grosnaja meteorite (type C3) was probably contaminated with paraffin wax during storage. Oro, Nooner and Olson, using gas chromatography and mass spectroscopy analysed organic materials extracted from 29 meteorites, with the following results:

(1) Chromatographic evidence for the presence of pristane and phytane was obtained for almost all the meteorites. In the majority of type C2 and in some of the type C3 chondrites evidence for norpristane was also obtained. Mass spectral analysis showed the presence of isoprenoids in the Borisknio (type C2), Mighei (type C2), Murray (type C2), Santa Cruz (type C2), and Mokoia (type C3) carbonaceous chondrites and also in the graphite nodule of the Odessa Iron meteorite.

(2) Samples of material from the interior and from other parts of the Orgueil meteorite showed that the hydrocarbons in this chondrite were quantitatively heterogeneous and qualitatively semi-homogeneous. Of the meteorites examined only the Orgueil had any odd over even carbon number preference in their *n*-alkane distribution. Even this observation was not recorded in all analyses on the Orgueil meteorite. The distribution of hydrocarbons in organic extracts of the Ivuna (type C1), Alais (type C1) and Warrenton (type C3) chondrites indicate possible bimodal distributions.

(3) The type C2 chondrites alone consistently showed similar hydrocarbon distributions and contents. The Mokoia (type C3), Lance (type C3), and the ordinary chondrites Saint Caprais and the graphite nodules from the Canyon Diablo and Odessa meteorites have hydrocarbon distributions similar to those in the type C2 chondrites.

(4) The hydrocarbon distributions in the various meteorites are quite similar to those found in Precambrian sediments, petroleum crudes and products derived from petroleum.

No specific cyclic alkanes have yet been identified in solvent extracts of meteorites, but gas–liquid chromatography–spectral analysis (gc–ms) gives some evidence for their presence. The results generally confirm the presence of cholestane-like compounds in the soluble organic matter from the Orgueil meteorite. But conclusive proof for the presence of cycloalkanes in carbonaceous chondrites is not yet available.

Aromatic hydrocarbons are generally present in meteorites only in very low concentrations but in a very few samples substantial amounts are present. A similar wide variety of aromatic hydrocarbons can also be produced by decomposition of biological materials (as in natural petroleum deposits) or by abiological synthesis (methane pyrolysis in a Fischer–Tropsch synthesis). Thus the range and type of aromatic hydrocarbons identified in the Orgueil and the Mokoia carbonaceous chondrites are very similar to those synthesized abiogenically by the pyrolysis of isoprene at 600–700°C. Even so, the possibility of a direct biological source for the aromatic hydrocarbons from carbonaceous chondrites cannot be excluded.

The general absence of soluble aromatic hydrocarbons in non-carbonaceous chondrites, the graphitic nodules of iron meteorites and in the majority of sedimentary and igneous terrestrial rocks might suggest that contamination of carbonaceous chondrites by these substances is not a serious problem. Also in the carbonaceous chondrites the aromatic hydrocarbons are generally present in very small amounts compared with the alkanes. This could of course simply reflect the greater ease of destruction of unsaturated organic matter including aromatics either by oxidation to simple substances or by hydrogen transfer processes involving alkanes. The last reactions are very likely and would lead to reactive and readily degraded alkenes.

$$\text{C}_6\text{H}_6 + R—CH_2—CH_3 \longrightarrow \text{C}_6\text{H}_8 + R—CH{=}CH_2$$

Infrared spectra of extracts of the Orgueil meteorite revealed absorption bands characteristic of carboxylic acids, and gas chromatography showed the presence of fatty acids with carbon numbers between C_{14} and C_{28}, even-numbered carbon atom components predominating. These results were confirmed by various workers, but no evidence for fatty acids in the ordinary Holbrook chondrite was obtained.

Since carbonaceous chondrites contain relatively large amounts of elemental sulphur, it is probable that sulphur compounds will be present in measurable concentrations.

Thiophenes, benzthiophenes, alkylthiophenes, alkylbenzthiophenes, dibenzthiophenes and possibly thiophenols have been identified in the Murray meteorite using mass spectrometry, but these organo-sulphur compounds were present in even less abundance than the low level of unsubstituted aromatic hydrocarbons.

Amino acids have also been identified, in various carbonaceous chondrites including the Orgueil, Mighei, Cold Bokkeveld and Murray meteorites. Seventeen amino acids were identified with amounts varying from 30 μg/g in the Mighei and Cold Bokkeveld to 112 μg/g in the Orgueil meteorite. The amino acids were present in both the free state and in combined forms which were easily hydrolysed. The most common of the amino acids identified were: serine, glycine, alanine, leucine and isoleucine, glutamic acid and threonine. Trace quantities of basic, aromatic and sulphur-containing amino acids were also detected.

The Murchison Meteorite

The Murchison meteorite (Fig. 11.06a and b) which fell at about 11.00 a.m. on 28th September 1969 near Murchison, Victoria, Australia, broke up during flight and scattered many fragments over an area of 14 square miles. The meteorite (with largest recorded stone of at least 2·5 kilograms), which contained about 2% carbon and 0·16 % nitrogen, has been classified as a type C2 carbonaceous chondrite.

Analysis of the nature of various organic components in a sample of the meteorite collected soon after its fall by a NASA team of investigators at Ames Research Center suggests that the identified amino acids and hydrocarbons are indigenous. The $\delta^{13}C$ values of +4·43 to +5·93 for solvent extractable organic material are significantly different from those of terrestrial organic matter, and the presence of amino acids, glycine, alanine, valine, proline, glutamic acid, 2-methylalanine and sarcosine were unequivocally established. Since there were almost equal mounts of the D- and L-enantiomers of valine, proline, alanine and glutamic acid the investigators suggest that this minimizes the possibility of terrestrial contamination and offers a possible extraterrestrial origin. Also the presence of 2-methyalanine and sarcosine which are not commonly found in terrestrial biological systems is presented as evidence of a possible abiogenic synthesis. Gas-chromatography–mass spectrometry analyses of the hydrocarbons show them to be similar to those obtained by laboratory abiogenic experiments and the investigators use the data to support the contention that the organic molecules identified in the Murchison meteorites are abiogenic and possibly extraterrestrial in origin.

Oró and co-workers (1971) at the University of Houston have examined the inner portions of a relatively large piece of the Murchison meteorite and after various extractions and derivative preparations they were able to identify a series of amino acids: glycine, alanine (by far

(a)

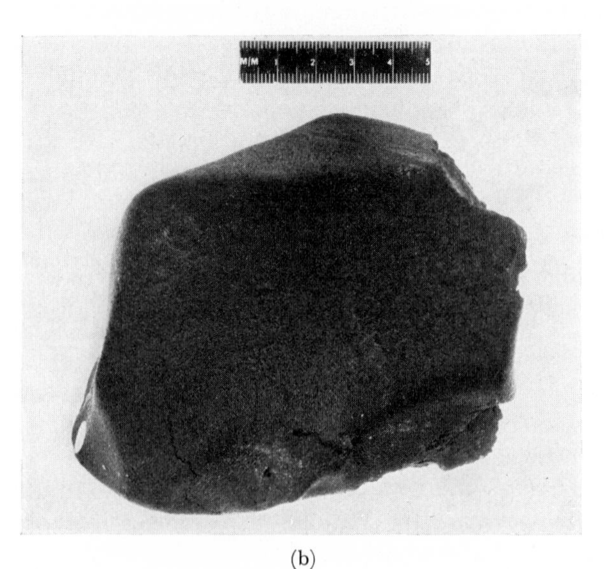

(b)

Fig. 11.06. The Murchison meteorite. Sample which is in the meteorite collection at the Natural History Museum. (a) Shows the meteorite surface which has been fused, probably during its entry into the earth's atmosphere. (b) Again shows some evidence of a fused surface, but to a less degree than the reverse side shows in (a). Photographs by kind permission of the British Museum (Natural History), London

the major components with 5·3 μg/g and 3·1 μg/g respectively), 2-methyl-alanine, amino-butyric acid, valine, glutamic acid, proline and some of the leucines. The results were in good agreement with the analyses carried out on the Murchison meteorite by the Ames Research Center Team. These workers suggest various explanations for the formation of amino acids in the meteorite and favour the "extraterrestrial abiotic synthesis", before, during or after the formation of the meteorite parent body. This is generally the most favoured suggestion for the presence of the amino acids, but what is often overlooked (and is very often the case with Precambrian Rock studies, see Chapters 9 and 10) is the method of extraction and subsequent treatment of the organic matter from the meteorite. After extraction of the organic portion with triply distilled water, the residue is treated with hydrochloric acid and then eluted from an exchange resin with ammonium hydroxide solution, followed by further acid treatment. These later treatments with ammonia and strong acid are not compatible with the identification of microgram quantities of glycine and alanine, and should be judged with some reservation.

Oró and co-workers also examined the Murchison meteorite for its hydrocarbon content and found aliphatic hydrocarbons ranging from C_{10} to relatively high molecular weight components. The most abundant series consisting of branched-chain saturated alkanes, with the monomethyl and dimethyl isomers predominating. The workers concluded that the Murchison meteorite, which seems to be one of the first examined in which the amount of terrestrial biological contamination is very small, contains substantial amounts of extractable organic compounds, which have spectra indicating the presence of compounds significantly different and more complex than those observed in other meteoritic studies. They further suggest that this indicates abiotic synthesis, extensive diagenesis and related chemical processes, or both.

It has been suggested that amino acids so far identified in the vast majority of meteorites are the result of terrestrial contamination. The amounts are so small that amino acids from human hands or from dust can contribute significantly to the amounts present in the extracts.

Similarly trace amounts (5–24 μg) of soluble carbohydrates have been detected in the Mighei and Cold Bokkeveld meteorites, especially mannose and glucose, with some evidence for the presence of arabinose. The problems of contamination are of the same order as with amino acids. The general problem of contamination is well illustrated by the following observations. Calvin and Vaughn noted that the ultraviolet absorption spectra of an aqueous extract of the Murray meteorite was pH sensitive and resembled the spectrum of cytosine, but the authors were careful to point out that this did not necessarily imply the presence of cytosine. The result, however, became interpreted by some reporters as having proved the presence of the pyrimidine base. Later work to duplicate this report and verify the presence of cytosine showed that very similar ultraviolet absorptions can be observed in the materials stripped from the ion exchange resins by hot eluants during the fractionation processes on the organic extract.

Vanadyl porphyrins have been identified in the Orgueil meteorite, but not in non-carbonaceous meteorites.

Attempts to establish optical activity in meteoritic organic compounds have led to disagreement. Optical activity was observed in the solvent extractable fraction of the Orgueil, but the results were subsequently criticized and it was claimed that the rotation was caused by instrument artefacts.

The Allende Meteorite

A meteorite fell near Pueblito de Allende, Chihuahua, Mexico, on 8th February 1969 and was collected shortly afterwards. The meteorite (type C3–C4) contained approximately 0·3% carbon. The organic matter present has been examined at Berkeley, where workers analysed the inner and outer parts separately. The surface layer contained traces (0·2 ppm) of soluble organic matter, but very little soluble organic matter could be detected in the inner part (0·002 ppm). Calvin reported that the organic matter present in the surface layer was of biological origin and cannot be other than terrestrial contamination. He further commented that the speed with which the contamination was acquired, and the diversity of the material, make doubtful any interpretation of such soluble organic materials that have been found in meteorites of unknown or at least considerably longer terrestrial history. The presence of a mono-saturated C_{18} fatty acid in the Allende meteorite is additional evidence for recent contamination, since such acids have only been found in recent terrestrial sediments.

Cronin and Moore (1971) have examined the Allende, Murray and Murchison meteorites for amino acid constituents and found that the Murray and Murchison (type C2 meteorites) specimens have a similar amino acid composition, whilst the Allende (type C3) meteorite was found to be essentially devoid of amino acids.

Breger and co-workers at the U.S. Geological Survey, together with Clarke at the Smithsonian Institution, Washington, have reported on the occurrence and significance of formaldehyde in the Allende carbonaceous chondrite. These workers examined various specimens of the meteorite and found that several showed unusual lustrous fusion crusts with examples of "antlers". These "antlers" are a kind of "mineralized capillary tube" and were found to be filled with a "colourless hydrocarbon material". Breger suggests that it is possible that organic material near the surface of the meteorite might have been pyrolysed to give products which passed into the interior of the meteorite as it entered the earth's atmosphere. On cooling, these pyrolysis products may have been drawn into the previous hot and evacuated hollow "antlers" to give filled "capillary-size fused tubes". Examination of the "colourless hydrocarbon material" showed that it was probably derived from formaldehyde, which is considered to occur as para-formaldehyde in the meteorite. This retention of formaldehyde is thought to be possibly unique in the Allende meteorite, and Breger and co-workers suggest that it could be the precursor of the various chain, ring and polyhydroxy-compounds present in meteorites and in space. These workers conclude their report with a very interesting speculation in which they postulate that certain organic compounds (amino acids, sugars, formaldehyde, hydrocarbons, etc.) that exist in space or may be formed on meteoritic surfaces or within meteorites can be distributed by those meteorites and, on landing on a planet, may serve as precursors of life. This, they comment, is a revision and rebirth of the well-developed and often-quoted panspermia theory of Arrhenius (1903), which suggested that life could be transmitted by means of spores carried by meteorites.

Examination of Indigenous Carbon Compounds

The extractable organic matter in carbonaceous chondrites has recently been re-examined by Kaplan and Smith (1970) and their results suggest that it can be considered to be indigenous.

Seven carbonaceous chondrites (Table 11.10) were extracted with organic solvents and the products separated. The $^{13}C/^{12}C$ ratio of the separated compounds and the remaining insoluble organic matter were measured.

The results (Table 11.10) indicate that the major proportion of the carbon compounds is present as *an acid-resistant insoluble organic substance.* In a type C3 chondrite (Mokoia meteorite) the insoluble organic matter amounts to 63% by weight of the original carbon content, whereas in types C1 and C2 meteorites it is somewhat less.

Three definite $\delta^{13}C$ groupings occur in the carbon compounds present in these particular carbonaceous chondrites:

(1) values of $\delta^{13}C$ greater than $+40$ per mil for the carbonate fraction;

(2) values of $\delta^{13}C$ between -14 and -18 per mil for the insoluble organic matter and probably the solvent extractable organic matter indigenous to the meteorite; and
(3) values of $\delta^{13}C$ between -4 and -8 per mil for the graphite and iron carbide in the meteorites.

These results indicate a non-random isotopic distribution of carbon in carbonaceous meteorites and show that the insoluble organic matter, and perhaps a small portion of the solvent extractable organic material, are indigenous to the meteorite.

Structured Morphological Objects in Carbonaceous Meteorites

In 1961 Claus and Nagy reported the presence of small spherical objects in samples of the Orgueil and Ivuna meteorites. They suggested that they were fossil micro-organisms and endeavoured to classify them into groups analogous to single-celled algae and the fossil group of hystrichospheres. The "organized elements" resembled spores of certain species of aquatic algae, but one form was morphologically dissimilar to any known terrestrial organism. Organized elements which were abundant in the Orgueil and Ivuna chondrites were absent in the Holbrook and Brunderheim meteorites. Similar but poorly defined structures were also found in the Murray and Mighei meteorites.

Claus and Nagy commented:

(1) The high concentration of organized elements in the Orgueil and Ivuna meteorites could only have developed in an aqueous environment, yet the samples were kept in dry museum storage and would have disintegrated if subjected to an aqueous medium. Furthermore, they were outdoors for only a few hours after they had fallen.
(2) The organized elements resemble those species of dinoflagellates or chrysomonads which live in water (sea, lakes) and do not occur in soil.
(3) The four types of organized elements were similar to, but not identical with, known terrestrial species. A fifth type was entirely dissimilar to any known terrestrial form.
(4) Terrestrial organisms which resemble the organized elements normally develop cysts (spores) under unfavourable conditions (such as a dry environment). Two or three of the organized elements in the Orgueil and in the Ivuna meteorites had structures resembling cysts.
(5) The same general type of organized elements were present in

> both Orgueil samples which fell in the temperate climate of southern France, and in the Ivuna meteorite which fell 74 years later in an arid, tropical region of central Africa. Consequently contamination with morphologically identical, locally derived, micro-organisms seem unlikely.

It was concluded that the organized elements may be micro-fossils indigenous to the meteorite.

These opinions aroused great opposition because the chemical composition of the meteorites did not resemble that of typical micro-fossil containing terrestrial sediments.

The lack of evidence of conditions within meteorites which might have been favourable for the support of living systems led Urey to suggest that life possibly evolved on one planetary object and then was transferred to a planetary object of primitive composition. Thus life may have been transported to the moon from the earth and in time returned to the earth as extra-terrestrial matter. This assumed that the moon might have a composition consistent with that of carbonaceous chondrites, but the evidence from the lunar explorations shows that carbonaceous meteorites are mineralogically dissimilar to the lunar rocks.

During 1962 additional finds of structured bodies were made by various other groups in studies with crushed fragments, thin sections and acid-resistant extracts of the Orgueil, Murray, Ivuna and Mighei carbonaceous chondrites. The ordinary stony-meteorites, Bruderheim and Holbrook, lacked objects. Later, Claus, Nagy and Hennessy also found structured bodies in the Alais and Tonk meteorites. Some of the micro-fossil objects had simple morphological structures, whilst others were more complex. Finch and Anders claim that some of these objects are recent pollen and spore contaminations and also that the structured bodies described by Claus and Nagy are mineral grains, hydrocarbon droplets as well as contamination.

From the Mighei chondrite (type C2), Timofeev extracted what he claimed to be fossilized objects similar to some algae, such as the "characteristic spores of water plants of the Late Cambrian Mire".

More recent examinations by Brooks and Muir of the structured material in carbonaceous chondrites has revealed organic box-like structures (Fig. 11.07a–d) with five or six sides. The basic symmetry of all the structures is hexagonal and those structures with a different number of sides, whether it be five or seven, are either imperfect or probably damaged during preparation.

The structures are hollow, and contain included mineral matter, and it is possible that they have been produced by condensation of organic material round a hexagonal mineral nucleus. They show no similarities to any known terrestrial biological organism.

There are several possible sources of the organic materials:

(1) they are biogenic;
(2) they were formed by Miller–Urey type reactions (Chapter 8);

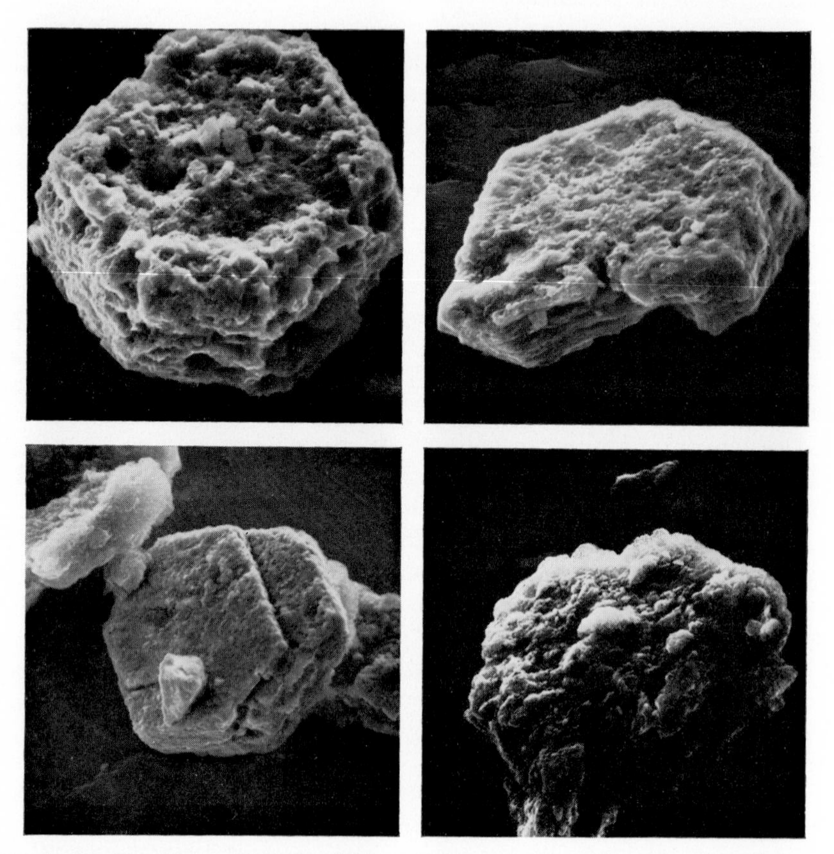

Fig. 11.07. Morphology of the insoluble organic matter from the Orgueil and Murray meteorites.
Organic residue from the Murray meteorite: Top left: Typical six-sided box-like structure which is almost perfectly hexagonal in form. Top right: Similar box-like structure, which is slightly damaged. Bottom left: Coating of organic material which covers a hexagonal mineral grain. Some of the natural cleavage of the mineral grain can be seen.
Organic residue from the Orgueil meteorite: Bottom right: Box-like structure. The hexagonal nature of this structure is less clear than that of the Murray specimens, but examination of many specimens showed a clear preponderance of five- and six-sided structures. J. Brooks and M. D. Muir (1971). *Grana* **11,** 9

(3) they are formed metastably by spontaneous reactions, especially Fischer–Tropsch type syntheses;
(4) they are derived from contamination by terrestrial biological matter.

Although it may be thought that further search for histological evidence for the presence of structured bodies of a possible micro-fossil nature in chondrites would provide the best evidence for life, this is not necessarily so. Such evidence is subjective, and not readily repeated. In micropaleontological studies on meteorites, unless chondrites should turn up with large quantities of micro-fossil contents which would prove beyond doubt that they are indigenous, organic in nature and of authentic origin, the main hope of reaching a decision on the question of the presence of extra-terrestrial life must rest mainly with the unstructured organic matter which the meteorites contain.

The Insoluble Organic Matter in Carbonaceous Chondrites

In previous studies on carbonaceous chondrites almost all the organogeochemistry has been concerned with the readily soluble organic fractions and the insoluble carbonaceous matter has been neglected. This is unfortunate since the extractable organic matter forms a minor part of the total. Up to 70% of the organic matter present in chondrites is in the form of an insoluble organic polymer and can be present in type C1 and C2 chondrites in 35 and 25 mg/g amounts respectively.

This insoluble organic matter is regarded as an indigenous component of the carbonaceous chondrites because:

(1) it is by far the major portion of the organic content;
(2) all the measured stable carbon isotope values on the insoluble organic matter in meteorites have values outside the terrestrial range for the equivalent types of complex polymeric material showing it to be indigenous to the meteorites;
(3) it was detected in the Orgueil and other meteorites a few weeks after the fall;
(4) because it is virtually impossible to explain the formation of up to 7% of such organic material (i.e. something like 2 oz in a piece of meteorite the size of a man's fist) by terrestrial contamination.

This question of quantity is extremely important in our view. There is no doubt that much of the work carried out on very minor constituents, although leading to precise molecular formulae of the single contained materials, is nevertheless virtually valueless since one can never be sure that the compounds have not been derived from a contaminating source.

The insoluble carbonaceous matter was isolated from the Orgueil meteorite by removing inorganic constituents with acid, and Electron Spin Resonance (ESR) Spectroscopy, and X-ray spectra results clearly showed that it was not graphite but was similar in many ways to kerogens found in terrestrial rocks.

The insoluble material in the Orgueil meteorite was analysed by Raia and corresponded to an empirical formula of $C_{90}H_{65}N_{18}O_9S_{3.6}$.

Spectroscopic evidence indicates that the insoluble material is polymeric, highly substituted, irregular and contains some sort of aromatic unsaturated skeleton. Vinogradov and Vdovykin, and Brooks and Shaw obtained infrared spectra on the material extracted from the Mighei and Orgueil and Murray meteorites respectively, which indicated the presence of a carbonyl substituted highly condensed unsaturated aromatic structure.

Phenolic acids were present in the hydrochloric acid hydrolysates of the Orgueil, the Murray, the Mokoia and the Felix meteorites, the most abundant being *m*-hydroxybenzoic acid. Other compounds tentatively identified included *p*-hydroxybenzoic acid, *p*-hydroxyphenylacetic acid and *p*-hydroxy-3-methybenzoic acid. Similar phenolic acids were also found by Briggs in the Mokoia and Haripura (type C2) meteorites.

Oxidative degradation of the material with ozone and hydrogen peroxide gave benzenetricarboxylic acid, benzenetetracarboxylic acid and benzenepentacarboxylic acid, adipic acid and perhaps a pyridine-carboxylic acid. These products suggested that the insoluble organic matter in the Orgueil was at least partially aromatic in nature. The insoluble organic matter (kerogen) from a Devonian Shale gave similar results.

Apart from this preliminary work by Bitz and Nagy the insoluble organic matter has been very much neglected. This is in many cases understandable since the soluble components are easier to study in that they may be obtained in a pure state and assigned precise molecular formulae. But with the problems (and thus invited criticisms) of contamination associated with soluble compounds it is extremely difficult to arrive at a useful conclusion about their presence and origin. Another criticism which is rarely aired is that if one is to provide acceptable evidence for biological origin, then the more complex the chemical is, the more likely it will be possible to identify it with a larger part of a differentiated biological unit and hence to increase enormously the value of the evidence.

Brooks and Shaw have previously shown that the structure of the

(a)

(b)

Fig. 11.08. The Orgueil meteorite. Sample of meteorite in the collection of meteorites at the Natural History Museum. Note in both (a) and (b) that the surface of the carbonaceous chondrite (type C1) is being gradually eroded away, due to the moisture of the atmosphere in the Museum. Photographs by kind permission of the British Museum (Natural History), London

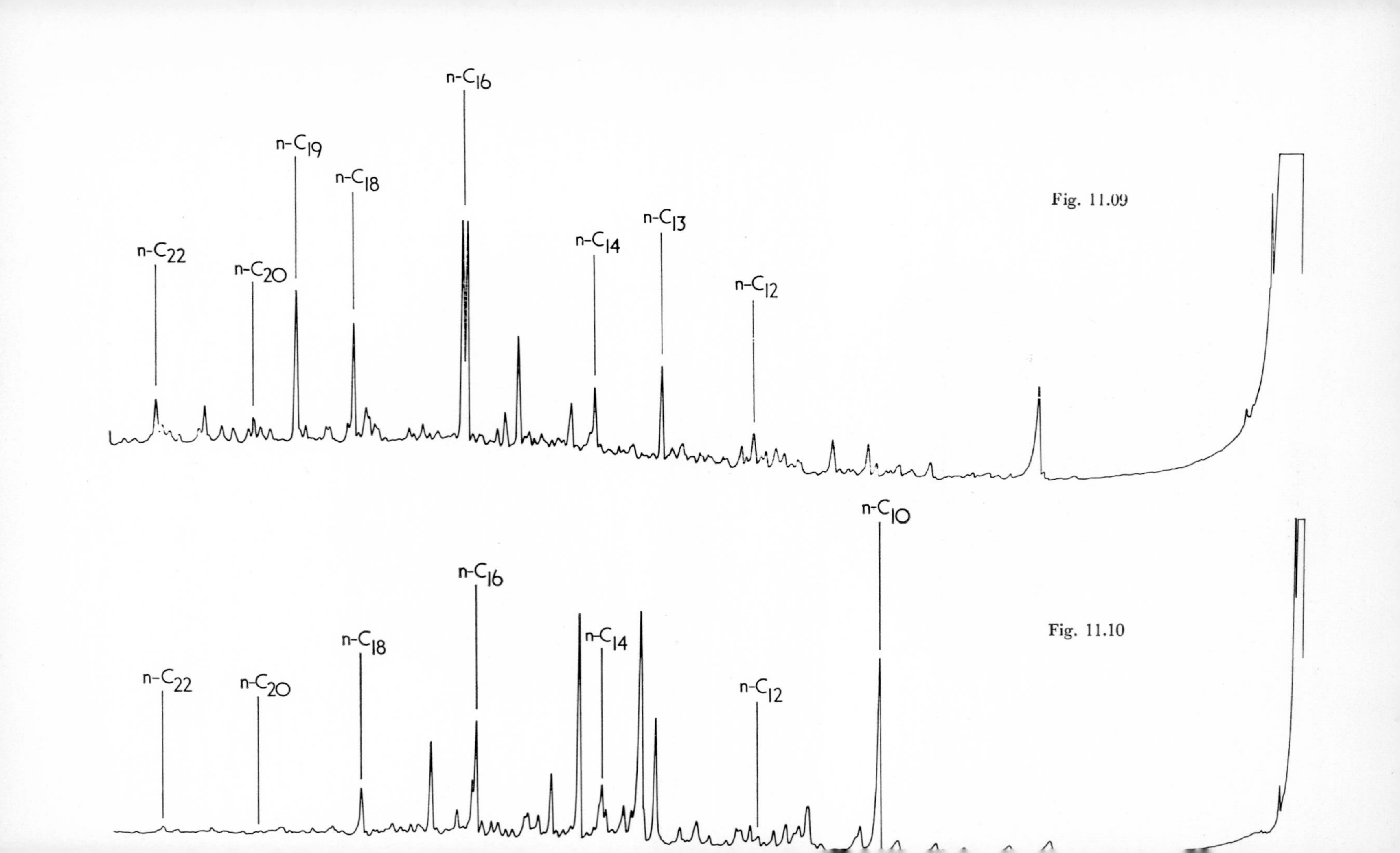

Fig. 11.09

Fig. 11.10

tough resistant material (sporopollenin), which makes up a major part of pollen and spore walls is chemically identical with much of the older kerogen derived from Pre-cambrian terrestrial sediments (Chapters 9 and 10). It seems a reasonable hypothesis therefore that the insoluble organic material in the carbonaceous chondrites might also (if of biogenic origin) be of a similar sporopollenin-type.

The Orgueil and Murray Carbonaceous Chondrites

On 14th May 1846, at 8 p.m., about twenty stones, the largest about the size of a football but most about the size of a cricket ball (Fig. 11.08 a and b), fell over an area of about two square miles near Orgueil in the province of Tarn-et-Garonne in southern France. A total amount of at least 10 kg was recovered and this material has been widely distributed throughout the world, but the largest amount (about 8 kg) is kept in the Museum d'Histoire Naturelle in Paris.

The Murray meteorite fell near the town of Murray in Calloway County, Kentucky, U.S.A., on 20th September 1950 at 1.33 a.m. Several stones fell, one penetrating the roof of a house, and a total of about 7 kg have been recovered. Most of the meteorite material is in the U.S. National (Smithsonian) Museum, Wellington.

Studies on the Insoluble Organic Matter present in the Orgueil and Murray Meteorites

The insoluble organic matter extracted from the Orgueil (3·5%) and Murray (4·4%), using standard geochemical techniques (see Chapter 9) was examined by Brooks and Shaw using infrared spectroscopy; potash fusion followed by thin layer chromatography of the products; oxidative degradation and capillary column gas chromatography of the products; acetolysis and elemental chemical analyses. The results of these studies were compared with similar examinations of sporopollenins derived from naturally occurring modern pollen grains and spores, some synthetic chemical analogues prepared by oxidative co-polymerization of carotenoids and carotenoid esters, some sporopollenins derived from fossil spores and related species, and with some artificially metamorphosed sporopollenin that had been heated with rock (see

Fig. 11.09

Fig. 11.09. II Mono-carboxylic acids (methyl ester derivatives) from ozone degradation products of the insoluble organic matter present in Murray meteorite

Fig. 11.10

Fig. 11.10. Mono-carboxylic acids (methyl ester derivatives) from ozone degradation products of a Dinoflagellate cyst concentrate

Chapter 9). The results of these comparative studies showed that all the substances belong to the same class of polymeric material, which has chemical properties similar to those of naturally occurring sporopollenin.

This common identity rests chiefly on the following set of criteria, which are characteristic of sporopollenin polymers:

(1) They have stability to non-oxidative chemical reagents including those strong hydrolytic reagents used to separate fossil pollen and spores, microfossils and insoluble organic matter from their inorganic environment.

(2) They are unsaturated and readily oxidized using ozone (and other reagents) to give characteristic soluble mono- and di-carboxylic acid products (Fig. 11.09), which compare qualitatively

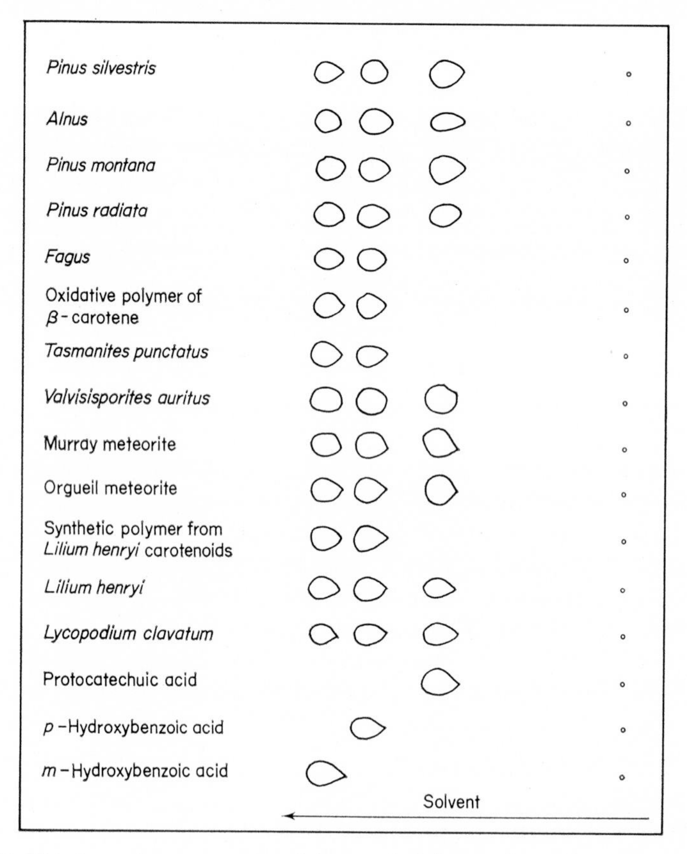

Fig. 11.11. Thin layer chromatography of the potash fusion products of some sporopollenins from pollen, spores, fossil spores, synthetic polymer and from the insoluble organic matter in meteorites

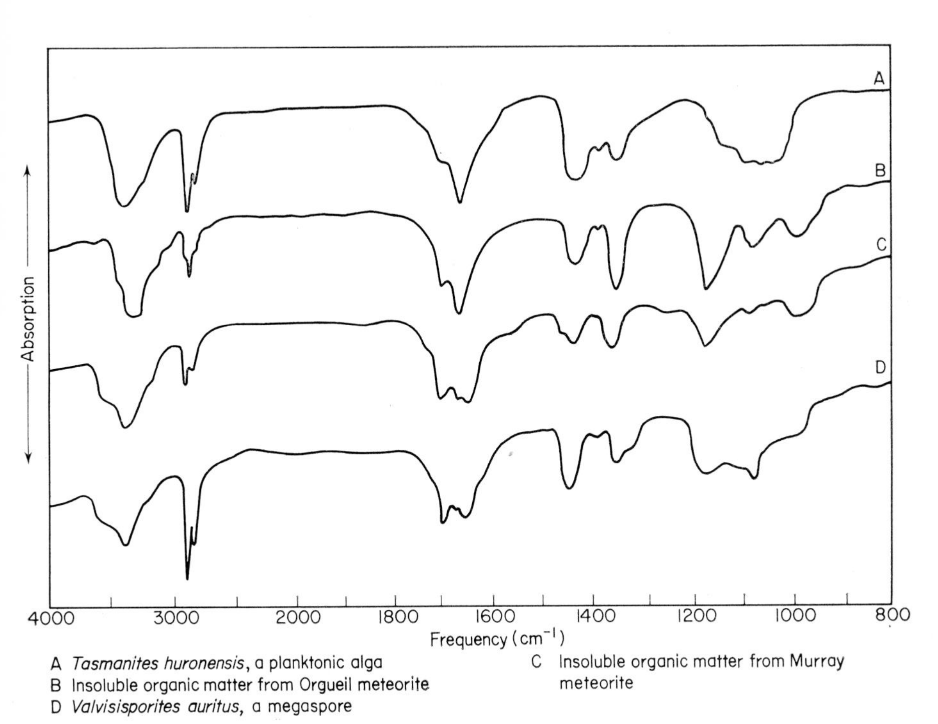

Fig. 11.12. The infrared spectra of some sporopollenins from fossil sources

and semi-qualitatively with products obtained from a wide range of known biological sporopollenin-containing materials (Fig. 11.10).

(3) They are stable to acetolysis (i.e. treatment with acetic-anhydride/concentrated sulphuric acid mixture).

(4) They give characteristic potash fusion products (*m*-hydroxybenzoic and *p*-hydroxybenzoic acids as the major products and trace amounts of protocatechuic acid) (Fig. 11.11).

(5) They have similar infrared spectra (Fig. 11.12), showing absorption bands corresponding to hydroxyl groups, carbon–hydrogen, carbon–carbon single and double bonds, carbon–oxygen bonds and carbon–methyl groups.

(6) They give similar pyrolysis–gas chromatograms and these show characteristic pattern changes when the pyrolysis temperature of the insoluble material is changed;
  pyrolysis temperature 770°C (Fig. 11.13a);
  pyrolysis temperature 610°C (Fig. 11.13b);
  pyrolysis temperature 480°C (Fig. 11.13c).

It is especially noteworthy that infrared spectra and pyrolysis–gas chromatograms of the insoluble organic matter extracted from the Orgueil and Murray meteorites are virtually indistinguishable from each other and from those of artificially metamorphosed sporopollenin derived from modern lower plants. The spectra and pyrolysis gas chromatograms are also very similar to those of various morphologically intact microfossil planktonic algal (e.g. *Tasmanites punctatus*), spore-like outer walls (Fig. 11.14) and megaspores from lower plants (e.g.

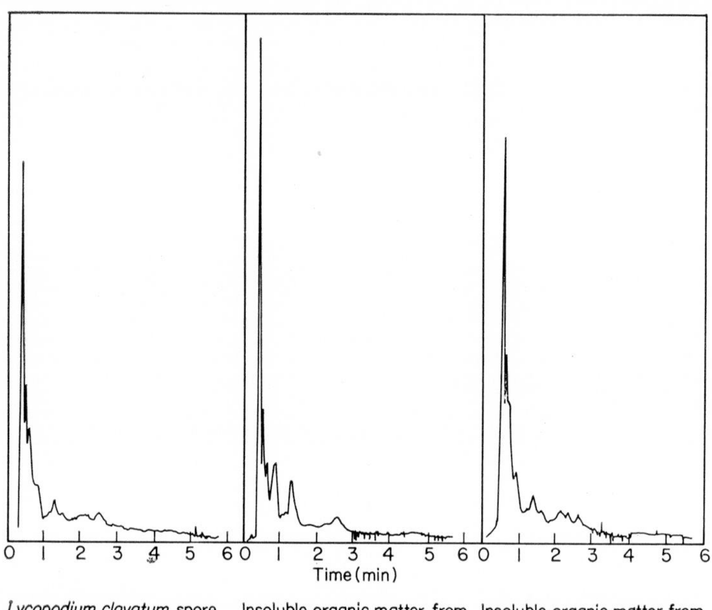

Fig. 11.13. (a) Pyrolysis–gas chromatograms of some insoluble organic matter. (Pyrolysis temperature 770°C)

Selaginella kraussiana). Also the spectra of oxidative degradation products from the insoluble matter present in the Murray meteorite (Fig. 11.09) are very similar to those produced by an identical chemical degradation of the insoluble organic material present in the resistant outer wall of lower plants (e.g. dinoflagellates) (Fig. 11.10).

Studies of this type led Brooks and Shaw (1969) to suggest that the insoluble organic matter present in carbonaceous chondrites might have a common chemical identity with the biological polymer, sporopollenin.

There is a possibility of course that sporopollenin-type polymers might have resulted from an abiogenic process.

This, however, seems unlikely if one considers the synthesis of sporopollenin polymers. Sporopollenin is an oxidative co-polymer of

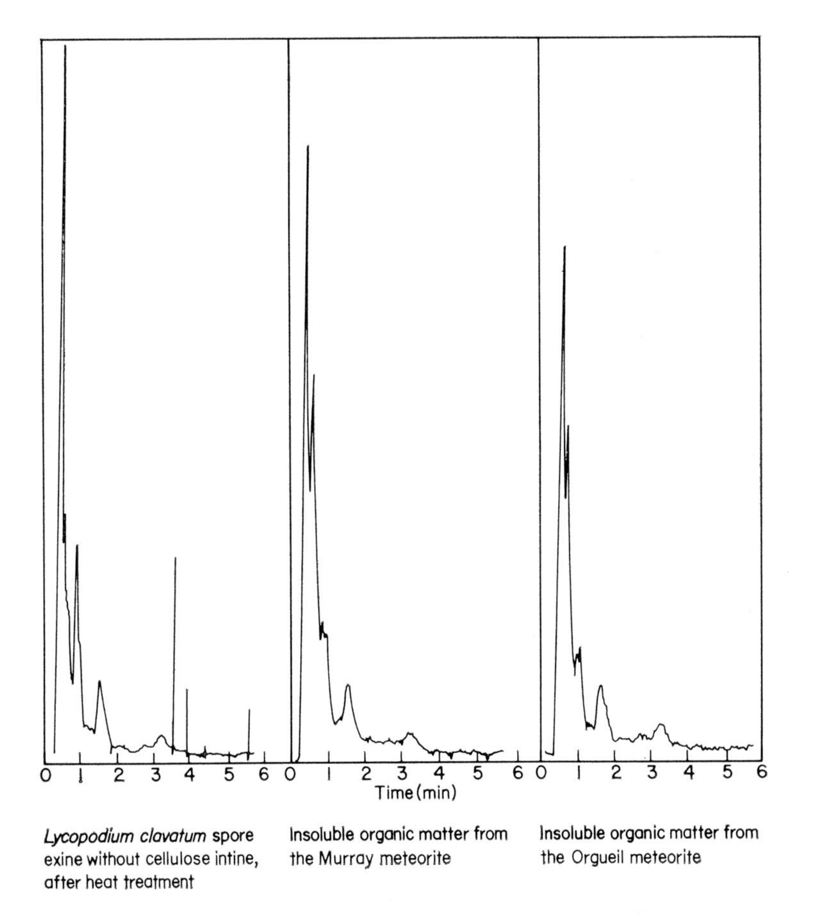

Fig. 11.13. (b) Pyrolysis–gas chromatograms of some insoluble organic matter. (Pyrolysis temperature 610°C)

carotenoids and related carotenoid esters, but paradoxically it will not survive in an oxidative environment. This suggests that sporopollenin can only be synthesized and survive in conditions where oxidation processes are under strict control. Such conditions are only to be found in living systems, which suggests that the insoluble organic polymeric material present in the Orgueil and Murray meteorites, with chemical properties similar to sporopollenin derived from modern and fossil higher and lower plants, may have had a biological origin.

In addition the normal products of heat-stimulated abiogenic reactions which lead to alkanes always strongly favour the production

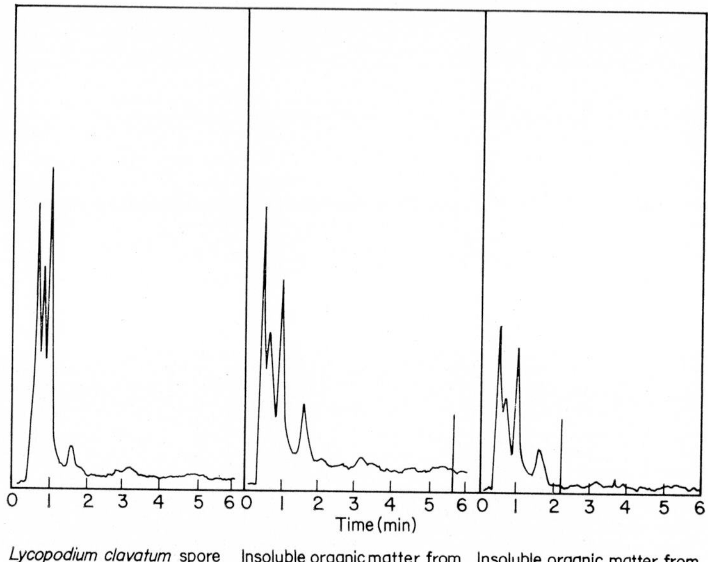

Fig. 11.13. (c) Pyrolysis–gas chromatograms of some insoluble organic matter. (Pyrolysis temperature 480°C)

of normal as distinct from branched chain compounds, since the latter are more readily degraded under such conditions.

i.e.

$$\begin{array}{c}\text{—C—C—}\\ | \\ \text{—C—C—C—}\end{array} \rightleftharpoons \begin{array}{c}\text{—C—C—}\\ + \\ \text{—C—C—C—}\end{array}$$

Although a reaction of this type is to some extent reversible the equilibrium lies very much to the right in favour of the non-branched materials. The isoprenoid nature of sporopollenin-like materials of the type mentioned therefore favours a biogenic origin.

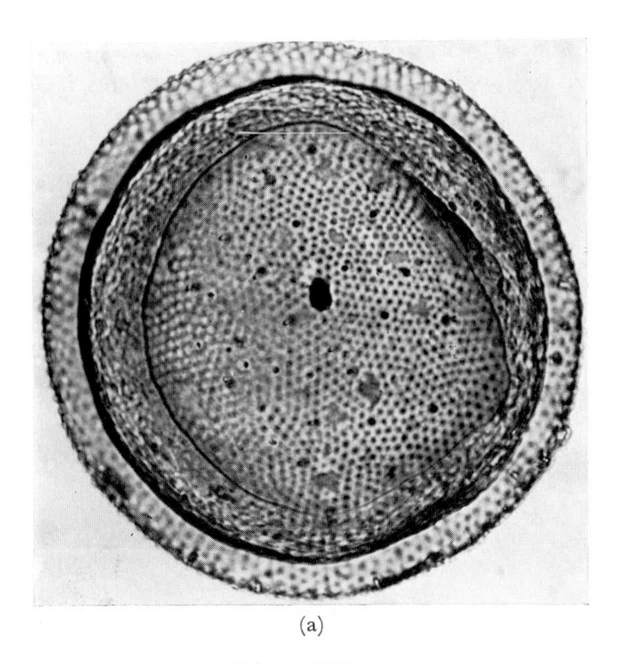

(a)

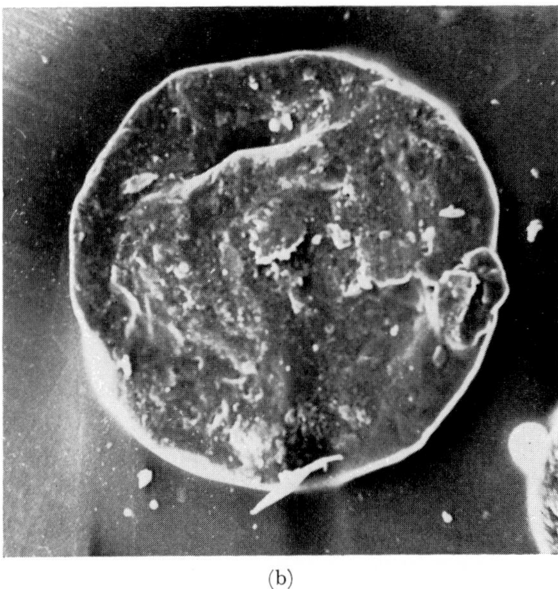

(b)

Fig. 11.14. Surface walls of Tasmanites. (a) Light microscopic photograph of *Crassosphaera* (Tasmanaceae) from the Lower Jurassic (about 180 million years old). Photograph by kind permission of M. G. Collett. (b) Scanning electron microscope photograph of *Tasmanite punctatus* from the Permian (about 240 million years old)

SUGGESTED FURTHER READING Chapter 11

Books

Cohen, E. (1894, 1903 and 1905). "Meteoritenkunde". Volumes I–III. E. Schweizerbart, Stuttgart.

Geiss, J. and Goldberg, E. D. (Eds) (1963). "Earth Science and Meteorites" North-Holland, Amsterdam.

Hey, M. H. (1966). "Catalogue of Meteorites". British Museum (Natural History), London.

Krinov, E. L. (1969). "Principles of Meteorites". Pergamon, Oxford.

Mamikunian, G. and Briggs, M. H. (Eds) (1965). "Current Aspects N Exobiology". Pergamon, Oxford.

Mason, B. (1962). "Meteorites". Wiley, New York.

Ponnamperuma, C. (Ed). (1972). "Exobiology". North-Holland, Amsterdam.

Wood, J. A. (1968). "Meteorites and the Origin of Planets". McGraw-Hill, New York.

Vdovykin, G. P. (1967). "Carbon Matter of Meteorites (Organic Compounds, Diamonds, Graphite)". Nauka Publ. Office, Moscow.

Articles

Anders, E. (1968). Chemical Processes in the Early Solar System, as inferred from Meteorites. *Accounts Chem. Res.* **1,** 289–299.

Arrhenius, S. (1908). Vorstellung vom Weltgebaude im Wandel der Zeiten. Akademische Verlags Gesellschaft Leipzig.

Arrhenius, S. (1910). L'Evolution des Mondes. La Vie dans l'Univers. Paris.

Bitz, M. and Nagy, B. (1966). Ozonolysis of Polymer-type Material in Coal, Kerogen and in the Orgueil Meteorite—a Preliminary Report. *Proc. natn. Acad. Sci. U.S.A.* **56,** 1383–1390.

Breger, I. V. *et al.* (1972). Occurrence and Significance of Formaldehyde in the Allende Carbonaceous Chondrite. *Nature, Lond.* **236,** 155–158.

Briggs, M. H. and Mamikonian, G. (1963). Organic Constituents of the Carbonaceous Chondrites. *Space Sci. Rev.* **1,** 647–682.

Brooks, J. and Muir, M. D. (1970). Morphology and Chemistry of the Organic Insoluble Matter from the Onverwacht Series Precambrian Chert and the Orgueil and Murray Carbonaceous Chondrites. *Grana* **11,** 9–14.

Brooks, J. and Shaw, G. (1969). Evidence for Extraterrestrial Life: Identity of Sporopollenin with Insoluble Organic Matter Present in the Orgueil and Murray Meteorites and also in Some Terrestrial Microfossils. *Nature, Lond.* **223,** 756.

Calvin, M. and Vaughn, S. K. (1960). Extraterrestrial Life: Some Organic Constituents of Meteorites and their Significance for Possible Extraterrestrial Biological Evolution. *In* "Space Research" (Ed. H. K. Bijl), Vol. 1, pp. 1171–1191. North-Holland, Amsterdam.

Clarke, R. S. *et al.* (1970). The Allende, Mexico, Meteorite Showers. Smithsonian Contributions to the Earth Sciences Number 5.

Claus, G. and Nagy, B. (1961). A Microbiological Examination of Some Carbonaceous Chondrites. *Nature, Lond.* **192,** 594–596.

Claus, G. and Nagy, B. (1962). Microfossils, new to Science, Resembling Algae and Flagellates, found in Meteorites. 1st International Congress on Palynology, Tucson, Arizona.

Cronin, J. R. and Moore, C. B. (1971). Amino Acid Analyses of the Murchison, Murray and Allende Carbonaceous Chondrites. *Science, N.Y.* **172,** 1327–1329.

Fitch, F. W. and Anders, E. (1963). Organized Element: Possible Identification in Orgueil Meteorite. *Science, N.Y.* **140,** 1097–1100.

Han, J. (1969). Organic Analysis on the Pueblito de Allende Meteorite. *Nature, Lond.* **222,** 364–365.

Hayes, J. M. (1967). Organic Constituents of Meteorites—a Review. *Geochim. cosmochim. Acta* **31,** 1395–1440.

Helmholtz, H. von. (1893). "Popular Lectures on Scientific Subjects". Longmans, Green and Company.

Jobbins, E. A. *et al.* (1966). The Barwell Meteorite. *Min. Mag.* **35,** 881–902.

Kaplan, I. R. and Smith, J. W. (1970). Endogenous Carbon in Carbonaceous Meteorites. *Science, N.Y.* **167,** 1367–1370.

Kvenvolden, K. *et al.* (1970). Evidence for Extraterrestrial Amino-acids and Hydrocarbons in the Murchison Meteorite. *Nature, Lond.* **228,** 923–926.

Lancet, M. S. and Anders, E. (1970). Carbon Isotope Fractionation in the Fischer–Tropsch Synthesis and in Meteorites. *Science, N.Y.* **170,** 980–982.

Levy, R. L. *et al.* (1970). Organic Analysis of the Pueblito de Allende Meteorite. *Nature, Lond.* **227,** 148–150.

Lovering, J. F., Le Maitre, R. W. and Chappell, B. W. (1971). Murchison C 2 Chondrite and its Inorganic Composition. *Nature, Lond.* **230,** 18 and 20.

Meinschein, W. G. (1963). Hydrocarbons in Terrestrial Samples and the Orgueil Meteorite. *Space Sci. Rev.* **2,** 653–679.

Meinschein, W. G., Nagy, B. and Hennessy, D. J. (1963). Evidence in Meteorites of Former Life. *Ann. N.Y. Acad. Sci.* **108,** 553–579.

Mueller, G. (1953). The Properties and Theory of Genesis of the Carbonaceous Complex within the Cold Bokkeveld Meteorite. *Geochim. cosmochim. Acta* **4,** 1–10.

Mueller, G. (1964). Interpretation of Microstructures in Carbonaceous Meteorites. *In* "Advances in Organic Geochemistry" (Eds U. Colombo and G. D. Hobson). Pergamon, Oxford.

Nagy, B. (1966). A Study of the Optical Rotation of Lipids Extracted from Soils, Sediments and the Orgueil Carbonaceous Meteorite. *Proc. natn. Acad. Sci. U.S.A.* **56,** 389–398.

Nagy, B. (1966). 20 Strange Meteorites, *Chemistry* **39,** 8–13.

Nagy, B. (1968). Carbonaceous Meteorites. *Endeavor* **27,** 81–86.

Nagy, B. (1968). Indication of Possible Biological Substances in Carbonaceous Meteorites. *J. Astronautical Sci.* **15,** 161–168.

Nagy, B., Meinschein, W. G. and Hennessy, D. J. (1961). Mass Spectroscopic Analysis of the Orgueil Meteorite. Evidence for Biogenic Hydrocarbons. *Ann. N.Y. Acad. Sci.* **93,** 27–35.

Nagy, B. and Claus, G. (1966). Mineralized Microstructures in Carbonaceous Meteorites. *In* "Advances in Organic Geochemistry" (Eds: U. Colombo and G. D. Hobson). Pergamon, Oxford.

Nagy, L. A., Kremp. G. O. W. and Nagy, B. (1969). Microstructures Approximating Hexagonal Forms (and of Unknown Origin) in the Orgueil Carbonaceous Meteorite. *Grana* **9,** 110–117.

Oró, J. *et al.* (1967). Organic Compounds in Meteorites I–II. *Geochim. cosmochim. Acta.* **31,** 1359–1394 and 1935–1948.

Oró, J. *et al.* (1971). Amino Acids, Aliphatic and Aromatic Hydrocarbons in the Murchison Meteorite. *Nature, Lond.* **230,** 105–106.

Pirie, N. W. (Ed) (1968). A Discussion on Anomalous Aspects of Biochemistry of Possible Significance in Discussing the Origins and Distributions of Life. *Proc. R. Soc. Lond. B.* **171.**

Raia, J. (1966). M.S. Thesis, Department of Chemistry, University of Houston. (Cited in J. M. Hayes, *Geochim. cosmochim. Acta.* **31,** 1395–1440).

Reed, S. J. B. (1969). Cosmic Metallurgy. *Sci. J.* December 69–74.

Staplin, F. L. (1965). Organic Remains in Meteorites. *In* "Current Aspects of Exobiology" (Eds G. Mamikunian and M. H. Briggs), 77–92. Pergamon, Oxford.

Studies, M. H., Anders, E. *et al.* (1968–1972). Origins of Organic Matter in Early Solar System, I–IV. A Series of Papers in *Geochim. cosmochim. Acta.*

Timofeer, B. V. (1963). Lebensspuren in Meteoriten. Resultate einer Microphytologischen Analyse. *Grana* **4,** 92–99.

Urey, H. C. and Lewis, J. S. (1966). Organic Matter in Carbonaceous Chondrites. *Science, N.Y.* **152,** 102–104.

Van Schmus, W. R. and Wood, J. A. (1967). A Chemical-petrologic Classification for the Chondritic Meteorites. *Geochim. cosmochim. Acta* **31,** 747–765.

Wiik, H. B. (1956). The Chemical Composition of Some Stony Meteorites. *Geochim. cosmochim. Acta* **9,** 279–289.

12

The Origin and Development of Living Systems— A Summary

INTRODUCTION

It is perhaps an unavoidable function of the human thought process to assume as almost axiomatic that large things such as the Universe must naturally and inevitably be built up from small things. Thus theories, whether concerned with the origin of the vast galaxies or of small units of primeval matter, always tend to assume some even smaller more primeval ancestry. Indeed this is a useful concept and one can outline a sequence of events in which the living system, especially man, is at the pinnacle of a sort of evolutionary pyramid of matter which has at its base the simplest of fundamental particles. This in a sense implies an ultimate goal, namely the conversion of all matter into the living system. An alternative rarely discussed but equally plausible concept is that the Universe began with a living system and we are now witnessing the conversion of this living system into simpler matter. Whether or not it is wise to have teleological concepts of the Universe is debatable, but it seems reasonably clear that there exists at least equilibrium relationships between the various types and states of matter in our known Universe. These relationships have been discussed in earlier chapters and may be briefly summarized as follows:

fundamental particles $\rightleftarrows$ atoms $\rightleftarrows$ molecules $\rightleftarrows$ complex molecules $\rightleftarrows$ living systems

These relationships have all been shown to be true in some context or other. The net conversion of molecules into the living system is certainly true on this planet, but of course it is under strict biological control. The reversibility of the final equation is only questioned when one considers the possibility of *de novo* abiogenic conversion of molecules into the living system.

Almost all our knowledge of the living system is confined to this planet and in this sense our information sources are to a very large

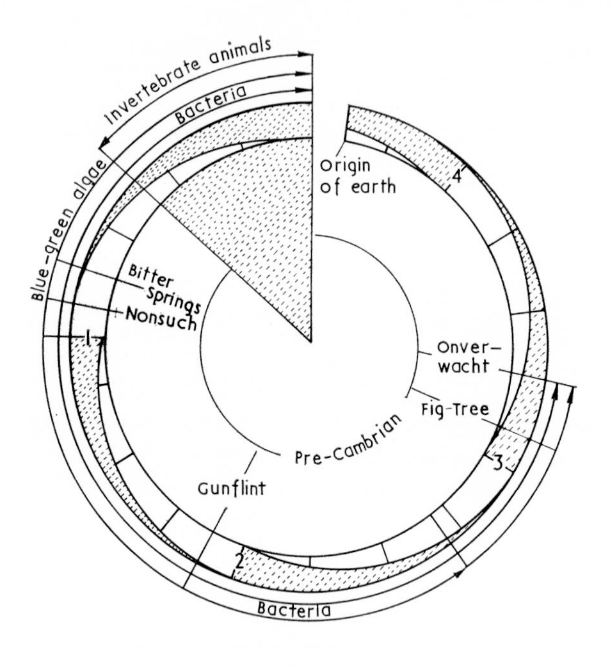

Fig. 12.01. The "Geological clock" (after J. W. Schöpf)

degree clearly restricted. Nevertheless there seems to be no reason why evidence should not be forthcoming which would give firm indications about the way in which life arose on this planet. Such evidence as there is, is largely to be found in the various residues, whether of biological, chemical or other nature which have been left by living systems in the various rock, especially sedimentary rock strata. It is surprising how much knowledge has in fact been gleaned from the relatively few studies of such material which, to date at least, have been carried out by very few workers who have little more than scratched the surface of the available geological and geochemcial evidence (Fig. 12.01) that must surely exist. It is perhaps only recently that the sort of physical techniques have become available that will allow a more profitable study of much of the materials of the rocks.

At the same time the rapid expansion of biological and biochemical knowledge in recent years has thrown much more light on the nature and chemical assemblage of the living system itself. Of special significance in this context is the well established universal nature of all the living processes. The chemistry and biochemistry of the genetic processes, of protein synthesis and structure and of enzyme occurrence and mechanism, all point to a fundamental origin for all the processes. The large morphological gaps which clearly exist between unicellular and differentiated multicellular organisms, between prokaryotic and eukaryotic organisms and between the unicellular organism and the virus in no way affect the concept of biochemical universality; indeed in many ways they help reinforce the concept by drawing attention to the manner in which the same type of processes operate in such widely differing biological forms. Thus the messenger function of a simple viral RNA reflects clearly the much more complex but completely analogous mRNA of a mammal. Further discoveries in these particular areas will undoubtedly make the subsequent search for living system residues all the easier and more profitable.

ORIGIN OF LIFE

It seems reasonably clear that the surface of the planet Earth at one time some 4×10^9 years ago was so hot no life could exist, nor for that matter could one expect even complex molecules such as nucleic acids and proteins to exist in any form remotely resembling that required for life assembly. Consequently at some time there had to be a time when life arose on a sterile planet and there are several possible ways in which this might have occurred.

(1) The living system arose on earth out of a neutral or reducing environment by the initial formation of various simple organic molecules and their subsequent conversion into more complex molecules. These in turn are seen to undergo a process of so-called chemical evolution, whereby not only are nucleic acids and proteins produced but complexes of the two types of macromolecules with each other and with other materials must arise to form ultimately a mutual and profitable partnership of organic and for that matter inorganic compounds which will possess properties of a simple living system. That is, to be able to survive in a given environment and reproduce without apparent

external aid, and moreover possess an inherent property of being capable of mutational and or other changes which will eventually allow more complex systems to arise.

(2) The living system came from an extraterrestrial source. There are several variations on this theme. Thus:

(a) A living system (possibly spore or other resistant cell) may have travelled in some as yet unknown manner through space from its home environment to land eventually on earth at a time when the planet's environment was receptive for the system. This is the so-called panspermia theory. This particular theory has been widely held for many years and has had many distinguished proponents. Thus Arrhenius introduced the radiopanspermia theory in which he visualized that some resistant micro-organisms (presumably spores) might be driven across the emptiness of space by radiation pressure to alight ultimately on some planet whose environment might favour survival and reproduction. There seems to be no physical objection to this concept, but there does seem to be doubt about the ability of a spore to survive the damaging radiations known to occur in space. However, recent work has suggested that many extant spores are a good deal more resistant both to high vacuum and to radiation than might have first been thought. Sporopollenin exines of modern spores offer substantial protection against radiation damage. An alternative includes the lithopanspermia theory which had support from both Helmholtz and Lord Kelvin, who believed that various rocks containing a suitable spore or seed of a micro-organism might occur throughout space. These rocks would have been produced by the explosion of a life-bearing body or by its collision with some other body. The seeds would be protected from radiation damage by their rocky envelopes, and equally during entry into the sterile planet's atmosphere they would escape heat damage because of the heat shield effect of the low conducting enveloping rock material.

(b) The living system may have come to earth as a contaminant of some long-since departed space traveller in much the same way as we are probably currently introducing some degree of microbiological contamination on to the moon, Mars, and perhaps to other planets of the solar system.

(c) The living system may have been deliberately introduced to the earth by that long since departed space traveller for reasons which are certainly not clear.

(d) The living system may indeed be truly a machine: a machine that was manufactured in some extraterrestrial laboratory for some specific purpose and introduced on to our planet either by accident or design. There is no particular reason why we should not regard ourselves even now as some form of rather (to us at least) sophisticated robot.

There are of course many more possible alternatives to these but many of them would find difficulty in accommodating the known facts. Thus it is not impossible that quite complex life was present on the planet at an early stage of its formation (perhaps of extra-terrestrial origin) and that this life died away leaving residues which have changed and evolved in a different manner throughout the ages. However, it seems reasonably clear from the evidence that up to a point the farther back one goes in time the more primitive is the living system, so that in the earliest of Precambrian rocks we can only detect remains of primitive algal-like organisms. More complex forms, one could argue, have been more fully destroyed, and although this may well be possible one would nevertheless expect some fingerprint impressions of such forms. We can therefore with reasonable confidence assume that the earth at an early stage contained living systems of a simple, probably unicellular, nature but whether prokaryotic or eukaryotic is far from clear.

(3) The living system arose in some as yet completely unknown or unforeseeable manner. Thus although we seem naturally to assume that all matter including living matter arose initially from small fundamental particles by a sort of physico-chemical evolutionary process, there is no conceptual difficulty in imagining that the Universe began as a living system and what we are now observing is the breakdown of that system into "fundamental particles". There are certainly many reasons for thinking that the Universe might oscillate between "life" and "death" since the living system alone makes the existence of the Universe meaningful. We will examine some of these various possibilities in more detail.

ABIOGENIC FORMATION OF THE LIVING SYSTEM

There is little doubt that the most widely held theory of the origin of the living system on this planet is the abiogenic synthesis or chemical evolutionary theory. The acceptance of this theory and its promulgation by many workers who have certainly not always considered all the facts in great detail has in our opinion reached proportions which could be regarded as dangerous. The arguments and evidence for the theory have been presented largely in Chapters 7 and 8.

The evidence falls essentially into two main parts. In the first part it has undoubtedly been shown (see Chapter 8) that under certain primitive conditions, such as a reducing atmosphere, an aqueous environment and reasonably moderate temperatures, mixtures of simple substances such as alkanes, water, carbon dioxide, nitrogen or ammonia when subjected to a wide variety of energy sources including sparks (lightning) heat, radioactivity and the like are readily converted into a wide variety of organic compounds. These compounds include all the small monomer building blocks required formally to produce the macromolecular and immediate precursors of all known living systems, namely amino acids (monomer units of proteins), carbohydrates, purines and pyrimidines (units of polysaccharides and of nucleic acids). We have pointed out already that claims for the abiogenicity of many of these reactions seem strange considering that they have been devised and executed by highly intelligent life forms. Nevertheless there seems little doubt that given the right conditions, and these perhaps need not be too critical, enough of the required monomer units could probably be formed. In this same section of evidence there have been many other much less convincing experiments. These include the various, largely unsuccessful, attempts to prepare useful polymers whether proteins, carbohydrates or nucleic acids by numerous *in vitro* polymerization procedures. Some very limited success has been achieved in the synthesis of protein-like substances in this way, but the formation of anything resembling a nucleic acid seems a long way off. However, given a protein (polymerase) with catalytic properties whereby nucleoside polyphosphates might be converted into nucleic acids, the inherent difficulties of producing a nucleic acid by an *in vitro* procedure could be overcome. Such a mechanism would, of course, have the advantage of relating the protein to the nucleic acid, and this is of great importance to the theory since the subsequent elaboration of a simple living system from the initially derived nucleoprotein requires tight relationships between the nucleic acids and the proteins. These relationships would presumably at a very early stage have to ape in many albeit primitive ways the relationships which exist in current living systems. In other words, the nucleic acid would have to code in some way for protein synthesis. The proteins synthesized would have to include RNA-polymerases and enzymes required for protein synthesis and not least enzymes required for energy production processes. The theory assumes also that at some time a choice of specific chiral molecules was made, and in addition an embracing membrane is produced which would equally require to be coded in the genetic material. One thing seems to be a constant feature of these theories and that is that all these happenings are going to take a great deal of materials and time.

This takes us to the second part of the theory. This section makes specific comments about the events which both preceded and followed the simple molecular synthesis. In particular it is assumed that the earth at an early stage in its formation had a reducing or at least neutral and a sufficiently low temperature atmosphere to allow the build-up and retention of large amounts of organic compounds of varying complexity. This concept is vital to the theory since in an oxidizing atmosphere most organic compounds would (it is assumed) be rapidly degraded to carbon dioxide, water and other simple molecules. The organic compounds produced would accumulate in the warm oceans over millions of years to form a "primitive organic soup" of enormous proportions. This organic soup would not be degraded (as it would today) into its ancestral units since there would be no available living metabolizing system to carry out such processes. The soup would consist, one assumes, of all the various sugars, amino acids, nucleic acid bases and many other compounds mentioned in Chapters 5 and 6 as prerequisities of the living system.

With the passage of time and with the subjection of all the various components to the variety of natural conditions that could arise on such a planet, from extremes of heat, radioactivity or other energy sources, to evaporation on sandy or other rocky surfaces ultimately presumably by chance, conditions arise which uniquely favour the conversion of amino-acids to proteins and of sugars and nucleic acid bases into nucleotides and into the derived various polyphosphate derivatives. With the passage of more time and the accumulation of more of these materials another unique situation again somehow arises whereby that particular mixture of proteins, nucleic acids and other materials get together to form a complete unified living system capable of self-duplication and mutation. From this simple system then arose all other living systems. The universality of the biochemistry of living systems would stem from the unique single precursor cell.

This in brief represents the main framework of the theory. What evidence is there in support of such a theory, and what chance is there of obtaining any evidence which might throw light on the geo-organic chemistry of those early days? At the moment the only source of evidence would seem to rest in the sedimentary rocks (Fig. 12.02) which were formed at the time and which over the subsequent aeons of time have remained in a few instances relatively little altered. These are the ancient Precambrian sediments, especially those remarkable formations the Fig-Tree and Onverwacht sediments of southern Africa. These rocks, as we have seen (Chapter 10), lie essentially on the igneous basement rocks which are about $4{\cdot}0 \times 10^9$ years old and which seem

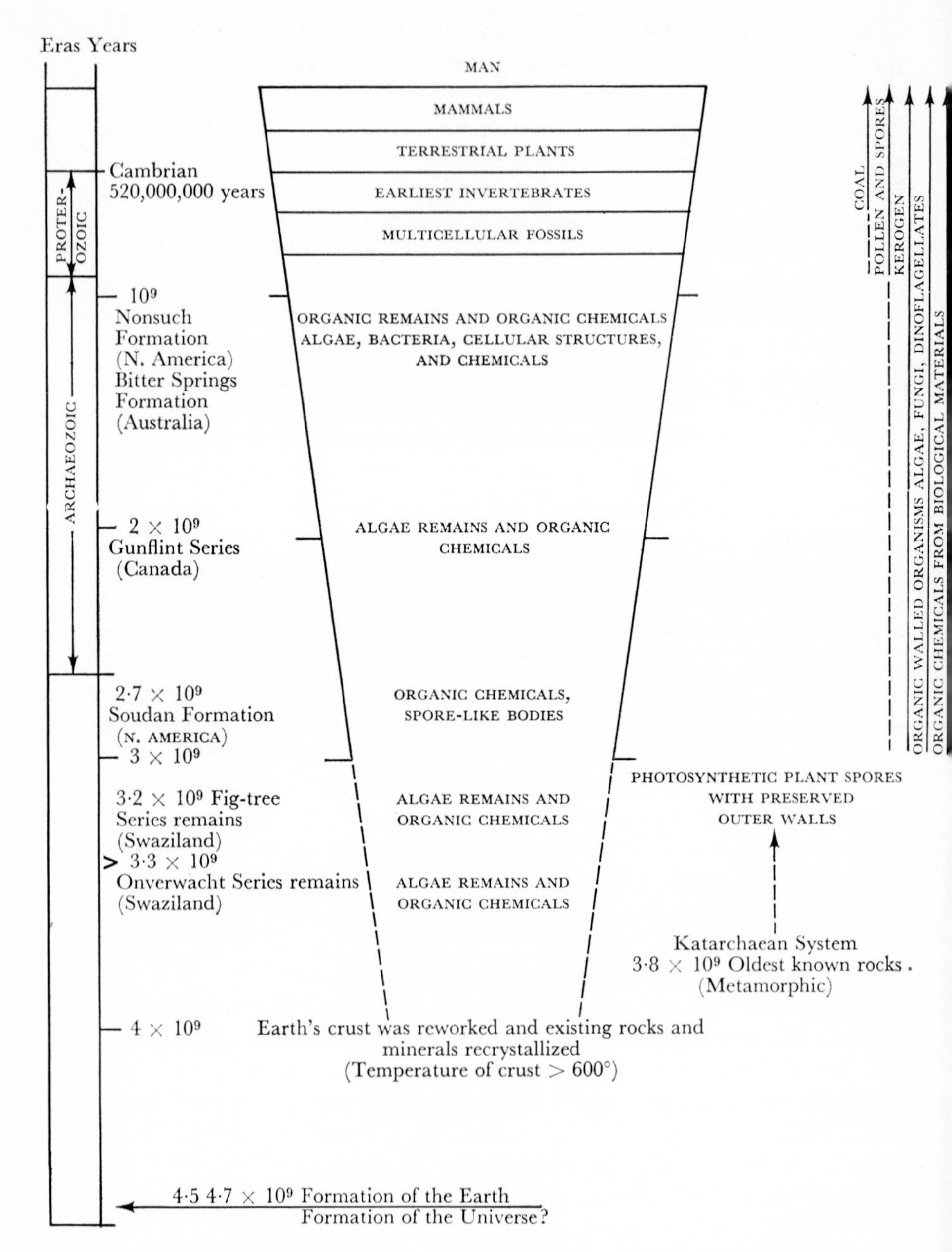

Fig. 12.02. Geological time-scale with reference to bio-organic geochemistry

reasonably clearly to represent basement throughout the whole planet, a time when the earth's crust was being reworked at temperatures almost certainly in excess of 600°C. Under such extremes of temperature not only would any living system be rapidly destroyed but, equally, even complex molecules such as proteins, polysaccharides, let alone nucleic acids, would not survive. So abiogenic formation of a living system would require to begin earlier than 4×10^9 years ago.

Now if we assume that a primitive organic soup was first formed as a precursor requirement for the living system and that it formed and accumulated organic matter over millions of years, then there should be evidence for such matter in the earliest of sediments. Thus the "primitive soup" would of necessity have to contain large amounts of highly nitrogenous matter, especially amino acids and purines and pyrimidines. Such compounds, let alone nucleotides or polymers thereof, are extremely readily absorbed on a variety of rock and clay particles. In a primeval ocean therefore we would expect these concentrated organic soups to deposit their organic matter in the form of massive sediments over millions of years. These sediments would undoubtedly undergo exposure to many extremes of temperature and pressure. When materials such as simple non-nitrogenous compounds are heated and compressed they will generally finish up as carbon, but highly nitrogenous materials when subjected to extremes of temperature form nitrogenous cokes—graphite-like materials containing large amounts of nitrogen. If there ever was a primitive soup, then we would expect to find at least somewhere on this planet either massive sediments containing enormous amounts of the various nitrogenous organic compounds, amino acids, purines, pyrimidines and the like, or alternatively in much metamorphosed sediments we should find vast amounts of nitrogenous cokes. In fact no such materials have been found anywhere on earth. Indeed to the contrary, the very oldest of sediments, ranging to something like $3{\cdot}7 \times 10^9$ years and consequently very close to the basement figure $4{\cdot}0 \times 10^9$ years mentioned earlier, contain organic matter very much of the sporopollenin-type (Chapter 10) and degradation products of such materials, especially a variety of alkanes and fatty acids and derivatives. Sediments of this type are extremely short of nitrogen. Some idea of the sparseness of nitrogen comes for example from the observation of Schöpf that extraction of Precambrian sediments with 0·5N ammonium acetate provided only nanomolecular (10^{-9}) amounts of amino acids, mainly glycine. Whether of course the glycine was derived from the ammonium acetate is another matter. Either way these and other experiments have served to emphasize the scarcity of either chemicals such as amino acids or of nitrogenous cokes in ancient sediments or in nearby metamorphosed rocks. There is, in other words, pretty good negative evidence that there never was a primitive organic soup on this planet that could have lasted but for a brief moment.

It is possible of course to assume that the various chemical and organizational changes which took place in those early days occurred over a very short period of time, so short in fact that no trace of the happenings remain. However, once one begins to make assumptions of this type one is biting into the very fundamental requirements of the abiogenic theory, namely the requirements of large amounts of organic matter and of time so that chance processes have neither the opportunity nor the time to operate successfully.

The other requirement of the abiogenic theory is that the primitive earth contained a reducing atmosphere. This would arise from the condensate of the various dust and gas particles which occurred in the formation of the solar system with hydrogen predominating in the primitive earth. Again there is no evidence for such a reducing atmosphere, although such evidence may be difficult to find since reduced minerals, for example, might have been subsequently reoxidized. The sort of "evidence" sometimes forwarded, namely the reduced iron beds, is somewhat absurd since the existence of living systems was well documented before such beds were deposited and the beds themselves are very likely reduced by living organisms.

EXTRATERRESTRIAL SOURCE OF THE LIVING SYSTEM ON EARTH

An objection to this particular theory which is frequently made is that it merely puts back the problem of the origin of the living system to some other planet or source. As far as this planet is concerned of course such objections are meaningless. We must be interested in the truths of matters and must not modify truths so that we can conveniently express our origins in ways which for some reason or other give us maximum satisfaction.

There is little doubt that this is, at the moment at least, the only theory which fits all the known facts. It does of course as a theory have the dubious advantage of distance, but apart from that there are a number of other points in its favour.

Firstly let us assume that life did in fact arrive from some other body in some as yet unknown manner whether by some form of panspermia, on an ancient astronaut's boot or by deliberate introduction. Clearly it would only survive if conditions on the surface of the planet were satisfactory. Such conditions undoubtedly occurred somewhere between about $4{\cdot}0 \times 10^9$ and $3{\cdot}7 \times 10^9$ years ago, since there is very good evidence for the presence of living systems in ancient Precambrian Onverwacht cherts of up to $3{\cdot}7 \times 10^9$ years old (Chapter 10). The living system would require a source of organic compounds on which it could feed and multiply, but it would only require very small amounts of such compounds; hence no primitive soup of massive proportions is required but only weak or even localized sources of organic substances. Such small amounts of organic compounds could undoubtedly arise in an oxidizing environment, so that we do not need to postulate the somewhat dubious reducing atmosphere. Indeed one could dispense with virtually all complex organic compounds so long as our living system could metabolize simple substances such as carbon dioxide, alkanes and so on. The living system might even at this point have powers of photosynthesis (see later).

Now one of the more remarkable observations which are revealed by the sediments from the earliest to latest Precambrian is the enormous length of time which elapsed before any sort of relatively complex life arose. Thus there is excellent evidence from something like $3{\cdot}8 \times 10^9$ years to about $1{\cdot}0 \times 10^9$ years for the existence on this planet of living systems, but always of a very simple algal-like nature. Even in the period $1{\cdot}0 \times 10^9$ to about $0{\cdot}5 \times 10^9$ years only slight evolution seems to have occurred with evidence for more organized tubular organisms, presumably still algal-like in character or possibly fungal or bacterial. It was not until the Cambrian that an extraordinary sudden outburst of evolutionary activity occurred with formation almost immediately (in this time context) of all the various complex living systems from higher plants and mammals or analogues known today. In other words, those primitive simple cells took something like 85% of our available time before they could mutate to produce differentiated complex living systems from unicellular simple organisms. And yet proponents of the chemical evolutionary theory will suggest that what appears a far more formidable task, namely the production of the living system *de novo* from organic chemicals, occupied a time span so short that no record remains—in other words, that the primitive soup was formed and converted into living systems over so short a time span that the residues of that system are too small to be detected.

It is invariably assumed and virtually a necessity of the chemical abiogenic theory that the simplest cell would be a prokaryotic cell since the initial formation of a eukaryotic cell requires organization and chance occurrences which seem beyond the wildest possible dreams. However, it is by no means certain that the simplest of cells first detected in the oldest of Precambrian sediments are in fact prokaryotic cells.

The evidence here is admittedly slim, but we, for example, have shown that these most primitive cells contain sporopollenin, and although we cannot be certain, nevertheless sporopollenin does seem to be characteristic of eukaryotic rather than prokaryotic cells. In other words, we have not to date found sporopollenin in any prokaryotic cell (bacterial spores have been examined) but it is present in many eukaryotic spores and appears to be associated with the sexual process. Thus in *Mucor mucedo* the asexual spores do not produce sporopollenin whereas the (±) zygote produced by fusion of the (+) and (−) forms contains a thin (*ca* 1% by weight) membrane of sporopollenin. We cannot of course be certain that this relationship will always hold, but at least it is an interesting pointer and suggests positive experiments. Moreover, there is fair evidence for oxygen-producing photosynthesis in the Onverwacht organisms and photosynthesis of this type is characteristic of eukaryotic cells.

It is important of course to bear in mind the fact that the particular morphological living system detected in ancient sediments may not necessarily bear much relationship to the original organism from which it was derived. In other words, it will almost certainly be the residue of a resistant organism, probably a spore, formed in an adverse environment when food or water was scarce or when temperature or atmospheres became untenable to the normal extant life form. It is under such conditions that an ultimate act of evolution occurs with formation of spores whereby the genetic material is allowed a final opportunity to preserve its identity and to commence new growth at some future time when the environment is more favourable. Thus the resistant (±) zygospores of the fungus *Mucor mucedo* (Fig. 12.03) are dark yellow brown spheroids bearing little or no resemblance to the normal filamentous form of the fungus or to its asexual non-resistant spores. In the same way pollen grains or the related megaspores of various higher plants bear no obvious relationship to the plant itself.

We must at all times be careful, therefore, to remember that the morphological bodies we find in the sediments which have invariably been termed "bacteria" or "algae", etc., must almost of necessity be spore or similar resistant remains and not necessarily relics of the original micro-organism's more normal form. The chemistry of such residues confirms this, since in general they appear to be composed almost entirely of the highly resistant chemical substance sporopollenin. As one would expect, and has been pointed out by White (see Chapter 9), polysaccharides such as cellulose will have been hydrolysed away within a relatively short time after deposition, the sedimentation process ensuring no lack of water for this purpose. Such processes would ensure that no remains of non-spore type organisms are likely to be detected in ancient sediments any more than they are in relatively recent sediments, unless substances such as sporopollenin can coat other tissues or cells of the organisms. This is certainly a possibility and has been discussed in Chapters 9 and 10.

It has been suggested that evolution has occurred by symbiosis of prokaryotic cells into one another to form eukaryotic cells. Margoulis envisages that the first eukaryotic cell began when a microbe capable only of anaerobic fermentation of glucose to pyruvate formed a symbiotic association with an aerobic smaller prokaryote which must have

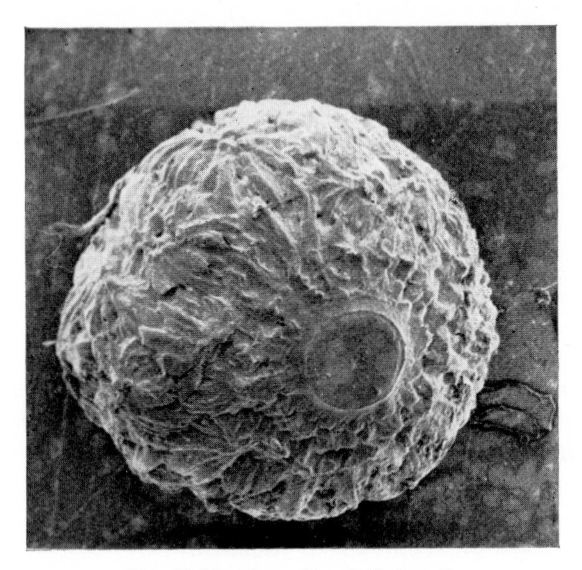

Fig. 12.03. *Mucor Mucedo* "sphere"

had the ability to produce cytochromes and to oxidize all foodstuffs into carbon dioxide. The original anaerobic organism (protoeukaryote) ultimately became the nucleus and cytoplasm and the smaller aerobic organism (protomitochondrion or protochloroplast) the mitochondria (or chloroplast) of an amoeba-like eukaryotic cell. Similar associations of prokaryotes are seen to give rise to other types of eukaryotes. Now although there are many well-documented examples of symbiosis between various organisms (e.g. *Cyanophora paradoxa* is a protozoan host which harbours blue green algal symbionts (*Cyanocyta*); the lichens are fungi with algal symbionts; various amoeba-containing symbiont bacteria occur and include *Pelomyxa palustris*), nevertheless it is pertinent to note that there appears to be no example of the occurrence of any symbiotic processes which led to conversion into new types of cells. It is possible of course that such conversions would be difficult to observe, but the rapid turnover in many symbiotic colonies examined might have been expected to throw up some suitable example by now. Indeed although one assumes that a simple prokaryotic cell such as a bacterium is simpler (more primitive) than a eukaryotic cell, and in basic structure it undoubtedly is, nevertheless bacteria frequently possess specialized properties, e.g. exogenous amino acid requirements, which give them a degree of sophistication more reminiscent of quite advanced and integrated living systems. In this sense one can make out a better case for the formation of the prokaryotic cell by degradation of a eukaryotic cell, and a virus particle could then be envisaged as an extension of such degradation. It is not impossible of course that both symbiotic and degradative processes may now occur, but it is certainly far from sure that symbiosis is a complete reasonable explanation of primary evolution, and indeed this particular theory does not seem to find much favour with most biologists.

The extraterrestrial origin therefore fits the facts. A simple living system arrived on planet Earth, perhaps a photosynthetic eukaryotic organism (algae). It survived on minimal amounts of organic compounds especially perhaps carbon dioxide, in a warm atmosphere containing large amounts of carbon dioxide. It increased and ultimately produced an oxygen-containing atmosphere, but with only limited mutation to more complex forms. It is possible that at a certain oxygen content in the atmosphere, which arose somewhere about $0{\cdot}5\text{–}1{\cdot}0 \times 10^9$ years ago, the accompanying modifications to the environment were such as to favour more complex mutational processes to provide integrated multicellular and multifunctional organisms. It would certainly be interesting to examine the effects of varying oxygen–carbon dioxide ratios on the abilities of extant eukaryotic organisms to undergo mutations of this type.

The evidence for an extraterrestrial origin for the living system would undoubtedly be greatly enhanced if it could be shown that any life has existed outside this planet, since after all it is not impossible that life is completely unique and has only ever occurred once in the whole history of the Universe. There are various possible sources of evidence for extraterrestrial life. They include the possibility of examining planetary bodies in our own solar system, investigation of the occurrence of any planetary bodies in other nearby solar systems and examination of such extraterrestrial material that might fall either on earth (meteorites) or perhaps more usefully on other planets and their satellites in the solar system.

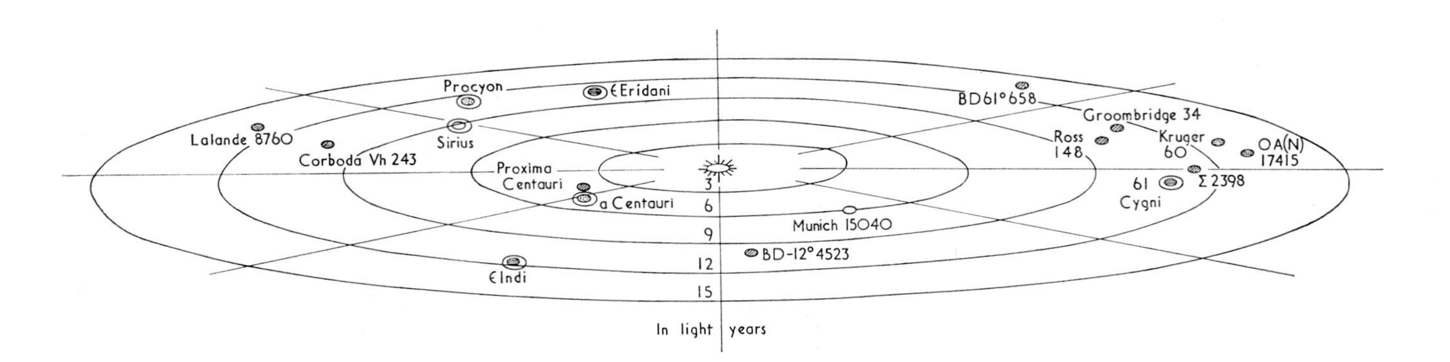

Surface Temperature

3000° K   5500° K   4500° K   10000° K   Naked Eye Stars

Fig. 12.04. The Sun's nearest neighbours

EVIDENCE FOR OTHER PLANETARY SYSTEMS

It is in many ways surprising how little effort appears to have been devoted by astronomers to an examination of nearby star systems (Fig. 12.04) with a view to obtaining evidence for planetary bodies. Certainly analysis of obvious radiation from various parts of the Universe would, in the short term at least, prove perhaps a more profitable pursuit than would attempts to seek for possibly non-existent planetary systems. Nevertheless one feels that far too little attention is being devoted to these types of studies. Indeed many of the triumphs claimed of modern astronomy owe as much if not more to nuclear and sub-atomic particle studies on earth as they do to more conventional astronomical studies. There seems to be little doubt, however, that search for other planetary systems could be much more fruitful if carried out in an environment free from the disturbing effects of our own atmosphere. It is hoped that it may soon be possible to have some form of extraterrestrial astronomical laboratory, either in an orbiting space laboratory or on the moon.

In spite of the difficulties there have been several observations of nearby stars which indicate that they are associated with large planetary type bodies. Thus Strand in 1944 found perturbations in the binary star system 61 Cygni B which is one of our nearest stars and distant some 11 light years. Of the two components of the system the brighter has a luminosity about one-nineteenth that of the Sun. The less bright component was found to be perturbed to an extent which indicated its proximity to a dark body some 15 times the mass of Jupiter, which is too small to be a star and is therefore apparently a planet. Very much smaller planets of earth size would have such small perturbing effects that they would not be normally detectable. There are other stars in which similar perturbations can be observed. They include 70 Ophiuchi with an associated body some 12 times the mass of Jupiter. There are other star systems relatively close to our own which are similar to the Sun in luminosity and spectral class and include Tau Ceti (10·2 light years distant) and Epsilon Eridani (10·5 light years distant), and indeed attempts (which commenced in 1960) to detect extraterrestrial messages from such sources were made by Drake from Greenbank, Virginia. He used an 85 ft radio-telescope and regularly observed any transmission from the vicinity of these sources at 21·1 cm, the wavelength at which hydrogen (the common Universe constituent) emits radio energy, since it was thought that an intelligent alien might equally consider this wavelength to be of fundamental importance. However, needless to say these observations failed to produce any intelligent messages, but it is an interesting and encouraging commentary on our changing attitudes that experiments such as this were considered worth while.

Perhaps the most immediate prospect of reaching a decision about an extraterrestrial living system lies in a close examination of the extraterrestrial material immediately available to us. This includes all the various types of meteorites especially the carbonaceous chondrites, samples of lunar rocks, perhaps in the near future Martian and other planetary rocks, and perhaps most usefully the carbonaceous meteorites which must have landed on the moon. We have seen in earlier chapters that there is a very close parallel between the pattern and type of organic matter which occurs in terrestrial, expecially in the ancient Precambrian sedimentary rocks, and that occurring in typical carbonaceous chondrites such as the Orgueil and the Murray meteorites. In each case the organic matter is present in the form of readily extractable (by organic solvents) material, much of which is of a hydrocarbon nature, and an insoluble polymeric-resistant substance readily oxidized to a mixture of organic acids. In the case of terrestrial sediments much of the insoluble organic matter is in the form of morphologically distinct spore remains (exines) and composed of the carotenoid–carotenoid ester polymer sporopollenin. The morphologically amorphous material which also exists in terrestrial sediments side by side with intact spores is chemically in no ways different—it is composed of the chemical substance sporopollenin and its lack of biological homogeneity is the result of physico-chemical abrasive processes which have occurred over the subsequent enormous time spans. Certainly such processes readily and similarly degrade even many modern types of sporopollenin including that derived from microspore (pollen) exines of Poplar, whereas that from other species (e.g. *Pinus*) is less susceptible to such abrasions. The various forms of sporopollenin, in other words, have different sensitivities to metamorphic processes and this is reflected in the occurrence of specific spores only in relatively late sediments.

The insoluble organic matter which is present in the above-mentioned carbonaceous meteorites is in all respects very much like that in the terrestrial sediments and there seems to be no reason to think that it too is not sporopollenin produced as a resistant outer exine by some long-since-extinct micro-organism. The amounts of such material present in both meteorites are so large (0·3%) as to exclude any possibility of contamination by terrestrial biological material although such comment would not apply to the far smaller amounts of soluble and readily extractable materials much of which, both in terrestrial and extraterrestrial materials, have been shown not to be indigenous to those materials and results of studies with these compounds can have only limited value. Certainly, however, the particular soluble components fully complement the sporopollenin structure from which for the most part they can be seen to arise by mild metamorphic processes.

The possible presence of living system remains in the carbonaceous chondrites certainly adds weight to the frequently made suggestion that the solar system at one time contained another planet which for some reason or other disintegrated, scattering its sediments and other rock materials throughout space, to be dragged ultimately into the asteroidal belt, and later into the earth's gravitational field.

THE LIVING SYSTEM AS A MACHINE

We have pointed out in an earlier chapter (7) that the most useful definition of the living system is that it is a biological machine, and this perhaps limited definition would generally brook no argument with most scientists. We certainly claim no originality for this concept which has been frequently expressed and almost regarded in a sense as an axiomatic part of modern molecular biology. However, there is perhaps a difference between regarding the living system as a machine with qualifications, as is normally implied in the above definition, and regarding it as a machine without any qualifications. Unlike a motor car, the living system can make itself. If we could look at a motor car factory in which all the people were invisible we should see what we might think was a living system creating and recreating itself over and over again, then suddenly (new model) mutating to a form which perhaps was more powerful or safer or possessing other useful features which might apparently seem to arise by some process of natural selection. Certainly a driven moving motor car in a way is every bit alive as the person driving, since the living system can clearly extend from the driver to the driven in a quite easily understood manner. Only when contact with the living system is removed do we now see the motor car as a dead inanimate object or as a sort of mechanical spore which might, however, on some future occasion spring to life. There is clearly a very close analogy between the car and the type of system we claim to be alive. The analogy may be much closer than one might think. The cell, in spite of its intricacies, is very much a collection of both physical and chemical nuts and bolts. Whether it was or is produced and subsequently perpetually driven by some little known celestial force either for pleasure or some other unknown purpose may not readily be open to objective study, but nevertheless it is a possibility that cannot be ignored.

Author Index

Numbers in italics indicate those pages where references appear in full

F

G

H

I

J

K

R

S

T

U

V

W

Subject Index

G

M

Q